THE PICKERING MASTERS

THE WORKS OF CHARLES DARWIN

Volume 11. *The Lepadidae*

THE WORKS OF CHARLES DARWIN

EDITED BY
PAUL H. BARRETT & R. B. FREEMAN
ADVISOR: PETER GAUTREY

VOLUME
11

A MONOGRAPH OF THE SUB-CLASS CIRRIPEDIA:
VOLUME I
THE LEPADIDAE

LONDON AND NEW YORK

First published 1992 by Pickering & Chatto (Publishers) Limited

Published 2016 by Routledge
2 Park Square, Milton Park, Abingdon, Oxon OX14 4RN
605 Third Avenue, New York, NY 10017

Routledge is an imprint of the Taylor & Francis Group, an informa business

Copyright © Taylor & Francis 1992

All rights reserved. No part of this book may be reprinted or reproduced or utilised in any form or by any electronic, mechanical, or other means, now known or hereafter invented, including photocopying and recording, or in any information storage or retrieval system, without permission in writing from the publishers.

Notice:
Product or corporate names may be trademarks or registered trademarks, and are used only for identification and explanation without intent to infringe.

Darwin, Charles, *1809–1882*
   The works of Charles Darwin.
   Vols. 11–20
   1. Organisms. Evolution – Early works
   I. Title II. Barrett, Paul H. (Paul Howard), *1917–1987* III. Freeman, R. B. (Richard Broke), *1915–1986* IV. Gautrey, Peter
575

ISBN 13: 9-781-85196-301-0 (hbk)

INTRODUCTION TO VOLUMES ELEVEN, TWELVE AND THIRTEEN

A Monograph of the Sub-Class Cirripedia, with Figures of all the Species. [Vol. I] *The Lepadidae; or, pedunculated Cirripedes.* 1851. [Vol. II] *The Balanidae (or sessile Cirripedes); the Verrucidae, etc., etc., etc.* 1854. Ray Society Publications No. 21 for 1851, and No. 25 for 1853. Freeman 339.

Darwin wrote on page 80 of his *Autobiography* 'in October, 1846, I began work on "Cirripedia"'. He first examined an aberrant form which he had collected on the coast of Chile. To understand its anatomy, he had to study that of other species, and from there was led into the taxonomy of all the living forms, as well as that of the British fossils. He spent eight years, between 1846 and 1854, studying the group, and had the results published in four volumes. The living forms were produced by the Ray Society in two volumes; these are printed here in three. The fossils were produced by the Palaeontographical Society in two slimmer volumes; these form Volume 14 of this series.

By 1846, Darwin had already shown, in published work, his competence in geology and vertebrate zoology. His work on the seashore when at Edinburgh had indicated that he knew his way around the marine invertebrates, but he had not yet studied any particular group in depth. Almost all working zoologists of the time were acknowledged experts in some group, and to be accepted by them he had to become one of their number. The cirripedes were ideal for this purpose, not too many in species, badly understood particularly so far as their internal anatomy was concerned, and with their nomenclature in a muddle. Whilst working on their taxonomy, he was also an active member of a Committee set up by the British Association for the Advancement of Science for formulating the rules of nomenclature under the chairmanship of Hugh Edwin Strickland.

To begin with, Darwin collected material from all over the world and found the work deeply rewarding, describing the animals in a letter of 1849 as 'beloved Barnacles'. However towards the end he was tired of the monotony and wrote in 1852 'I hate a Barnacle as no man ever did before'. He never did any more formal taxonomy for the rest of his life. Darwin already had the outlines of his theory of evolution in mind and the disciplines which he perfected here, as well as the acceptance by his colleagues, were to stand him in good stead. These two volumes were illustrated with 40 plates, a few with some coloured figures, by George Brettingham Sowerby, the second of the name. The work is still useful today, as well as forming an ideal introduction to formal taxonomy.

PUBLISHER'S NOTE

The publisher deeply regrets the deaths of Professor Paul Barrett and Richard Freeman. Professor Barrett and Mr Freeman had completed their work on the introductions and texts and Peter Gautrey has kindly agreed to act as advisor for the rest of the series.

The Lepadidae

A MONOGRAPH

ON THE SUB-CLASS

CIRRIPEDIA,

WITH

FIGURES OF ALL THE SPECIES.

BY

CHARLES DARWIN, F.R.S., F.G.S.

THE LEPADIDÆ;

OR,

PEDUNCULATED CIRRIPEDES.

LONDON:
PRINTED FOR THE RAY SOCIETY.
MDCCCLI.

PREFACE

My duty in acknowledging the great obligations under which I lie to many naturalists, affords me most sincere pleasure. I had originally intended to have described only a single abnormal cirripede, from the shores of South America, and was led, for the sake of comparison, to examine the internal parts of as many genera as I could procure. Under these circumstances, Mr J. E. Gray, in the most disinterested manner, suggested to me making a monograph on the entire class, although he himself had already collected materials for this same object. Furthermore, Mr Gray most kindly gave me his strong support, when I applied to the Trustees of the British Museum for the use of the public collection; and I here most respectfully beg to offer my grateful acknowledgements to the Trustees, for their most liberal and unfettered permission of examining, and when necessary, disarticulating the specimens in the magnificent collection of cirripedes, commenced by Dr Leach, and steadily added to, during many years, by Mr Gray. Considering the difficulty in determining the species in this class, had it not been for this most liberal permission by the Trustees, the public collection would have been of / no use to me, or to any other naturalist, in systematically classifying the cirripedes.

Previously to Mr Gray suggesting to me the present monograph, Mr Stutchbury, of Bristol, had offered to intrust to me his truly beautiful collection, the fruit of many years' labour. At that time I refused this most generous offer, intending to confine myself to anatomical observations; but I have since accepted it, and still have the entire splendid collection for my free use. Mr Stutchbury, with unwearied kindness, further supplied me with fresh specimens for dissection, and with much valuable information. At about the same period, Mr Cuming strongly urged me to take up the subject, and his advice had more weight with me than that of almost any other person. He placed his whole magnificent collection at my disposal, and urged me to treat it as if it were my own: whenever I told him that I thought it necessary, he permitted me to open *unique* specimens of great value, and dissect the included animal. I shall always feel deeply honoured by the confidence reposed in me by Mr Cuming and Mr Stutchbury.

I lie under obligations to so many naturalists, that I am, in truth, at a loss how to express my gratitude. Mr Peach, over and over again, sent me fresh specimens of several species, and more especially of *Scalpellum vulgare*, which were of invaluable assistance to me in making out the singular sexual relations in that species. Mr Peach, furthermore, made for me observations on several living individuals. Mr W. Thompson, the distinguished natural historian of Ireland, has sent me the / finest collection of British species, and their varieties, which I have seen, together with many very valuable MS. observations, and the results of experiments. Professor Owen procured for me the loan of some very interesting specimens in the College of Surgeons, and has always given me his invaluable advice and opinion, when consulted by me. Professor E. Forbes has been, as usual, most kind in obtaining for me specimens and information of all kinds. To the Rev. R. T. Lowe I am indebted for his particularly interesting collection of cirripedes from the Island of Madeira – a collection offering a singular proof what treasures skill and industry can discover in the most confined locality. The well-known conchologist, Mr J. G. Jeffreys, has sent for my examination a very fine collection of British specimens, together with a copious MS. list of synonyms, with the authorities quoted. To the kindness of Messrs Mc Andrew, Lovell Reeve, G. Busk, G. B. Sowerby, Sen., D. Sharpe, Bowerbank, Hancock, Adam White, Dr Baird, Sir John Richardson, and several other gentlemen, I am greatly indebted for specimens and information: to Mr Hancock I am further indebted for several long and interesting letters on the burrowing of cirripedes.

Nor are my obligations confined to British naturalists. Dr Aug. Gould, of Boston, has most kindly transmitted to me some very interesting specimens; as has Professor Agassiz other specimens collected by himself in the Southern States. To Mr J. D. Dana, I am much indebted for several long letters, containing original and valuable information on points connected with the anatomy of the / Cirripedia. to Mr Conrad I am likewise indebted for information and assistance. Both the celebrated Professors, Milne Edwards and Müller, have lent me, from the great public collections under their charge, specimens which I should not otherwise have seen. To Professor W. Dunker, of Cassel, I am indebted for the examination of his whole collection. I have, in a former publication, expressed my thanks to Professor Steenstrup, but I must be permitted here to repeat them, for a truly valuable present of a specimen of the *Anelasma squalicola* of this work. I will conclude my thanks to all the above British and foreign naturalists, by stating my firm conviction, that if a person wants to ascertain how much true kindness exists among the disciples of Natural History, he should undertake, as I have done, a monograph on some tribe of animals, and let his wish for assistance be generally known.

Had it not been for the Ray Society, I know not how the present volume could have been published; and therefore I beg to return my most sincere thanks to the council of this distinguished institution. To Mr G. B. Sowerby, Jnr, I am under obligations for the great care he has taken in making preparatory drawings, and in subsequently engraving them. I believe naturalists will find that the ten plates here given are faithful delineations of nature.

In monographs, it is the usual and excellent custom to give a history of the subject, but this has been so fully done by Burmeister, in his *Beiträge zur Naturgeschichte der Rankenfüsser*, and by M. G. Martin St Ange, in his *Mémoire sur l'Organisation des Cirripèdes*, that it / would be superfluous here to repeat the same list of authors. I will only add, that since the date, 1834, of the above works, the only important papers with which I am acquainted, are:

1. Dr Coldstream 'On the Structure of the Shell in Sessile Cirripedes', in the *Encyclopaedia of Anatomy and Physiology*.

2. Dr Lovén 'On the *Alepas squalicola*' (*Ofversigt of Kongl. Vetens., &c.* Stockholm, 1844, p. 192), giving a short but excellent account of this abnormal cirripede.

3. Professor Leidy's very interesting discovery (*Proceedings of the Academy of Natural Sciences*, Philadelphia, vol. iv, No. I, Jan. 1848), of eyes in a mature Balanus.

4. Mr A. Hancock's Memoir (*Annals of Natural History*, 2nd series, Nov. 1849), on his *Alcippe lampas*, the type of a new order of cirripedes.

5. Mr Goodsir's Paper (*Edinburgh New Philosoph. Journal*, July 1843), on the 'Larvae in the First Stage of Development in Balanus'.

6. Mr C. Spence Bate's valuable paper on the same subject, lately published (Oct. 1851), in the *Annals of Natural History*; and lastly, M. Reinhardt has described, in the *Copenhagen Journal of Natural History*, Jan. 1851, the *Lithotrya Nicobarica*, and has discussed its powers of burrowing into rocks.

I have given the specific or diagnostic characters, deduced from the external parts alone, in both Latin and English. As I found, during the progress of this work, that a similarly abbreviated character of the softer internal parts, was very useful in discriminating the species, I have inserted it after the ordinary specific character.

In those cases in which a genus includes only a single / species, I have followed the practice of some botanists, and given only the generic character, believing it to be impossible, before a second species is discovered, to know which characters will prove of specific, in contradistinction to generic, value.

In accordance with the rules of the British Association, I have faithfully endeavoured to give to each species the first name attached to it, subsequently to the introduction of the binomial system, in 1758, in the tenth edition[1] of the *Systema Naturae*. In accordance with the rules I have rejected all names before this date, and all MS. names. In one single instance, for reasons fully assigned in the proper place, I have broken through the great law of priority. I have given much fewer synonyms than is usual in conchological works; this partly arises from my conviction that giving references to works, in which there is not any original matter, or in which the plates are not of a high order of excellence, is absolutely injurious to the progress of Natural History, and partly, from the impossibility of feeling certain to which species the short descriptions given in most works are applicable; thus, to take the commonest species, the *Lepas anatifera*, I have not found a single description (with the exception of the anatomical description by M. Martin St Ange) by which this species can be certainly discriminated from the almost equally common *Lepas Hillii*. I have, however, been fortunate in having / been permitted to examine a considerable number of authentically named specimens (to which I have attached the sign [!] used by botanists), so that several of my synonyms are certainly correct.

The Lepadidae, or pedunculated cirripedes, have been neglected under a systematic point of view, to a degree which I cannot quite understand: no doubt they are subject to considerable variation, and as long as the internal surfaces of the valves and all the organs of the animal's body, are passed over as unimportant, there will occasionally be some difficulty in the identification of the several forms, and still more in settling the limits of the variability of the species. But I suspect the pedunculated cirripedes have, in fact, been neglected, owing to

[1] In the rules published by the British Association, the 12th edition (1766), is specified, but I am informed by Mr Strickland that this is an error, and that the binomial method was followed in the 10th edition.

their close affinity, and the consequent necessity of their being included in the same work with the sessile cirripedes; for these latter will ever present, I am fully convinced, insuperable difficulties in their identification by external characters alone.

I will here only further remark, that in the introduction I have given my reasons for assigning distinct names to the several valves, and to some parts of the included animal's body; and that in the introductory remarks, under the general description of the Lepadidae, I have given an abstract of my anatomical observations. /

CORRIGENDA AND ADDENDA

Page 42, 43, I should have added, that the number of the segments in the cirri increases with the age of the specimen; but that the relative numbers in the different cirri keep, as far as I have seen, nearly constant; hence the numbers are often given in the descriptions.

Page 156, in a footnote, I have alluded to a new genus of sessile cirripedes, under the name of Siphonicella, I now find that this species has been called, by Professor Steenstrup, *Xenobalanus globicipitis*. /

MONOGRAPH ON THE CIRRIPEDIA

INTRODUCTION

I should have been enabled to have made this volume more complete, had I deferred its publication until I had finished my examination of all the other known cirripedes; but my work would thus have been rendered inconveniently large. Until this examination is completed, it will be more prudent not to discuss, in detail, the position of the Lepadidae among the Cirripedia, or of these latter in the great class of Crustacea, to which they now, by almost universal consent, have been assigned. I may, however, remark that I believe the Cirripedia do not approach, by a single character, any animal beyond the confines of the Crustacea: where such an approach has been imagined, it has been founded on erroneous observations; for instance, the closed tube within the stomach, described by M. Martin St Ange (to whose excellent paper I am greatly indebted), as indicating an affinity to the annelides, is, I am convinced, nothing but a strong epithelial lining, which I have often seen ejected with the excrement. Again, a most distinguished author has stated that the Cirripedia differ from the Crustacea:

1. In having 'a calcareous shell and true mantle'; but there is no essential difference, as / shown by Burmeister, in the shells in these two classes; and cirripedes certainly have no more claim to a mantle than have the bivalve entomostraca.

2. 'In the sexes joined in one individual'; but this, as we shall see, is not constant, nor of very much weight, even if constant.

3. 'In the body not being ringed'; but if the outer integument of the thorax of any cirripede be well cleaned, it will be seen (as was long ago shown by Martin St Ange), to be most distinctly articulated.

4. 'In having salivary glands'; but these glands are, in truth, the ovaria.

5. 'In the liver being formed on the molluscous type'; I do not think this is the case, but I do not quite understand the point in question.

6. 'In not having a head or organs of sense'; this is singularly erroneous: Professor Leidy has shown the existence of eyes in the mature cirripede; the antennae, though preserved, certainly become functionless soon after the last metamorphosis; but there exist other organs of sense, which I believe serve for smelling and hearing: and lastly, so far from there being no head, the whole of the cirripede externally visible, consists exclusively of the three anterior segments of the head.

The subclass, Cirripedia, can be divided into three orders; the first of which, mainly characterized by having six pair of thoracic cirri, includes all common cirripedes: these latter may be divided into three families – the Lepadidae, or pedunculated cirripedes, the subject of the present memoir; the Verrucidae containing the single genus Verruca or Clisia; and, lastly, the Balanidae, which consist of two very distinct subfamilies, the Balaninae and Chthamalinae. Of the other two orders above alluded to, one will, I believe, contain the remarkable burrowing genus Alcippe, lately described by Mr Hancock, and a second burrowing genus, or rather family, obtained by me on the coast of South America. The third order is highly singular, and differs as much from all other cirripedes as does a Lernaea from other crustaceans; it has a suctorial mouth, but is destitute of an anus; it has / not any limbs, and is as plainly articulated as the larva of a fly; it is entirely naked, without valves, carapace, or capitulum, and is attached to the cirripede, in the sack of which it is parasitic, by *two* distinct threads, terminating in the usual larval, prehensile antennae. I intend to call this cirripede, Proteolepas. I mention it here for the sake of calling attention to any parasite at all answering to this description.

Although the present volume is strictly systematic, I will, under the general description of the Lepadidae, give a very brief abstract of some of the most interesting points in their internal anatomy, and in the metamorphoses of the whole class, which I hope hereafter to treat, with the necessary illustrations, in detail. I enter on the subject of the metamorphoses the more readily, as by this means alone can the homologies of the different parts be clearly understood.

On the names given to the different parts of cirripedes

I have unwillingly found it indispensable to give names to several valves, and to some few of the softer parts of cirripedes. The accompanying figure of an imaginary Scalpellum includes every valve; the two most important valves of Lepas are also given, in which the direction of the lines of growth and general shape differ from those of Scalpellum as much as they do in any genus. The names which I have imposed will, I hope, be thus acquired without much difficulty.

Whoever will refer to the published descriptions of recent and fossil Cirripedia, will find the utmost confusion in the existing nomenclature: thus, the valve named in the woodcut the *scutum*, has been designated by various well-known naturalists as the 'ventral', the 'anterior', the 'inferior', the 'ante-lateral', and the 'latero-inferior' valve; the first two of these titles have, moreover, been applied to the rostrum or rostral valve of sessile cirripedes. / The *tergum* has been called the 'dorsal', the 'posterior', the 'superior', the 'central', the 'terminal', the 'postero-lateral', and the 'latero-superior' valve. The *carina* has received the first two of these identical epithets, viz. the 'dorsal' and the 'posterior'; and likewise has been called the 'keel-valve'. The confusion, however, becomes far worse, when any individual valve is described, for the very same margin which is anterior or inferior in the eyes of one author, is the posterior or superior in those of another; it has often happened to me that I have been quite unable even to conjecture to which margin or part of a valve an author was referring. Moreover, the length of these double titles is inconvenient. Hence, as I have to describe all the recent and fossil species, I trust I may be thought justified in giving short names to each of the more important valves, these being common to the pedunculated and sessile cirripedes.

The part supported by the peduncle, and which is generally, though not always, protected by valves, I have designated the *capitulum*.

The title of *peduncle*, which is either naked or squamiferous, requires no explanation; the scales on it, and the lower valves of the capitulum, are arranged in whorls, which, in the Latin specific descriptions, I have called by the botanical term of verticillus.

I have applied the term *scutum* to the most important and persistent of the valves, and which can generally be recognized by the hollow giving attachment to the adductor scutorum muscle, from the resemblance which the two valves taken together bear to a shield, and from their office of protecting the front side of the body. From the protection afforded by the two *terga* to the dorso-lateral surface of the animal, these valves have been thus called. The term *carina*[1] is a mere

---

[1] In the carina of fossil species of Scalpellum, I have found it necessary to distinguish different parts, viz., A, the tectum, of which half is seen; B, the parietes; and C, the intra-parietes.

translation / of the name already used by some authors, of keel-valve.

The *rostrum* has been so called from its relative position to the carina or keel. There is often a *sub-carina* and a *sub-rostrum*.

The remaining valves, when present, have been called *Latera*; there is always one large upper one inserted between the lower halves of the scuta and terga, and this I have named the upper latus or latera; the other latera in Pollicipes are numerous, and require no special names; in Scalpellum, where there are at most only three pair beneath the Upper Latera, it is convenient to speak of them (*vide* Woodcut, I) as the *carinal*, *infra-median*, and *rostral latera*.

As each valve often requires (especially among the fossil species) a distinct description, I have found it indispensable to give names to each margin. These have mostly been taken from the name of the adjoining valve (see fig. I). In Lepas, Pollicipes, etc., the margin of the scutum adjoining the tergum and upper latus, is not divided (fig. II) into two distinct lines, as it is in Scalpellum, and is therefore called the tergo–lateral margin. In Scalpellum (fig. I) these two margins are separately named tergal and lateral. The angle formed by the meeting of the basal and lateral or tergo–lateral margins, I call the baso–lateral angle; that formed by the basal and occludent margins, I call, from its closeness to the rostrum, the rostral angle. In Pollicipes the carinal margin of the tergum can be divided into an upper and lower carinal margin; of this there is only a trace (fig. I) in Scalpellum.

That margin in the scuta and terga which opens and *shuts* for the exsertion and retraction of the cirri, I have called the occludent margin. In the terga of Lepas (fig. III) and some other genera, the occludent margin is highly protuberant and arched, or even formed of two distinct sides.

Occasionally, I have referred to what I have called the / *primordial valves*: these are not calcified; they are formed at the first exuviation, when the larval integuments are shed: in mature cirripedes they are always seated, when not worn away, on the umbones of the valves.

The membrane connecting the valves, and forming the peduncle, and sometimes in a harder condition replacing the valves, I have often found it convenient to designate by its proper chemical name of *chitine*, instead of by horny, or other such equivalents. When this membrane at any articulation sends in rigid projections or crests, for the attachment of muscles or any other purpose, I call them, after Audouin, *apodemes*. For the underlying true skin, I use the term *corium*.

The animal's body is included within the capitulum, within what I call the *sack* (see Pl. IV, figs 2 and 8'a, and Pl. IX, fig. 4). The body consists of the *thorax* supporting the cirri, and of an especial enlargement, or downward prolongation of the thorax, which includes the stomach, and which I have called the *prosoma* (Pl. IX, fig. 4n). The *cirri* are composed of two arms or *rami*, supported on a common segment or support, which I call the *pedicel*. The *caudal appendages* are two little projections, either uni- or multi-articulate (Pl. IV, fig. 8'a), on each side of the anus, and just above the long proboscis-like penis. On the thorax and prosoma, or on the pedicels of the cirri, there are in several genera, long, thin, tapering filaments, which have generally been supposed to serve as branchiae; these I call simply *filaments*, or *filamentary appendages* (Pl. IX, fig. 4g–l). The mouth (fig. 4b) is prominent, and consists of *palpi* soldered to the *labrum*; *mandibles*, *maxillae*, and *outer maxillae*, these latter serve as an under lip; to these several organs I sometimes apply the title used by entomologists, of 'trophi'. Beneath the outer maxillae, there are either two simple orifices or tubular projections; these, I believe, serve as organs of smell, and have hence called them the *olfactory orifices*. Within the sack, there are often two sheets of ova (Pl. IV, fig. 2b), these I call (after Steenstrup, and / other authors) the *ovigerous lamellae*; they are united to two little folds of skin (Pl. IV, fig 2f), which I call the *ovigerous fraena*.

From the peculiar curved position which the animal's body occupies within the capitulum, I have found it far more convenient (not to mention the confusion of nomenclature already existing) to apply the term rostral instead of ventral, and carinal instead of dorsal, to almost all the external and internal parts of the animal. Cirripedes have generally been figured with their surfaces of attachment downwards, hence I speak of the lower or basal margins and angles, and of those pointing in an opposite direction as the upper; strictly speaking, as we shall presently see, the exact centre of the usually broad and flat surface of attachment is the anterior end of the animal, and the upper tips of the terga, the posterior end of that part of the animal which is externally visible; but in some cases, for instance in Coronula, where the base is *deeply concave*, and where the width of the shell far exceeds the depth, it seemed almost ridiculous to call this, the anterior extremity; as likewise does it in Balanus to call the united tips of the terga, lying deeply within the shell, the most posterior point of the animal, as seen externally.

I have followed the example of botanists, and added the interjection

[!] to synonyms, when I have seen an authentic specimen bearing the name in question.

Every locality, under each species, is given from specimens ticketed in a manner and under circumstances appearing to me worthy of full confidence – the specific determination being in each case made by myself. /

CLASS: CRUSTACEA
SUB-CLASS: CIRRIPEDIA

FAMILY: LEPADIDAE

Cirripedia pedunculo flexili, musculis instructo: scutis[1] musculo adductore solummodô instructis: valvis caeteris, siquae adsunt, in annulum immobilem haud conjunctis.

Cirripedia having a peduncle, flexible, and provided with muscles. Scuta[1] furnished only with an adductor muscle: other valves, when present, not united into an immovable ring.

Metamorphoses; larva, first stage, p. [9] 8; larva, second stage, p. [13] 11; larva, last stage, p. [14] 12; its carapace, ib.; acoustic organs, p. [15] 13; antennae, ib.; eyes, p. [16] 14; mouth, p. [17] 15 thorax and limbs, p. [18] 15; abdomen, p. [19] 16; viscera, ib.; immature cirripede, p. [20] 17; homologies of parts, p. [25] 21.

Description of mature Lepadidae, p. [28] 23; capitulum, ib.; peduncle, p. [31] 26; attachment, p. [33] 28; filamentary appendages, p. [38] 32; shape of body, and muscular system, p. [39] 32; mouth, ib.; cirri, p. [42] 33; caudal appendages, p. [43] 36; alimentary canal, [44] 36; circulatory system, p. [46] 38; nervous system, ib.; eyes, p. [49] 40; olfactory organs, p. [52] 43; acoustic [?] organs, p. [53] 44; male sexual organs, p. [55] 45; female organs, p. [56] 46; ovigerous lamellae, p. [58] 48; ovigerous fraena, ib.; exuviation, p. [61] 50; rate of growth, ib.; size, ib.; affinities of family, p. [64] 52; range and habitats, p. [65] 53; geological history, p. [66] 54.

Metamorphoses. I will here briefly describe the metamorphoses, as far as known, common to all Cirripedia, but more especially in relation to the present family. I may premise, that since Vaughan Thompson's capital discovery of the larvae in the last stage of development in Balanus, much has been done on this subject: this same author subsequently published[2] in the *Philosophical Transactions*, an account of the larvae of Lepas and Conchoderma (Cineras) in the first stage; and seeing how

---

[1] The meaning of this and all other terms is given in the Introduction, at pp. 3–7.
[2] *Philosophical Transactions*, 1835, p. 355, Pl. vi.

totally distinct they were from the larva of the latter stage in Balanus, he erroneously attributed the difference to / the difference in the two families, instead of to the stage of development. Burmeister[3] first showed, and the discovery is an important one, that in Lepas the larvae pass through two totally different stages. This has subsequently been proved by implication to be the case in Balanus, by Goodsir,[4] who has given excellent drawings of the larva in the first stage; and quite lately, Mr C. Spence Bate, of Swansea, has made other detailed observations and drawings of the larvae of five species in this same early stage, and has most kindly permitted me to quote from his unpublished paper.[5] I am enabled to confirm and generalize these observations, in all the cirripedes in the order containing the Balanidae and Lepadidae.

The ova, and consequently the larvae of the Lepadidae, in the *first stage*, whilst within the sack of the parent, vary in length from 0.007 to 0.009 in Lepas, to 0.023 of an inch in Scalpellum: my chief examination of these larvae has been confined to those of *Scalpellum vulgare*; but I saw them in all the other genera. The larva is somewhat depressed, but nearly globular; the carapace anteriorly is truncated, with lateral horns; the sternal surface is flat and broad, and formed of thinner membrane than the dorsal. The horns just alluded to are long in Lepas and short in Scalpellum; their ends are either rounded and excessively transparent, or, as in Ibla, furnished with an abrupt, minute, sharp point: within these horns, I distinctly saw a long filiformed organ, bearing excessively fine hairs in lines, so exactly like the long plumose spines on the prehensile antennae of the larvae in the last stage; that I have not the least doubt, that these horns are the cases in which antennae are in process of formation. Posteriorly / to them, on the sternal surface, near each other, there are two other minute, doubly curved, pointed horns, about 0·004 in length, directed posteriorly; and within these I again saw a most delicate articulated filiformed organ on a thicker pedicel: in an excellent drawing, by Mr C. S. Bate, of the larva of a Chthamalus (*Balanus punctatus* of British authors), after having kept alive and moulted once, these organs are distinctly shown as articulated antennae (without a case), directed

---

[3] *Beiträge zue Naturgeschichte der Rankenfüsser*, 1834. Mr J. E. Gray, however, briefly described, in 1833 (*Proceedings, Zoological Society*, October), the larva in the first stage of Balanus; in this notice the anterior end of the larva is described as the posterior.
[4] *Edinburgh New Philosophical Journal*, July 1843, Pls iii and iv.
[5] This will appear in the October number (1851) of the *Annals of Natural History*.

forwards: hence, before the first moult in Scalpellum, we have two pair of antennae in process of formation. Anteriorly to the bases of these smaller antennae is seated the heart-shaped eye (as I believe it to be), 0·001 in diameter, with apparently a single lens, surrounded, except at the apex, by dark-reddish pigment-cells. In some cases, as in some species of Lepas, the larvae, when first excluded from the egg, have not an eye, or a very imperfect one.

There are three pairs of limbs, seated close together in a longitudinal line, but some way apart in a transverse direction: the first pair always consists of a single spinose ramus, it is not articulated in Scalpellum, but is multi-articulate in some genera; it is directed forwards. The other two pair have each two rami, supported on a common haunch or pedicel: in both pair, the longer ramus is multi-articulate, and the shorter ramus is without articulations, or with only traces of them: the longer spines borne on these limbs (at least, in Scalpellum and Chthamalus), are finely plumose. The abdomen terminates, a little beyond the posterior end of the carapace, in a slightly upturned horny point; a short distance anteriorly to this point, a strong, spinose, forked projection depends from the abdominal surface.

Messrs V. Thompson, Goodsir, and Bate, have kept alive for several days the larvae of Lepas, Conchoderma, Balanus, Verruca, and Chthamalus, and have described the changes which supervene between the first and third exuviations. The most conspicuous new character is the / great elongation of the posterior point of the carapace into an almost filiform, spinose point in Lepas, Conchoderma, Chthamalus, and Balanus, but not according to Goodsir, in one of the species of the latter genus. The posterior point, also, of the abdomen becomes developed in Balanus (Goodsir) into two very long, spear-like processes, serrated on their outer sides; in Lepas and Conchoderma, according to Thompson, into a single, tapering spinose projection; and in Chthamalus, as figured by Mr Bate, the posterior bifid point, as well as the depending ventral fork, increase much in size. Another important change, which has been particularly attended to by Mr Bate, is the appearance of spinose projections and spines (some of which are thick, curved, and strongly plumose, or, almost pectinated along their inner sides) on the pedicels and lower segments of the shorter rami of the two posterior pairs of limbs.

The mouth in its earliest condition alone remains to be described; in *S. vulgare*, it is seated on a very slight prominence, in a most remarkable situation, namely, in a central point between the bases of the three pairs of legs. I traced by dissection the oesophagus for some little way, until lost in the cellular and oily matter filling the whole animal, and it was directed anteriorly, which is the direction that might have been expected, from the course followed by the oesophagus in the larva in the last stage, and in mature cirripedes. Mr A. Hancock has called my attention to a prosbosciformed projection on the under side of the larva of *Lepas fascicularis*, when just escaped from the egg. Mr Bate has described this same proboscis in Balanus and Chthamalus, and states the important fact, that it is capable of being moved by the animal; and, lastly, I have seen it in an Australian Chthamalus, and in Ibla, of remarkable size. This proboscis, which is always directed posteriorly (like the mouth in the mature animal), certainly answers to the mouth as made out by dissection in Scalpellum; and I believe I saw, as has Mr Bate, a terminal orifice: it certainly does / not possess any trophi. In Ibla (in which the larva is large enough for dissection), the base of the proboscis arises posteriorly to the first pair of legs, and the orifice at the other end reaches beyond or posteriorly to the point, where the mouth in Scalpellum opens, namely between the middle pair of legs. The mouth being either so largely prosbosciformed or seated only on a slight eminence, in two genera so closely allied as Ibla and Scalpellum, and (judging from Mr Thompson's figures, and from what I have seen myself), in the species of the same genus Lepas, is a singular difference: in the cases in which, at first, the proboscis is absent, it would probably soon be developed. I cannot but suppose that the inwardly directed spines on the bases of the two posterior legs, which are so rapidly developed, serve some important end, namely, as organs of prehension for the larvae, like the mandibles and maxillae of mature cirripedes, for seizing their prey, and conveying it to their movable mouths, conveniently seated for this purpose.

The first pair of legs answers, as I believe from reasons hereafter to be assigned, to the outer pair of maxillipods in the higher Crustacea; and the other four legs to the first two pair of thoracic limbs in these same Crustacea; this being the case, the highly remarkable position of the mouth in the larva, either between the bases of the two posterior pair of legs, or at least posteriorly to the first pair, together with the probable functions of the spiny points springing from the basal segments of the two pair of true thoracic limbs, forcibly bring to mind the anomalous structure of the mouth being situated in the middle of the underside of the thorax, in Limulus – that most ancient of crustaceans, and therefore one likely to exhibit a structure now embryonic in other orders. I will only further remark, that I suspect that the truncation of the anterior end of the carapace, has been effected by the segments having been driven inwards, and consequently, that the larger antennae within the lateral horns, though standing more in front than the / little approximate pair, are normally the posterior of the two pair. According to Milne Edwards, the posterior pair are normally seated outside the anterior pair, and this is the case with those within the lateral horns.

Larva in the second stage. Notwithstanding the considerable changes, already briefly given, which the larva undergoes during the first two or three exuviations after leaving the egg, all these forms may be conveniently classed under the first stage. The larva in the second stage is known only from a single specimen described, figured, and found by Burmeister,[6] adhering to seaweed in the midst of other larvae of Lepas in the last stage. In its general shape and compressed form, it seems to come nearer to the last than to the first stage. It has only three pair of legs, situated much more posteriorly on the body than in the first stage, and all directed posteriorly; they are much shorter than heretofore, and resemble rather closely those of the last stage, with the important exception that the first pair has only one ramus. It is this circumstance which leaves no doubt on my mind, that we here have the three pair of limbs, of the first stage, metamorphosed. The body is prolonged some way behind these limbs, and ends in a blunt, rounded point, in which, probably, are developed the three posterior pair of legs and the abdomen of the larva in the last stage. The mouth is now seated some way anteriorly to the limbs, is large and prosboscifomed, and is, I presume, still destitute of trophi. There are now two closely approximate eyes, but as yet both are *simple*. The smaller pair of antennae has disappeared. The whole animal was attached to the seaweed by a (I presume, pair of) 'fleischigen Fortsatz', which Burmeister considers as the prehensile antennae, to be presently described, in an early state of development. I have little doubt that this is correct, for in an abnormal cirripede of another order, in which the larva appears in the *first* stage with prehensile antennae, the eggs have two great projecting horns including / these organs, and attached by their tips, through some unknown means, to the sack of the parent, apparently in the same manner as Burmeister's larva was attached to the seaweed. I will only further remark on the

[6] *Beiträge zur Naturgeschichte der Rankenfüsser*, p. 16, Tab. i, figs 3, 4.

larva of this second stage, that its chief development since the first stage, has been towards its anterior end. The next great development, to be immediately described, is towards the posterior end of the animal.

Larva, last stage. My chief examination has been directed, at this stage of development, to the larvae of *Lepas australis*, which are of unusual size, namely, from 0·065 to even almost 0·1 of an inch in length; I examined, however, the larvae of several other species of Lepas, of Ibla and of Balanus, with less care, but sufficiently to show that in all essential points of organization they were identical; this, indeed, might have been inferred from the similarity of the larval prehensile antennae, preserved in the bases of all mature cirripedes, and which I have carefully inspected in almost every genus. The larvae in this final stage, in most of the genera, have increased many times in size since their exclusion from the egg; for instance, in *Lepas australis*, from 0·007 to 0·065, or even to 0·1 of an inch. They are now much compressed, nearly to the shape of a cypris or mussel-shell, with the anterior end the thickest, the sternal surface nearly or quite straight, and the dorsal arched. Almost the whole of what is externally visible consists of the carapace; for the thorax and limbs are hidden and enclosed by its backward prolongation; and even at the anterior end of the animal, the narrow sternal surface can be drawn up, so as to be likewise enclosed. As in several stomapod crustaceans, the part of the head bearing the antennae and organs of sense, in front of the mouth, equals, or even exceeds in length, and more than exceeds in bulk, the posterior part of the body, consisting of the enclosed thorax and abdomen. I will now briefly describe, in the following order, the carapace, the organs of sense, mouth, thorax and limbs, abdomen, and internal viscera. /

The form of the *carapace* has been sufficiently described; it consists of thick chitine membrane, marked with lines, and sometimes with stars and other patterns; it is obscurely divided into two halves by a line or suture along part of the dorsal margin; these halves or two valves are drawn together by an adductor muscle, in the same relative position as in the mature cirripede. The part overhanging and enclosing the thorax is lined by an excessively delicate membrane, obviously homologous with the lining of the sack in the mature animal, and is nothing but a duplicature of the carapace, rendered very thin from being on the under or protected side: a layer of true skin or corium, probably double, separates these two folds.

Acoustic organs. On the borders of the carapace, at the anterior end, on the sternal surface, there are two minute orifices, in *L. australis* 0·002 in diameter, sometimes having a distinct border round them; the membrane of the carapace on the inside is prolonged upwards and inwards in two short funnel-shaped tubes, lodged in closed sacks of the corium: within these sacks on each side a delicate bag is suspended, and hangs in the mouth of the above funnel; at the upper end a large nerve could be distinctly seen to enter the bag: I cannot doubt that this is a sense-organ; from its position and from the animal not feeding (as we shall presently see), I conclude that it is an acoustic organ.

Antennae. These are large and conspicuous; they are attached very obliquely on the sternal surface, a little way from the anterior end of the carapace, beyond which, when exserted, they extend;[7] they can (at least in Ibla) / be retracted within the carapace. They consist of three segments: the first or basal one is much larger than the others, and apparently always has a single spine on the outer distal margin. The second segment consists either of a large, thin, circular, sucking disc, or is hoof-like (Pl. V, figs 5, 10, 11, 12); in all cases it is furnished with one or more spines (seven very long ones in Lepas), on the exterior-hinder margin. The third and ultimate segment is small; it is articulated on the upper surface of the disc, and is directed rectangularly outwards; it is sometimes notched, and even shows traces of being bifid; it bears about seven spines at the end; some of these spines are hooked, others simple, and in *Lepas* and *Conchoderma*, two or three are very long, highly flexible, and plumose, a double row of excessively fine hairs being articulated on them. I can hardly doubt that these latter spines (within which the purple corium could be seen to enter a little way), floating laterally outwards, serve as feelers. The antennae, at first, are well furnished with muscles. They serve, in Lepas, according to Mr King, and in Balanus, according to Mr Bate, and as I saw myself in another unnamed order, for the purpose of

---

[7] Mr J. D. Dana, who has examined these organs in the larvae of Lepas, informs me in a letter, that in his opinion they 'correspond with the inferior antennae, the superior being wanting, as in most Daphnidae'. He continues – 'I know of no case in which the inferior are obsolete when the superior are developed; but the reverse is often true.' In position these antennae certainly correspond to the inferior and central pair of the larva in the first stage, which belong, as it would appear, to the first segment of the body; but judging from the drawing by Burmeister of the larva in the second stage, I am, in some respects, more inclined to consider that they correspond to the larger pair seen within the lateral horns of the carapace in the first stage.

walking, one limb being stretched out before the other; but their main function is to attach the larva for its final metamorphosis into a cirripede. The disc can adhere even to so smooth a surface as a glass tumbler.[8] The attachment is at first manifestly voluntary, but soon becomes involuntary and permanent, being effected by special and most remarkable means, which will be most conveniently described in a later part of this introduction. I will here only state that I traced with ease the two cement ducts running from two large glandular bodies, to within the antennae up to the discs.

Eyes. Close behind the basal articulations of the antennae, the sternal surface consists of two approximate, elongated, narrow, flat pieces, or segments. These / Burmeister considers as the basal segments of the antennae: as they are not cylindrical, I do not see the grounds for this conclusion: their posterior ends are rounded, and the membrane forming them is reflected inwards, in the form of two, forked, horny apodemes, together resembling two letters, UU, close together; these project up, inside the animal, for at least one third of its thickness from the sternal to the dorsal surface. The two great, almost spherical eyes in *L. australis*, each $\frac{1}{50}$th of an inch in diameter, are attached to the outer arms, thus, ●UU●, in the position of the two full stops. Hence the eyes are included within the carapace. Each eye consists of eight on ten lenses, varying in diameter in the same individual from $\frac{1}{2000}$ to $\frac{3}{2000}$th of an inch, enclosed in a common membranous bag or cornea and thus attached to the outer apodemes. The lenses are surrounded halfway up by a layer of dark pigment-cells. The nerve does not enter the bluntly pointed basal end of the common eye, but on one side of the apodeme. The structure here described is exactly that found according to Milne Edwards, in certain Crustacea. In specimens *just attached*, in which no absorption has taken place, two long muscles with transverse striae may be found attached to the knobbed tips of the two middle arms of the two ●UU●, and running up to the antero-dorsa surface of the carapace, where they are attached; other muscles (without transverse striae) are attached round the bases, on both sides of both forks. The action of these muscles would inevitably move the eyes, but I suspect that their function may be to draw up the narrow deeply folded, sternal surface, and thus cause the retraction of the great prehensile antennae within the carapace.

[8] Rev. R. L. King. *Annual Report of R. Institution of Cornwall*, 1848, p. 55.

Mouth. This is seated in exactly the same position as in the mature cirripede, on a slight prominence, fronting the thoracic limbs, and so far within the carapace, that it was obviously quite unfitted for the seizure of prey; and it was equally obvious, that the limbs were natatory, and incapable of carrying food to the mouth. / This enigma was at once explained by an examination of the mouth, which was found to be in a rudimentary condition and absolutely closed, so that there would be no use in prey being seized. Underneath this slightly prominent and closed mouth, I found all the masticatory organs of a cirripede, in an immature condition. The state of the mouth will be at once understood, if we suppose very fluid matter to be poured over the protuberant mouth of a cirripede, so as to run a little way down, in the shape of internal crests, between the different parts, and in the shape of a short, shrivelled, certainly closed tube, a little way (0·008 of an inch in *L. australis*) down the oesophagus. Hence, the larva in this, its last stage, cannot eat; it may be called a *locomotive Pupa*;[9] its whole organization is apparently adapted for the one great end of finding a proper site for its attachment and final metamorphosis.

Thorax and limbs. The thorax is much compressed, and consists of six segments, corresponding with the six pair of natatory legs; the anterior segments are much plainer (even the first being distinctly separated by a fold from the mouth), than the posterior segments, which is exactly the reverse of what takes place in the mature cirripede; in the latter, the first segment is confounded with the part bearing the mouth. The epimeral elements of the thorax are distinguishable; the sternal surface is very narrow, and is covered with complicated folds and ridges. The six pair of legs are all close, one behind the other, and all are alike in having a haunch or pedicel of two segments, directed forwards, bearing two arms or rami, each composed of two segments, the outer ramus / being a little longer than the inner one. On the lower segments in both rami of all the limbs, there is a single spine. In all the limbs, the obliquely truncated summit of the

---

[9] M. Dujardin has lately ('Comptes Rendus', Feb. 5, 1850, as cited in *Annals of Nat. History*, vol. v, p. 318), discovered that the 'Hypopi are Acari with eight feet, without either mouth or intestine, and which, being deprived of all means of alimentation, fix themselves at will, so as to undergo a final metamorphosis, and they become Gamasi or Uropodi.' Here, then, we have an almost exactly analogous case. M. Dujardin asks – 'Ought, therefore, the Hypopi to be called larvae, when, under that denomination, have hitherto been comprised animals capable of nourishing themselves?'

terminal segment of the inner ramus bears three very long, beautifully plumose spines: in the first pair, the summit of the outer ramus bears four, and in the five succeeding pair, six similar spines. This difference, small as it is, is interesting, as recalling the much greater difference between the first and succeeding pairs, in the first and second stage of development. The terminal segments of all the rami, bearing the long plumose spines, are directed backwards. The limbs and thorax are well furnished with striated muscles. The animal, according to Mr King, swims with great rapidity, back downwards. The limbs can be withdrawn within the carapace.

Abdomen and caudal appendages. The abdomen is small, and its structure might easily be overlooked without careful dissection of the different parts: it consists of three segments; the first can be seen to be distinct from the last thoracic segment, bearing the sixth pair of limbs, only from the fold of the epimeral element, and from its difference in shape; the second segment is very short, but quite distinct; the third is four or five times as long as the second, and bears at the end two little appendages, each consisting of two segments, the lower one with a single spine, and the upper one with three, very long, plumose spines, like those on the rami of the thoracic limbs. The abdomen contains only the rectum and two delicate muscles running into the two appendages, between the bases of which the anus is seated.

Internal viscera. Within the body, in front of the mouth, it was easy to find the stomach (with two pear-shaped caeca at the upper end), running first anteriorly, and then curving back and reaching the anus by a long rectum, difficult to be followed: it appeared, however, to me, that this stomach had more relation to the young cirripede, of which every part could now generally be traced, than to the larva, with its closed and rudimentary / mouth: the fact, however, of its being prolonged to the anus, which is in a different position in the larva and mature state, shows that the stomach serves, at least, as an excretory channel. Besides the stomach, the several muscles already alluded to, and much pulpy and oily matter, the only other internal organs consist of two long, rather thick, gut-formed masses, into the anterior ends of which the cement ducts running from the prehensile antennae could be traced. These masses are formed of irregular orange balls, about 0·001 of an inch in diameter, made up of rather large cells, so to have a grape-like appearance, held together by a transparent pale yellowish substance, but apparently not enclosed in a membrane: these masses lie rather obliquely, and approach each other at their anterior ends; they extend from above the compound eyes, to the caeca of the stomach to which they cohere, but in young specimens, they extend some way beyond the caeca, between the folds of the carapace. The two cement ducts, at the points where they enter these bodies, expand and are lost; at this point, also, the little orange-coloured masses of cells have the appearance of being broken down into a finer substance. Within the cement ducts I saw a distinct chord of rather opaque cellular matter. We shall presently see, that these gut-formed masses are the incipient ovaria.

The young cirripede within the larva. Several times I succeeded in dissecting off the integuments of the lately attached larva, and in displaying the young *Lepas australis* entire. The following description applies to the cirripede in this state; but for convenience sake, I shall occasionally refer to its condition when a little more advanced. I may premise, and the fact in itself is curious, that the bivalve-like shell of the larva, together with the compound eyes, is first moulted, and some time afterwards, the inner lining of the sack, together with the integuments of the thorax and of the natatory legs: hence, I often found specimens, which externally seemed to have perfected their metamorphoses, but which, within their / sacks, retained all the characters of the natatory larva. According to Mr King, the larva of Lepas throws off its external shell five days after becoming attached. Whilst the young Lepas is closely packed within the larva, the capitulum, as known by the five valves, about equals in length the peduncle. The peduncle occupies the anterior half of the larva; when fully stretched, it becomes narrower and slightly longer than the capitulum; the separation between the capitulum and peduncle is almost arbitrary in the mature animal, and corresponds with no particular line in the larva. Even at this early period, the muscles of the peduncle are quite distinct. No vestige is preserved in the outer integument, of the sternal and dorsal sutures of the larval carapace; but in the corium of the peduncle, three coloured marks which occur near the eyes, and two little curled marks which occur near the acoustic orifices of the larva, are all preserved for some time after maturity. The compound eyes, as we have seen, are attached to apodemes, springing from the sternal surface of the larval carapace, and are consequently cast off with it: whilst the young cirripede is packed within the larva, the outer integument of its peduncle necessarily forms a deep transverse fold passing over the eyes and apodemes, and this, as we shall presently see, plays an important part in the future position of the animal. The antennae are not moulted with the carapace, but left cemented to the surface of attachment; their muscles are converted into sinewy fibres, the corium after a short period is absorbed, and they are then preserved in a functionless condition. No trace of the two acoustic sacks can be perceived in the corium of the young cirripede, excepting the coloured marks above alluded to.

In the young capitulum, the five valves stand some way apart from each other; they are elegant objects under the microscope; they are not calcified, but consist exclusively of chitine; they are rather thick, composed of an outer membrane lined by hexagonal prisms, / quite unlike any other membrane in the animal. These valves, which I have called *primordial valves*, resemble pretty closely in shape the valves of the mature animal; the fork of the carina, however, is indicated only by a slight constriction above the lower end. After the exuviation of the larval integuments, and when calcification commences, the first layer of shell is deposited under, and then round these primordial valves. The latter, in well preserved old specimens, may often be detected on the umbones of the scuta, terga, and carina, but not on the umbones of any other valves.

The *mouth* seems one of the earliest parts developed: in the youngest larva dissected, I could make out at least points corresponding with each organ; and, at the period when the young cirripede could be dissected out of its larval envelopes, their general details were quite plain. The labrum, however, had not become bullate. The mouth, as we have seen, is formed under the rudimentary mouth of the larva, and at the same relative spot occupied by the probosciformed mouth of the larva in the second stage. Thus far, in the young cirripede and larva, there has been no great change in the relative positions of the parts: the rudimentary eyes, however, of the former are developed posteriorly to (or above, as applied to a cirripede), the cast-off compound eyes of the larva; but the position of the mouth, of the antennae, and of the several coloured marks in the corium, prove to demonstration, the correspondence in both of part to part. The case is rather different with what follows.

The *cirri* are developed at first of considerable length, so that the young animal may soon provide itself with food; in *Lepas australis* they are of great length, the sixth pair consisting of seventeen or eighteen obscure segments. The extreme tips of the twenty-four rami of the six pair of cirri, are formed within the twenty-four, corresponding, little, bi-segmental rami of the six pair of natatory legs; but as the cirri are many times longer than these legs, they occupy in a bundle the / whole thorax of the larva; no part whatever of the thorax of the cirripede is formed within the thorax of the larva, but (together with the pedicels of the anterior cirri) within the cephalic cavity. As a consequence of this, the longitudinal axis of the thorax of the young cirripede lies almost transversely to the longitudinal axis of the larva; and the cirripede, from this transverse position of its thorax, comes to be, as it were, internally, almost cut in twain, and the sack thus produced. As soon as the young cirripede is free and can move itself, the cirri are curled up, and the thorax is advanced towards the orifice of the capitulum, its longitudinal axis resuming the position of approximate parellelism to the longitudinal axis of the whole body, which it had in the larval condition. The reader will, perhaps, understand what I mean, if he will look at the mature cirripede (figured in Pl. IX, fig. 4). In this, he will see that the body or thorax is united to the peduncle only by a small part below the mouth; on the other hand, if he imagines the whole bottom of the body (as high up as the letter h) united and blended into the peduncle, he will see the state in which these parts exist in the larva. Now, let him greatly shorten the cirri, so as to resemble the natatory legs of the larva, and then imagine a young cirripede, with cirri *of full length*, formed within the old one, he will see that the new thorax supporting the cirri will have to be developed in an almost transverse position – the animal consequently being internally almost separated into twain.

Of the internal organs, whilst the cirripede is still within the larva, I have already mentioned the stomach with its pair of caeca: from the retracted position of the thorax and rudimentary abdomen, and consequently of the anus, compared with these parts in the larva, the alimentary canal is not above half its former length. There is, as yet, no trace of the filaments supposed by some to act as branchiae, at the base of the first pair of cirri. Nor could I perceive a trace of the testes or / vesiculae seminales: the penis is represented by a minute, apparently imperforate projection. I have already briefly described the pair of large, gut-formed bodies in the larva, into the anterior ends of which the cement-ducts ran, and evidently derived their slightly opaque, cellular contents. At a very early age, before the young cirripede can be distinctly made out, the posterior ends of these gut-formed bodies are absorbed, so as not to pass beyond the caeca of the stomach. When the young cirripede is plainly developed within the larva, these bodies in a relatively reduced condition are still distinct near the caeca, and at the opposite or anterior end (i.e. lower, in the position in which cirripedes are usually figured), they have branched out into a sheet of delicate inosculating tubes; these could be traced by every stage, until, in the young perfected cirripede, they filled the peduncle as ordinary ovarian tubes. In the larva, the two gut-formed bodies or incipient ovaria keep of equal thickness from one to the other end, but in the mature cirripede, the ovarian tubes in the peduncle and the small, glandular, grape-like masses, near the stomach-caeca, are connected only by a delicate tube; this I failed in tracing in specimens in the very immature condition of those now under description.

The larva fixes itself with its sternal surface parallel and close to the surface of attachment, and the antennae become cemented to it: if the cirripede, after its metamorphosis, had remained in this position, the cirri could not have been exserted, or only against the surface of attachment; but there is a special provision, that the young cirripede shall immediately assume its proper position at right angles to the position which it held whilst within the larva, namely with its posterior end upwards. This is effected in a singular manner by the exuviation of the great compound eyes, which we have seen are fastened to the outer arms of the double ●UU●-like, sternal apodemes: these together with the eyes stretch transversely across, and internally far up into, / the body of the larva; and, as the whole has to be rejected or moulted, the membrane of the peduncle of the young cirripede has necessarily to be formed with a wide and deep inward fold, extending transversely across it; this when stretched open, after the exuviation of the larval carapace and apodemes, necessarily causes the sternal side of the peduncle to be longer than the dorsal, and, as a consequence, gives to the young cirripede its normal position, at right angles to that of the larva when first attached.

I may here state, that I have examined the larvae in this the final or perfect stage in four species of Lepas, in *Conchoderma virgata*, *Ibla quadrivalvis*, and, though rather less minutely, in *Balanus balanoides*, and I find all essential points of organization similar. With the exception of diversities in the proportional sizes of the different parts, and in the patterns on the carapace, the differences, even in the arrangement of the spines on the limbs and antennae, are less than I should have anticipated.

I have in this abstract treated the metamorphoses at greater length than I should otherwise have done, on account of the great importance of arriving at a correct homological interpretation of the different parts of the mature animal. In Crustacea, according to the ordinary view, there are twenty-one segments; of these I can recognize in the cirripede, on evidence as good as can generally be obtained, all with the exception of the four terminal abdominal segments; these do not occur in any species known to me, in any stage of its development. If that part of the larva in front of the mouth, bearing the eyes, the prehensile antennae, and in an earlier stage two pair of antennae, be formed, as is admitted in all other Crustacea, of three segments, then beyond a doubt, from the absolute correspondence of every part, and even every coloured mark, the peduncle of the Lepadidae is likewise thus formed. The peduncle being filled by the branching ovarian tubes is no objection to this view, for I am / informed on the high authority of Mr J. D. Dana,[10] that this is the case with the cephalo-thorax in some true crustaceans, for instance, in Sapphirina. To proceed, the mouth, formed of mandibles, maxillae, and outer maxillae, correspond with the fourth, fifth, and sixth segments of the archetype crustacean. Posteriorly to the mouth, we come, in the larva, to a rather wide interspace without any apparent articulation or organ, and then to the thorax, formed of six segments, bearing the six pair of limbs, of which the first pair differs slightly from the others. The thorax is succeeded by three small segments, differently shaped, with the posterior one alone bearing appendages; these segments, I cannot doubt, from their appearance alone, and from their apparent function of steering the body, are abdominal segments. If this latter view be correct, the thoracic segments are the six posterior ones of the normal seven segments, and there must be two segments missing between the outer

---

[10] This distinguished naturalist has given his opinion in the *American Journal of Science*, March, 1846, that 'the pedicel of Anatifa corresponds to a pair of antennae in the young'; although the peduncle or pedicel is undoubtedly thus terminated, even in mature individuals, I think it has been shown that it is the whole of the anterior part of the larva in front of the mouth, which is directly converted into the peduncle. Professor E. Forbes, in his *Lectures*, and Professor Steenstrup, in his *Untersuchungen über das vorkommen des Hermaphroditismus in der Natur*, ch. v, have considered the peduncle as a pair of fused legs. Lovén has taken, judging from a single sentence, the same view of the homologies of the external parts as I have done; in his description of *Alepas squalicola* (*Ofversigt of Kongl. Vetens.*, etc., Stockholm, 1844, pp. 192–4), he uses the following words: 'Capitis reliquae partes, ut in Lepadibus semper, in *pedunculum mutatae et involucrum*', etc.; his involucrum is the same as the capitulum of this work.

maxillae and first thoracic pair of legs, which latter on this view springs from the ninth segment. Now, in a very singular cirripede, already alluded to under the name of Proteolepas, the two missing segments are present, the mouth being actually succeeded by eight segments, and these by the three usual abdominal segments – every segment in the body being as distinct as in an annelid: hence in Proteolepas, adding the three segments for the mouth and three for the carapace, we have altogether / seventeen segments, which, as I stated, is the full number ever observed in any cirripede, the four missing ones being abdominal, and, I presume, the four terminal segments. That the cavity in which the thorax is lodged, in the larva and therefore in the mature cirripede, is simply formed by the backward production of the carapace, does not require any discussion. The valves have no homological signification.

As we have just seen that the first pair of natatory legs is borne on the ninth segment of the body, so it must be with the first pair of cirri, which consequently correspond to the outer maxillipods (the two inner pair of maxillipods or pied-machoires being here aborted) of the higher Crustacea, and hence their difference from the five posterior pair, which correspond with the five, ordinary pair of ambulatory legs in these same Crustacea. The part of the body, which I have called the prosoma, that is the protuberant, non-articulated, lower part of the thorax (Pl. IX, fig. 4*n*), is a special development, either of the ninth segment, bearing the first pair of cirri, or of the segments corresponding with the organs of the mouth. The three abdominal segments of the larva are represented in the mature cirripede, in the order containing the Lepadidae, only by a minute, triangular gusset, let in between the V-shaped tergal arches of the last thoracic segment: in this gusset, small as it is, is seated the anus, and on each side the caudal appendages, often rudimentary and sometimes absent. In another order, I may remark (including, probably, the Alcippe of Mr Hancock), the cirri, of which there are only three pair, are abdominal.

I feel much confidence, that the homologies here given are correct. The cause of their having been generally overlooked arises, I believe, from the peculiar manner, already described, in which the animal, during its last metamorphosis, is internally almost intersected: even for some little time after discovering that the larval antennae were always embedded in the centre of the surface of attachment, I did not perceive, that this was the anterior / end of the whole animal. The accompanying woodcut gives at a glance, a view of the homologies of

THE LEPADIDAE

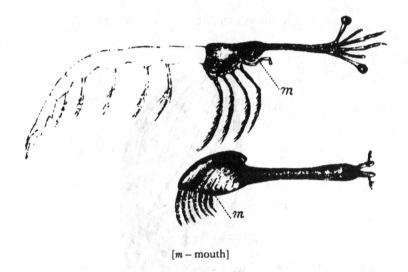

[*m* – mouth]

the external parts: the upper figure (from Milne Edwards) is a stomapod crustacean, Leucifer of Vaughan Thompson, and the abdomen, which we know becomes in cirripedes, after the metamorphosis, rudimentary, and therefore does not fairly enter into the comparison, is given only in faint lines: the lower figure is a mature Lepas, with the antennae and eyes, which are actually present in the larva, retained and supposed to have gone on growing. All that we externally see of a cirripede, whether pedunculated or sessile, is the three anterior segments of the head of a crustacean, with its anterior end permanently cemented to a surface of attachment, and with its posterior end projecting vertically from it.

Capitulum. I will now proceed to a general description of the different parts and organs in the Lepadidae. The capitulum is usually much flattened, but sometimes broadly oval in section. It is generally formed of five or more valves, connected together by very narrow or broad strips of membrane; sometimes the valves are rudimental or absent, when the whole consists of membrane. When the valves are numerous, and they occasionally exceed / a hundred in number, they are arranged in whorls, with each valve generally so placed as to cover the interval between the two valves above. Of all the valves, the scuta are the most persistent; then come the terga, and then the carina; the rostrum and

NOMENCLATURE OF THE VALVES

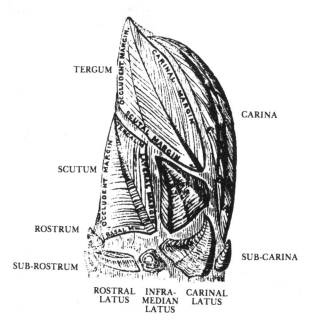

Fig. 1   Capitulum

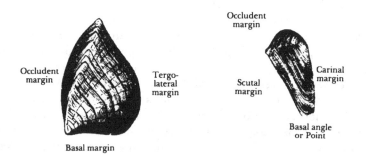

Fig. 2   Scutum of Lepas

Fig. 3   Tergum of Lepas

latera occur only in Scalpellum and Pollicipes, and in a rudimentary condition in Lithotrya, and, perhaps, in the fossil genus Loricula. The valves are formed sometimes of chitine (as in Ibla and Alepas), but usually of shell, which varies from transparency to entire opacity. The shell is generally white, occasionally reddish or purple; exteriorly, the valves are covered by more or less persistent, generally yellow, strong membrane. The scuta and terga are always considerably larger than the other valves: in the different genera the valves differ so much in shape that little can be predicated of them in common; even the direction of their lines of growth differs – thus, in Lepas and some allied genera, the chief growth of the scuta and of the carina is upwards, whereas in Pollicipes and Lithotrya, it is entirely downwards; in Oxynaspis, and some species of Scalpellum, it is both upwards and downwards. Even in the same species, there is often very considerable variation in the exact shape of the valves, more especially of the terga. The adductor muscle is always attached to a point not far from the middle of the scuta, and it generally has a pit for its attachment. In several genera, namely, Paecilasma, Dichelaspis, Conchoderma, and Alepas, the scuta show a tendency to be bilobed or trilobed. The valves are placed either at some distance from each other, or close together; but their growing margins very rarely overlap each other, though this is sometimes the case with their upper, free, tile-like apices; in a few species the scuta and terga are articulated together, or united by a fold. The membrane connecting the valves, where they do not touch each other, is like that forming the peduncle, and is sometimes brilliantly coloured crimson-red; generally, it appears blueish-grey, from the corium being seen through. Small pointed / spines, connected with the underlying corium by tubuli, are not infrequently articulated on this membrane: the tubuli, however, are often present where there are no spines. To allow of the growth of the capitulum, the membrane between the valves splits at each period of exuviation, when a new strip of membrane is formed beneath, connected on each side with a fresh layer of shell – the old and outer slips of membrane disintegrating and disappearing: when there are many valves, the line of splitting is singularly complicated. This membrane consists of chitine,[11] and is

---

[11] Chitine is confined to the Articulata. It was Dr C. Schmidt (Contributions, etc., being a Physiologico-Chemical investigation: in Taylor's *Scientific Memoirs*, vol. v), who discovered that the membrane connecting the valves and forming the peduncle, and the tissues of the internal animal, were composed of this substance. But Dr Schmidt says that the valves in Lepas are composed of 3·09 of albuminates,

composed of numerous fine laminae. After the valves have been placed in acid, a residue, very different in bulk in different genera, is left, also composed of successive laminae of chitine. It appears to me that each single lamina of calcified chitine, composing the shell, must once have been continuous with a non-calcified lamina in the membrane connecting the several valves: at the line where this change in calcification supervenes, the chitine generally assumes some / colour, and becomes much harder and more persistent; and as the whole valve is formed of component laminae thus edged (the once continuous laminae of non-calcified chitine connecting the valves, having disintegrated and disappeared) the surfaces of the valves are generally left covered by a persistent membrane, constituted of these edgings: this membrane has been called the epidermis. In some genera, as in Lepas, this so-called epidermis is seldom preserved, excepting on the last zone of growth: in Scalpellum and Pollicipes it usually covers the whole valves. It appears to me that the laminae of chitine, and of calcified chitine composing the valves, are both formed not by secretion, but by the metamorphosis of an outer layer of corium into these substances.

Within the capitulum is the sack, which, together with the upper internal part of the peduncle, encloses the animal's body. The sack is lined by a most delicate membrane of chitine, under which there is a double layer of corium; this double layer is united together by short, strong, transverse bundles of fibres, branched at both ends:[12] in some

---

and 96·81 of incombustible residue; I cannot but think that the existence of the albuminates is an error caused by Dr Schmidt's belief that the Cirripedia were intermediate between Crustacea and Mollusca, in the shells of which latter, the animal basis consists of albuminates. For after placing the valves of Lepas and Pollicipes in cold acid, I found that the membrane left could *not be dissolved* in boiling caustic potash, but could, though slowly (and without change of colour), in boiling muriatic acid; and these are the main diagnostic characters of chitine, compared with albuminous substances. I may add, that Schmidt was also induced to consider the shells of Cirripedia as having the same nature with those of Mollusca, from finding that in the above 96·81 of incombustible matter, 99·3 consisted of carbonate and only 0·7 of phosphate of lime; but Dr Schmidt's own analyses prove how extremely variable the proportions of these salts are in the Crustacea, as the following instance shows:

	Lobster	*Squilla*
Phosphate of lime	12·06	47·52
Carbonate of lime	87·94	52·48

And, therefore, it is not very surprising that Cirripedia should have still less phosphate of lime in their shells, than has a lobster compared with a squilla.

[12] I am much indebted to Mr Inman of Liverpool for having kindly sent me excellent specimens illustrating this structure.

genera, the ovarian tubes extend between these two layers. We have seen, under the head of the metamorphoses, that the delicate tunic lining the sack is simply a duplicature of the thick membrane and valves forming the capitulum, the whole being the posterior portion of the carapace of the larva slightly modified.

Peduncle. Its length varies greatly in different species, and even in the same species, according to the situation occupied by the individual; its lower end is sometimes pointed, but generally only a little narrower than the upper end. In outline, the peduncle is usually flattened, but sometimes quite cylindrical. It is composed of very strong, generally thick, transparent membrane, rarely coloured reddish, and often penetrated by numerous tubuli. The underlying corium is sometimes / coloured in longitudinal bands. At each period of growth a new and larger integument is formed under the old one, which gradually disintegrates and disappears; the extreme lower point is often deserted by the corium, and ceases to grow, whilst the whole upper part still continues increasing in diameter: in length the chief addition is made (as is clearly seen in those genera having calcified scales), round the upper margin, at the base of the capitulum. The surface of the membrane is either naked or superficially clothed with minute, pointed, articulated spines, or it is penetrated by calcified scales or styles (in Ibla alone formed of chitine), which pass through it to the corium, and are added to at their bases, like the valves, at each period of growth. In Lithotrya alone the scales of the peduncle are moulted together with the connecting membrane. These scales on the peduncle are generally placed symmetrically in whorls, with each scale corresponding with the junctions of two scales, both above and below. Except in *Scalpellum ornatum* and the fossil *Loricula pulchella*, they are very small compared with the valves of the capitulum. When the scales are symmetrical, new ones are first formed only round the summit of the peduncle, and only those in the few uppermost whorls continue to grow or to be added to at their bases; afterwards membrane is deposited under them. The shelly matter of the scales resembles that of the valves, and the manner of growth is the same; tubuli generally run to and through them from the corium. From the continued enlargement of the membrane of the peduncle, the scales come to stand, in the lower portion, some way apart. In Ibla, new horny styles are formed indifferently in all parts of the peduncle. In some species of Pollicipes, the calcareous styles are not symmetrical or symmetrically arranged; and besides those first formed round the top of the peduncle, there are other and larger ones formed near its base. Lastly, in Lithotrya we have a row of calcareous discs or an irregular, basal cup, formed in the same / manner as the valves of the capitulum: in this genus alone (as already stated), the calcified scales are moulted, and here alone their edges are serrated.

The peduncle is lined within by three layers of muscles, longitudinal, transverse, and oblique, all destitute of the transverse striae, characteristic of voluntary muscles; they run from the bottom of the peduncle to the base of the capitulum, as in Lepas, or halfway up it, as in Conchoderma; in Alepas alone they surround the whole capitulum up to its summit. In Lithotrya there are two little, fan-like, transverse muscles (involuntary), extending from the basal points of the terga to a central line on the underside of the carina. The gentle swaying to and fro movements, and the great power of longitudinal contraction – movements apparently common, as I infer from facts communicated to me by Mr Peach, to all the Pedunculata – are produced by these muscles. The interior of the peduncle is filled up with a great mass of branching ovarian tubes; but in Ibla and Lithotrya, the upper part of the peduncle is occupied by the animal's body.

Means of attachment. If the peduncle be very carefully removed (Pl. IX, fig. 7 and Pl. I, fig. 6*b*), from the surface of attachment, quite close to the end, but not at the actual apex, the larval prehensile antennae can always be found: these have been sufficiently described for our present purpose under the head of the Metamorphoses; but I may add, that the diagnostic differences between them in the several genera are briefly given, for a special purpose, in a discussion on the sexes of Scalpellum at the end of that genus. We have seen in the larva, that the cement ducts, with their opaque cellular contents, can be traced from within the discs of the antennae to the anterior or lower ends of the two gut-formed bodies, which it can be demonstrated are the incipient ovaria.

In mature cirripedes these ducts can be followed, in a slightly sinuous course, along the muscles on each side / within the peduncle, till they expand into two small organs, which I have called cement glands. These glands are found with great difficulty, except in *Conchoderma aurita*, where they are placed on each side under the inner layer of corium, at the bottom of the sack, so as to be just above the top of the peduncle; they resemble in shape a retort (Pl. IX, fig. 3). In *Pollicipes mitella* and *polymerus* they lie halfway down the peduncle, close together, and apparently enclosed within a common membrane; in these two species the broad end of the gland is bent towards the neck of the retort. In Scalpellum the position is the same, but the shape is more globular. In Ibla the structure is more simple, namely, a tube slightly enlarged, running downwards, bent a little upwards, and then resuming its former downward course, the lower portion forming the duct. The gland contains a strongly coherent, pulpy, opaque, cellular mass, like that in the cement ducts; but in some instances, presently to be mentioned, this cellular mass becomes converted within either the ducts or gland, or within both, into transparent, yellow, tough cement. Generally in Conchoderma, Pollicipes, and Scalpellum, two ovarian tubes, but in one specimen of *Conchoderma aurita*, three tubes, and in Ibla one tube could be seen running into or forming the gland; of the nature of the tubes there could not be the least doubt, for at a little distance from the glands they gave out branches (Pl. IX, fig. 3), containing ova in every state of development. In some specimens as in that figured of *Conchoderma aurita*, the ovarian tube on one side of the gland is larger than on the other, and has rather the appearance of being deeply embedded in the gland than of forming it; but, in other specimens, the two ovarian tubes first formed a little pouch, into which their cellular contents could be clearly seen to enter; and then this pouch expanded into the gland; thus quite removing a doubt which I had sometimes felt, whether the ovarian tube was not simply attached to or embedded in the gland, without any further connection. By dissection / the multiple external coats of the gland and ovarian tubes could be seen to be continuous. The cellular contents of the tubes passed into the more opaque cellular contents of the gland, by a layer of transparent, pulpy, pale, yellowish substance. There appeared in several instances to be a relation, between the state of fulness and condition of the contents of the gland, and of the immediately adjoining portions of the ovarian tubes. In one specimen of *Pollicipes mitella* it was clear that the altered, tough, yellow, transparent, non-cellular contents of the two glands and ducts, had actually invaded for some little distance, the two ovarian tubes which ran into them, thus showing the continuity of the whole. From these facts I conclude, without hesitation, that the gland itself is a part of an ovarian tube specially modified; and further, that the cellular matter, which in the ovarian tubes serves for the development of the ova, is, by the special action of the walls of the gland, changed into the opaquer cellular matter in the ducts, and this again subsequently into that tissue or substance, which cements the cirripede to its surface of attachment.

As the individuals grow and increase in size, so do the glands and cement-ducts; but it seems often to happen, that when a specimen is immovably attached, the cementing apparatus ceases to act, and the cellular contents of the duct become converted into a thread of transparent tough cement; the investing membrane, also, of the ducts, in Conchoderma sometimes becomes hard and mamillated. I have already alluded to the case of a Pollicipes, in which both glands and ducts, and even a small portion of the two adjoining ovarian tubes, had become thus filled up. As in sessile cirripedes, at every fresh period of growth a new cement gland is formed, it has occurred to me, that possibly in Pollicipes something similar may take place. In sessile cirripedes, the old cement-glands are all preserved in a functionless condition, adhering to the membranous or calcareous basis, each new larger one attached to that last formed, and / each giving out cement-ducts, which, bifurcating in the most complicated manner, pass outside the shell and thus attach it to some foreign body.

The cement, removed from the outside of a cirripede, consists of a thin layer of very tough, bright-brown, transparent, laminated substance, exhibiting no structure under the highest powers, or at most a very fine dotted appearance, like a mezzotinto drawing. It is of the nature of chitine; but boiling caustic potash has rather more effect on it than on true chitine; and I think boiling nitric acid rather less effect. In one single instance, namely, in Coronula, the cement comes out of the four orifices of the two bifurcating ducts, in the shape of distinct cells, which, between the whale's skin and the basal membrane, arrange themselves so as to make a circular, continuous slip of cement; then the cells blend together, and are converted into transparent, structureless cement. Cementing tissue or membrane would, perhaps, have been a more correct title than cement; but, in ordinary cases, its appearance is so little like that of an organized tissue, that I have for this reason, and for brevity-sake, preferred the simple term of cement.

In the larva the cement always escapes through the prehensile antennae; and it thus continues to do throughout life in most or all of the species of Lepas, Conchoderma, Dichelaspis and Ibla. In the first two of these genera, the cement escapes from the borders of the lower side of the disc or penultimate segment of the antennae, and can be there seen radiating out like spokes, which at their ends divide into finer and finer branches, till a uniform sheet of cement is formed, fastening the antennae and the adjoining part of the peduncle down to the surface of attachment. In *Dichelaspis Warwickii* and *Scalpellum*

Peronii, the cement, or part at least, comes out of the ultimate segment of the antennae, in the shape of one tube, within another tube of considerable diameter and length. In *Scalpellum vulgare*, and probably in some of the other / species, which live attached to corallines, the cement soon ceases to debouch from the antennae, but instead, bursts through a row of orifices on the rostral margin of the peduncle (Pl. IX, fig. 7), by which means this margin is symmetrically fastened down to the delicate, horny branches of the zoophyte. In Pollicipes, the two cement-ducts, either together or separately (Pl. IX, fig. 2, 2a'), wind about the bottom of the peduncle in the most tortuous course, at each bend pouring out cement through a hole in the membrane of the peduncle. In Ibla the lower part of the peduncle is internally filled by cement, and thus rendered rigid. In *Lepas fascicularis* a vesicular ball of cement surrounding the peduncle is thus formed (Pl. I, fig. 6), and serves as a float! All these curious, special adaptations are described under the respective genera. How the cement forces its way through the antennae, and often through apertures in the thick membrane of the peduncle, I do not understand. I do not believe, though some appearances favoured the notion, that the duct itself debouches and divides, at least this is not the case in Corunula, but only that the internal chord of cellular matter thus acts and spreads itself out; nor do I understand how, when the antennae and immediately adjoining parts are once cemented down, any more cement can escape; yet this must take place, as may be inferred from the breadth of the cemented, terminal portion of the peduncle in Lepas and Conchoderma; and from the often active condition in old individuals of the cementing organs.

I have entered on this subject at some length (and I wish I had space for more illustrations), from its offering, perhaps, the most curious point in the Natural History of the Cirripedia. It is the one chief character of the subclass. I am well aware how extremely improbable it must appear, that part of an ovarian tube should be converted into a gland, in which cellular matter is modified, so that instead of aiding in the development of new beings, it forms itself into a tissue or substance, which leaves the / body[13] in order to fasten it to a foreign

[13] The protrusion of the egg-bearing pouches in Cyclops and its kindred genera, outside the body, offers a feeble analogy with what takes place in cirripedes. Professor Allman (*Annals of Natural History*, vol. xx, p. 7), who has attended to the subject, says that the external egg-bearing pouches are 'a portion of the membrane of the true ovaries:' if the membrane of these pouches had been specially made adhesive, the analogy would have been closer.

support. But on no other view can the structure, clearly seen by me both in the mature cirripede and in the larva, be explained, and I feel no hesitation in advancing it. I may here venture to quote the substance of a remark made by Professor Owen, when I communicated to him the foregoing facts, namely, that there was a new problem to solve – new work to perform – to attach permanently a crustacean to a foreign body; and that hence no one could, *a priori*, tell by what singular and novel means this would be effected.

Filamentary appendages. These have generally been considered to act as branchiae; they occur at the bases of the first pair of cirri in Lepas, Alepas, Conchoderma, and in three species of Pollicipes: in Conchoderma there are similar appendages attached to the pedicels of the cirri (Pl. IX, fig. 4*g–k*); and in the above three species of Pollicipes there is a double row of them on the prosoma: their numbers differ in different species (in some there being none) of the same genus, and even in different individuals of the same species; they are entirely absent in the majority of the genera. These facts would indicate that they are not of high functional importance; and they seem so generally occupied by testes (Pl. IV, fig. 5), that I suspect their function is quite as much to give room for the development of these glands, as to serve for respiratory purposes. With the exception of the four above-named genera, the mere surface of the body and of the sack must be sufficient for respiration: in *Conchoderma aurita* the two great expansions of surface, afforded by the folded, tubular, ear-like projections, aid, as I believe, towards this end. /

The shape of the body varies, owing to the greater or less development of the lower part of the prosoma, the greater or less distance of the first from the second pair of cirri, and of the mouth from the adductor scutorum muscle (Pl. IX, fig. 4, and Pl. IV, 8*a*′). In all the genera, the body is much flattened. I may here mention a few particulars about the muscular system. One of the largest muscular masses is formed by the adductor scutorum, and by the muscles which surround in a double layer (the fasciae being oblique to each other) the whole of the upper part of the prosoma. From under the adductor, a pair of delicate muscles runs to the basal edge of the labrum, so as to retract the whole mouth, and two other pair to the integument between the mouth and the adductor, so as to fold it: again, there are other delicate muscles in some (for instance in *Lepas Hillii*) if not in all the Lepadidae, crossing each other in the most singular loops, and serving apparently to fold the membrane between the occludent edges of the scuta. Within the prosoma there is a strong adductor muscle, running straight from side to side, for the purpose, as it appears, of flattening the body. The thorax, on the dorsal and ventral surfaces, is well furnished with straight and oblique muscles (without striae), which straighten and curl up this part of the body. The muscles running into the pedicels of the cirri, cross each other on the ventral surface of the thorax; the muscles within the rami are attached to the upper segments of the pedicels. Finally, I may remark that the whole of the body and the cirri are capable of many diversified movements.

Mouth. This is prominent, and almost probosciformed (Pl. IX, fig. 4*b*), and in the abnormal Anelasma (Pl. IV, fig. 2*d*), quite probisciformed – such, also, was its character in the larval condition. In outline, it is either subtriangular, or oval with the longer axis transverse; the whole is capable, as well as the separate organs, of considerable movement, as I have seen in living sessile cirripedes. It is composed (Pl. V, fig. 2) of a labrum, / swollen or bullate, often to such an extent as to equal in its longitudinal axis the rest of the mouth; of palpi soldered to the labrum; of mandibles, maxillae, and outer maxillae, the latter serving as a lower lip. These organs have only their upper segments free, but there are traces, clearly seen in the mandibles (Pl. X, fig. 1*a*, *b*), of their being formed of three segments. The two lower segments are laterally united, and open into each other, the prominence of the mouth being thus caused: this condition appears to me curious, and is, to a certain limited extent, intermediate between those articulated animals which have their trophi soldered into a proboscis, and those furnished with entirely free masticatory or prehensile organs. The palpi adhere to the corners of the labrum; and I call them palpi only from seeing that they spring laterally from above the upper articulation of the mandibles. The prominence of the mouth, measured from the basal fold by which the whole is separated from the body, is much greater on the half formed by the labrum and mandibles, than on the other half facing the cirri. The trophi surround a cavity – the supra-oesophageal cavity – in the middle of which, between the mandibles is seated the orifice of the oesophagus. The oesophagus is surrounded by long, fine, muscular fasciae, radiating in all directions, opposing the constrictor muscles, and is capable of violent swallowing movements – constriction after constriction being seen to run down its whole course: there are also some fine muscles attached to the membrane forming the supra-oesophageal cavity. The trophi serve merely for the prehension of prey, and not for mastication.

The *labrum*, as stated, is always bullate or swollen; and sometimes the upper exterior part forms, as in Ibla (Pl. IV, fig. 8a, c), and Dichelaspis, an overhanging blunt point. The object, I suspect, of this bullate form is to give, in the upper part, attachment to longer muscles running to the lateral surfaces of the mandibles, and lower down to the oesophagus. The crest close over the / supra-oesophageal cavity, is generally furnished with small, often bead-like teeth. The *Palpi* are small, their apices never actually touching each other; they are more or less blunt, not differing much in shape in the different genera (Pl. X, figs 6–8), and clothed with spines. They are not capable of movement; their function seems to be to prevent prey, brought by the cirri, escaping over the labrum; I infer this from finding in *Anelasma* and in the male of Ibla, which have the cirri functionless, that the palpi are rudimentary.

The *mandibles* (Pl. X, figs 1–5) have from two to ten strong teeth in a single row; where the number exceeds five, several of the teeth are small; the inferior angle is generally pectinated with fine spines; in Lithotrya (fig. 2), the interspaces between the teeth are also pectinated. in the same individual there is not infrequently one tooth, more or less, on opposite sides of the mouth. Internally, the mandibles are furnished on their outer and inner sides with several ligamentous apodemes in Lithotrya roughened with points (Pl. X, fig. 2), for the attachment of the muscles; of these (fig. 1), there is a chief depressor and elevator, attached at their lower ends to near the basal fold of the mouth, and a lateral muscle, attached to the broad basal end of the palpi, and serving, apparently, to oppose the edge of mandible to mandible. The *maxillae* in the different genera (Pl. X, figs 9 to 15) differ considerably in outline; they are generally about half the size of the mandibles; at the upper corner, there are always two or three spines larger than the others, and often separated from them by a notch; the rest of the spinose edge is straight, or irregular, or step-formed, or with the lowest part projecting, or with one or two narrow prominences bearing fine spines. All these spines, quite differently from the teeth of the mandibles, are articulated on the edge of the organ, and stand in a double row. At a point corresponding with the upper articulation of the mandibles, a long, thin, narrow, rigid apodeme, projects inwards (fig. 10), and running / down nearly parallel to the thin, outer, flexible membrane of the mouth, is attached to the corium, and thus serves as a support to the whole organ. This apodeme is embedded in muscles (Pl. X, fig. 10); there are other large muscles attached to the inner side of the organ, and again others running laterally towards the mandibles. The apodeme, of course, is moulted with the integuments of the mouth. The *outer maxillae* (Pl. X, figs 16, 17) serve as a lower lip; they are thicker than the other trophi; they have their inner surfaces clothed with spines, sometimes divided into an upper and lower group, and occasionally separated by a deep notch: there are often long bristles outside. They are furnished with at least two muscles; in sessile cirripedes I have seen that they are capable of a rapid to and fro movement, and I have no doubt that their function is to brush any small creature, caught by the cirri, towards the maxillae, which are well adapted to aid in securing the prey, and to hand it over to the mandibles, by them to be forced down the oesophagus. On the exterior face of the outer maxillae, above a trace of an upper articulation, either two small orifices or two large tubular projections can always be discovered; and these, as will presently be mentioned, I believe to be olfactory organs.

Cirri. The five posterior pair are seated close to each other and equidistant; the first pair is generally seated at a little distance, and sometimes at a considerable distance from the second pair. The first pair is the shortest; the others, proceeding backwards, increase gradually in length. The rami of each pair are either equal in length or slightly unequal: those of the first pair are oftenest unequal. The number of segments in the posterior cirri is sometimes very great; in one species of Alepas, there were above sixty segments in one ramus, the other ramus being in this unique case (Pl. X, fig. 28) small and rudimentary. The pedicels consist of two segments, a lower, longer, and upper short one (fig. 18*c*, *d*). In the usual arrangement of the spines on the segments of the three posterior pair of cirri, there are (figs 26, 27) from three to six pair of / long spines on the anterior face, with generally some minute spines (occasionally forming a tuft) intermediate between them: on the dorsal surface, in the uppermost part of each segment, there is a tuft of short spines generally mingled with some longer, finer ones: on the inner side of each segment, on the upper rim, there are generally a few extremely minute and short spines. From the increase of these latter and of the intermediate spines, the antero–lateral faces of the segments of the first cirrus, and of the lower segments of the anterior ramus of the second cirrus (Pl. X, fig. 25), are almost always thickly paved with brush-like masses of spines. The lower segments of the anterior ramus of the third cirrus is generally, though not always, thus paved: these paved segments are much broader than the others. The posterior rami of the second and third cirri are often in some slight degree paved, though in other cases they resemble the three posterior pair of cirri. The two segments of the pedicels have bristles on their anterior faces, essentially arranged on the same plan as on the segments of the rami: the bristles are generally not so symmetrically arranged on the pedicels of the second and third cirri, as on the three posterior pair. There are some exceptions to the foregoing general rules: in the posterior cirri of *Alepas cornuta*, there is only one pair of long spines to each segment (fig. 28); in *Dichelaspis Lowei*, there are eight pair; in *Lepas fascicularis*, in old specimens, the segments are paved with a triangular brush of spines; the upper segments in *Paecilasma eburnea* support small oblong brushes; and, lastly, in *Paecilasma fissa* (fig. 29), and *crassa*, the spines form a single circle round each segment, interrupted on the two sides. These spines are often doubly serrated or plumose: many of them on the protuberant segments of the first three pair of cirri, are sometimes coarsely and doubly pectinated.

Caudal appendages. These are present (Pl. X, figs 18 to 24) seated on each side of the anus, in all the genera, except in Conchoderma, Anelasma, and *Scalpellum / villosum*; they consist of a very small single segment, destitute of spines in Lepas, and spinose in Paecilasma, Dichelaspis, Oxynaspis, Scalpellum, and some species of Pollicipes; they consist of several segments in Alepas, Ibla, Lithotrya, and in some species of Pollicipes. In the latter genus, some species have their caudal appendages multiarticulate, though so obscurely articulated, that the passage (fig. 22) from several to one segment is seen to be easily effected. When the appendage consists of many articulations, it is generally about as long as the pedicel of the sixth cirrus; but in *Ibla quadrivalvis*, it is four times as long. The segments are narrow, slightly flattened, much tapering; each (fig. 24) is surmounted by a ring of short spines, which are generally longest on the apex of the terminal segment. I could never trace muscles into these appendages.

Alimentary canal. The oesophagus is of considerable length: it is formed of strong, transparent, much folded membrane, continuous with the outer integuments, and moulted with them: it is surrounded by corium, and as already stated, by numerous muscles: at its lower end it expands into a bell, with the edges reflexed, and sometimes sinuous: this bell lies within the stomach, and keeps the upper broad end expanded. According to the less or greater distance of the mouth from the adductor muscle, the oesophagus runs in a more or less parallel course to the abdominal surface between the first and succeeding pairs of cirri, and enters the stomach more or less obliquely. In Ibla alone, it passes exteriorly to, and over the adductor scutorum muscle. The stomach lies in a much curved, almost doubled course; it is often a little constricted where most bent; it is broadest at the upper end, and here, in Lepas and Conchoderma, there are some deep branching caeca; in the latter of these two genera, the whole surface is, in addition, pitted in transverse lines. The stomach is coated by small, opaque, pulpy, slightly arborescent glands, believed to be hepatic; these are arranged in longitudinal lines, in all / the genera, except in Alepas, in which they are transverse and reticulated: the whole stomach is thus coated. There is, also, a coating of excessively delicate, longitudinal and transverse muscles without striae. The rectum varies in length, extending inwards from the anus to between the bases of the second and fifth pair of cirri: it is narrow, and formed of much folded transparent membrane, resembling the oesophagus, continuous with the outer integuments, with which it is periodically moulted. The anus is a small longitudinal slit, in the triangular piece of membrane representing the abdomen, let in between the last thoracic tergal arches, as already mentioned under the head of the Metamorphoses; it lies almost between the caudal appendages, and opens on the dorsal surface. Within the stomach, there can generally be plainly seen, in accordance with the period of digestion when the specimen was taken, a thin, yet strong, perfectly transparent epithelial membrane, not exhibiting under the highest power of the microscope any structure: it enters the branching caeca, and extends from the edge of the bell of the oesophagus to the commencement of the closed rectum, and consequently terminates in a point: it consists of chitine, like the outer integuments of the animal, and by placing the whole body in caustic potash, I have dissolved the outer coats of the stomach, and seen the bag open at its upper end, perfectly preserved, floating in the middle of the body, and full of the debris of the food. In most of the specimens which I have examined, preserved in spirits of wine, this epithelial lining was some little way distant and separate from the coats of the stomach; and hence was thought by M. Martin St Ange to be a distinct organ, like the closed tube in certain annelids. Occasionally, I have seen one imperfect epithelial bag or tube within another and later formed one. Digestion seems to go on at the same rate throughout the whole length of the stomach; if there be any difference, the least digested portions lie in the lower and narrower part. The prey, consisting generally of crustacea, infusoria, / minute spiral univalves, and often of the larvae of cirripedes, is not triturated: when the nutritious juices have been absorbed, the rejectamenta are cast out through the anus, all kept together in the epithelial bag, which is excluded like a model of the whole stomach, with the exception of that part coated by the bell of the oesophagus. I have sometimes thought that the bag was formed so strong, for the sake of thus carrying out the excrement entire, so as not to befoul the sack. I believe Lepas can throw up food by its oesophagus; at least, I found in one case, many *half-digested* small crustaceans in the sack, and others of the same kind in the stomach.

Circulatory system. I can add hardly anything to what little has been given by M. Martin St Ange: like others, I have failed, as yet, in discovering a heart. The whole body is permeated by channels, which have not any proper coat: there is one main channel along the ventral surface of the thorax, dividing and surrounding the mouth, and giving out branches which enter the inner of the two channels in each cirrus: as Burmeister has shown, there are also two channels in the penis. There are two dorso-lateral channels in the prosoma, which are in direct connection with the great main channel, running down the rostral (i.e. ventral) side of the peduncle. This latter main channel branches out in the lower part, and transmits the fluid through the ovarian tubes, whence, I believe, it flows upwards and round the sack, re-entering the body near the sides of the adductor scutorum muscle. The main rostral channel (or artery?) in the uppermost part of the peduncle, has a depending curtain, which, I think, must act as a valve, so as to prevent the circulating fluid regurgitating into the animal's body during the contractions of the peduncle.

Nervous system and organs of sense. In most of the genera, there are six *main* ganglia, namely, the supra-oesophageal, and five thoracic ganglia; but in *Pollicipes mitella* there are only four thoracic ganglia. Of these, the first thoracic or infra-oesophageal ganglion is considerably / the largest and most massive; it is squarish, or oval, or heart-shaped; it presents no trace of being formed by the union of two lateral ganglia.

Two great nerves spring from its underside A (represented in the woodcut on page 41, by dotted lines), and run straight down among the viscera in the prosoma: these nerves are about as large as those forming the collar and those running to the second ganglion; hence, six great nerves meet here, two in front, two behind, and two on the underside. At the anterior end, over the junction with the collar chord, three equal-sized nerves rise on each side, with a fourth, smaller one, outside; these go to the trophi and to the two olfactory sacks. At the posterior end, on each side, a pair of nerves branch out rectangularly, one of which *a*, goes to the first cirrus, and there divides into two branches; of these, the upper runs up the cirrus, and the lower one downwards. The other nerve *b*, proceeding on each side from this first thoracic ganglion, runs to the muscles beneath the basal articulation of the first cirrus. The collar surrounding the oesophagus is generally very long, sometimes equalling the whole thoracic chord; at a middle point, a small branch is sent off, and at the anterior end *e, e*, close to the supra-oesophageal ganglia, double or treble fine branches run to the true ovaria, lying close to the upper end of the stomach. The four (or only three) other thoracic ganglia, when viewed as transparent bodies, are seen to be solid; but in some of the genera, as in Conchoderma, the outline plainly shows, that each consists of a lateral pair fused together. The second thoracic ganglion B is rather small; it is either close to the first, as in *Pollicipes mitella* and *Lepas fascicularis*, or far distant, as in Ibla. The third C and fourth are of about the same size with the second: these three ganglia send large branches to the second, third, and fourth pair of cirri: other minute branches spring from their undersides, and from the intermediate double chords. The fifth ganglion is larger and longer than the three preceding ones, and gives off / nerves to the fifth and sixth pair of cirri; it is clearly formed by the union of the fifth, with what ought to have formed a sixth ganglion. The two nerves going to the sixth cirrus give off on their inner sides, each a great branch to the penis. In *Pollicipes mitella*, in which there are only four instead of five thoracic ganglia, it is evident from the outline and position of the nerves going to the fourth pair of cirri, that the fourth ganglion is fused into the fifth, itself, as we have just seen, normally composed of two consecutive ganglia. In this Pollicipes there is other evidence of concentration in the nervous system, for none of the ganglia show signs of being formed of lateral pairs; the second is close to the first; and the abdominal double chord is in part separated by a mere cleft; lastly, as we shall immediately see, the same remark is applicable to the supra-oesophageal ganglia.

The latter D alone remain to be described; they present far more diversity in shape than do the thoracic ganglia; they are almost always seen in outline to be laterally distinct, and usually resemble two pears with their tapering ends cut off and united; in a transverse line they are as long as the infra-oesophageal ganglion, but are much less massive. In *Lepas fascicularis* D, they are pear-shaped; in *Pollicipes mitella* they are globular, and separated by a third globular ganglion, which I believe is the ophthalmic ganglion, presently to be described; in *Pollicipes spinosus*, however, the ophthalmic ganglion is, as usual, placed in advance of the supra-oesophageal ganglion, which latter, in this one species, shows no sign of being formed of a lateral pair fused together. In *Alepas cornuta* the supra-oesophageal ganglion consists of two quite distinct ganglia, elongated in the longitudinal axis of the body, and separated from each other by the whole width of the mouth; the chord which unites them is of the same thickness as the rest of the collar. In all the genera, from the front of each of the two supra-oesophageal ganglia, a pair of nerves, / *f, f*, united and together as large as the collar nerve, rises, and can be traced running unbranched, in a nearly straight line, for a length equalling the whole rest of the nervous chord, so as to supply the peduncle and the inside of the capitulum or sack. At the inner ends of these two same ganglia, from a central point where they are united, a little central branch runs in front to the adductor scutorum and other adjoining muscles; and still smaller fibrils run behind to the oesophageal muscles.

Ophthalmic ganglia and eyes. Owing to Professor Leidy's[14] discovery of eyes in a Balanus, I was led to look for them in the Lepadidae. Extending from the front of the two supra-oesophageal ganglia, two chords may be seen in *Lepas fascicularis* (of which a rude diagram is here given), to run into two small, perfectly distinct oval/ ganglia E, which are not united by any transverse commissure. From the opposite ends of these two ganglia smaller nerves run, and, bending inwards at right angles, enter, beyond the middle, an elongated F, almost black, eye, composed of two eyes united together. Although in outline the eye appears single, two lenses can be distinctly seen at the end, directed upwards and towards the ganglia; two pigment capsules can also be

[14] *Proceedings of the Academy of Natural Sciences*, Philadelphia. No. i, vol. iv, Jan. 1848.

Diagram of the anterior portion of the nervous system in *Lepas fascicularis*. A. First thoracic or infra-oesophageal ganglion. B. Second thoracic. C. Third thoracic ganglion. D. Supra-oesophageal ganglion. E. The two ophthalmic ganglia. F. Double eye. *a.* Nerve going to first cirrus; *b*, to the muscles below the first cirrus; *c*, to the second cirrus; *d*, to the third; *e*, nerves running to the ovaria; *f*, double nerves supplying the sack and peduncle.

distinguished; these are deep and cup-formed, and of a dark reddish-purple. The following measurements will show the proportions of the parts in a specimen of the *Lepas fascicularis* having a capitulum 4/10ths of an inch in length.

Double eye length	26/6000	Supra-oesophageal ganglion, transverse or longest axis of both together	126/6000
Double eye width	13/6000		
Diameter of single lens	6/6000	Supra-oesophageal ganglion, longitudinal axis of	43/6000
		Infra-oesophageal ganglion, transverse axis of	120/6000
Ophthalmic ganglion length	16/6000	Infra-oesophageal ganglion, longitudinal axis of	114/6000
Ophthalmic ganglion breadth	11/6000		

In *Conchoderma aurita* the ophthalmic ganglia are much smaller, and nearer to the supra-oesophageal ganglion, than in *L. fascicularis*. In

Alepas cornuta the ophthalmic chords run towards each other from the two distant and separate supra-oesophageal ganglia; and the ophthalmic ganglia (instead of being quite separate, as in *L. fascicularis*), are united by their front ends, and the two eyes instead of standing some way in front, with nerves running to them, are embedded on the double ophthalmic ganglion; the pigment capsules here, also, have the shape of mere saucers, and are joined back to back, with the two lenses projecting far out of them. In neither sex of Ibla could I perceive that the eye was double. In *Pollicipes spinosus* the ophthalmic ganglion stands in front of the single supra-oesophageal ganglion, and shows no signs of being formed of a lateral pair; the eyes themselves, however, differently from, in all the foregoing cases, are, though approximate, quite distinct. In *Pollicipes / mitella* I did not see the eyes; but the ophthalmic ganglion consists, as I believe, of a single globular one, placed exactly between the two globular, supra-oesophageal ganglia, all three being of nearly equal size. Professor Leidy does not mention the ophthalmic ganglia; hence I infer that in Balanus, which is a more highly organized cirripede, they are fused into the supra-oesophageal ganglion.

In all the genera, the double eye is seated deep within the body; it is attached by fibrous tissue to the radiating muscles of the lowest part of the oesophagus, and lies actually on the upper part of the stomach; consequently, a ray of light, to reach the eye, has to pass through the exterior membrane and underlying corium connecting the two scuta, and to penetrate deeply into the body. In living sessile cirripedes, vision seems confined to the perception of the shadow of an object passing between them and the light; they instantly perceived a hand passed quickly at the distance of several feet between a candle and the basin in which they were placed.

As the infra-oesophageal ganglion sends nerves to the trophi and to the first pair of cirri, it must correspond to the segments, from the fourth to the ninth inclusive, of the archetype crustacean. The state of the supra-oesophageal and ophthalmic ganglia appears to me very interesting: I do not believe that in any *mature* ordinary crustacean, the first or ophthalmic ganglion can be shown to be distinct from the two succeeding ganglia, or to be itself composed of a pair laterally distinct. The ganglia, corresponding with the second and third segments of the body, which should normally support two pair of antennae, are in the Lepadidae united together; but laterally they are generally distinct in outline, and are actually separate in Alepas: the supra-oesophageal ganglion shows also its double nature, by giving rise to a pair of large double nerves, evidently corresponding with the two pair of antennular nerves in ordinary crustaceans. The embryonic condition of the whole supra-oesophageal portion of the nervous system in the Lepadidae, corresponds with the / rudimentary state of the only organ of sense supplied by it, namely, the eye, which in size and general appearance has retrograded to the state in which it was in, during the first stage of development of the larva; I have used the term embryonic, because, in the embryos of ordinary Crustacea, all the ganglia are at first longitudinally distinct, and laterally quite separate. The conclusion at which we before arrived from studying the metamorphoses, namely, that the whole peduncle and capitulum consisted of the first three segments of the head, is beautifully supported by the structure of the nervous system, in which these parts are seen to be supplied with nerves exclusively from the supra-oesophageal ganglion: now in ordinary crustacea the supra-oesophageal ganglion sends nerves to the eyes and the two pair of antennae corresponding, as is known by embryological dissections, to the first three segments of the body. Moreover, it is asserted that the carapace which covers the thorax in crustacea, is not formed by the development of the first segment; and this, likewise, may be inferred to be the case with the peduncle and capitulum in the Lepadidae, as the nerves of the ophthalmic ganglia go exclusively to the eyes. Finally, I may remark that in Pollicipes, looking to the whole nervous system, the state of concentration nearly equals that in certain macrourous decapod crustaceans, for instance the *Astacus marinus*, of which a figure is given by Milne Edwards.

Olfactory organs. In the outer maxillae, at their bases where united together, but above the basal fold separating the mouth from the body, there are, in all the genera, a pair of orifices (Pl. X, fig. 16); these are sometimes seated on a slight prominence, as in Lithotrya, or on the summit of flattened tubes (Pl. X, fig. 17), projecting upwards and towards each other, as in Ibla, Scalpellum, and Pollicipes. In Ibla these tubular projections rise from almost between the outer and inner maxillae. It is impossible to behold these organs, and doubt that they are of high functional importance to the animal. The orifice leads / into a deep sack lined by pulpy corium, and closed at the bottom. The outer integument is inflected inwards (hence periodically moulted), and becoming of excessive tenuity, runs to near the bottom of the sack, where it ends in an open tube: so excessively thin is this inflected membrane, that, until examining Anelasma, I was not quite certain that I was right in believing that the outer integument did not extend over the whole bottom. I several times saw a nerve of considerable size entering and blending into a pulpy layer at the bottom of the sack of corium; but I failed in tracing to which of the three pair of nerves, springing from the front end of the infra-oesophageal ganglion, it joined. I can hardly avoid concluding, that this *closed* sack, with its naked bottom, is an organ of sense; and, considering that the outer maxillae serve to carry the prey entangled by the cirri towards the maxillae and mandibles, the position seems so admirably adapted for an olfactory organ, whereby the animal could at once perceive the nature of any floating object thus caught, that I have ventured provisionally to designate the two orifices and sacks as olfactory.

Acoustic [?] *Organs.* A little way beneath the basal articulation of the first cirrus (Pl. IX, fig. 4d, and Pl. IV, fig. 2e), on each side, there may be seen a slight swelling, and on the under side of this, a transverse slit-like orifice, 1/20th of an inch in length in Conchoderma, but often only half that size. In Ibla this orifice is seated lower down (Pl. IV, fig. 8a', e), between the bases of the first and second cirri, which are here far apart: in *Alepas cornuta* it is placed rather nearer to the adductor scutorum muscle, namely, beneath the mandibles. The orifice leads into a rather deep and wide meatus; the external integument is turned in for a short distance, widening a little, and then ends abruptly. The meatus, enlarging upwards, is lined by thick pulpy corium, and is closed at the upper end; from its summit is suspended a flattened sack of singular and different shapes in the different genera. This, the so-called acoustic sack of *Conchoderma virgata*, *l* is figured (Pl. IX, fig. 6). The deep and wide notch faces towards the posterior end of the animal; the inferior lobe, thus almost cut off, is flattened in a different plane from the upper part; the lobe is lodged in a little pouch of corresponding form, leading from the open meatus in which the upper part is included. In *Conchoderma aurita*, the top of the acoustic sack is narrower and more constricted, the whole more rounded, and the lobe more turned down. In *Lepas fascicularis* the notch is not so deep or wide, and the lobe larger. In *Ibla Cumingii* the sack is of the shape of a vase, with one corner folded over. In *Scalpellum vulgare* it is small, oval, with the lower end much pushed in, and furnished with a little crest. Lastly, in *Pollicipes mitella* it is simply oval. In all cases the sack is empty, or contains only a little pulpy matter: it consists of brownish, thick, and remarkably elastic tissue, formed, apparently, of transverse little pillars, becoming fibrous on the outside, and with their inner ends appearing like hyaline points. The mouth of the acoustic sack (removed in the drawing) is closed by a tender diaphragm, through which I saw what I believe was a moderately sized nerve enter; I have not yet succeeded in tracing this nerve. The first pair of cirri seem, to a certain extent, to serve as antennae, and therefore the position of an acoustic organ at their bases, is analogous to what takes place in crustacea; but there are not here any otolites, or the siliceous particles and hairs, as described by Dr Farre, in that class. Nevertheless, the sack is so highly elastic, and its suspension in a meatus freely open to the water, seems so well adapted for an acoustic organ, that I have provisionally thus called it. In the larva, as I have shown, a pouch, certainly serving for some sense, I believe for hearing, is seated in quite a different position at the anterior end of the carapace. I may mention that I found sessile cirripedes very sensitive of vibrations in objects adjoining them, though not, apparently, of noises in the air or water. In a group of specimens, I could not touch one even most delicately / with a needle, without all the adjoining ones instantly withdrawing their cirri; it made no difference if the one touched had its operculum already closed and motionless.

Reproductive system – male organs. All the Cirripedia which I have hitherto examined, with the exception of certain species of Ibla and Scalpellum, are hermaphrodite or bisexual.[15] I shall so fully describe the sexual relations of the several species of these two genera, under their respective headings, and at the end of the genus of Scalpellum, that I will not here give even an abstract of the grounds on which my

---

[15] I am compelled to differ greatly from the account given by Prof. Steenstrup of the reproductive system in the Cirripedia, in his *Untersuchungen über das Vorkommen des Hermaphroditismus*, ch. v, 1846; a translation of which I have seen, owing to the great kindness of Mr Busk. Mr Goodsir has described (*Edin. New Phil. Journal*, July, 1843), what he considers the male of Balanus; but I have seen this same parasitic creature charged with ova, including larvae! From the resemblance of the larvae to the little crustacean described by Mr Goodsir, in the same paper, as a distinct parasite, I believe the latter to be the male of his so-called male Balanus, and that all belong to the same species, allied to Bopyrus. This genus, as is well known, is parasitic on other Crustacea; and it is a rather interesting fact thus to find, that this new parasite which is allied to Bopyrus, in structure, is likewise allied to it in habits, living attached to Cirripedia, a sub-class of the Crustacea.

firm belief is based, that the masculine power of certain hermaphrodite species of Ibla and Scalpellum, is rendered more efficient by certain parasitic males, which, from their not pairing, as in all hitherto known cases, with females, but with hermaphrodites, I have designated *complemental males*.

The male organs have been well described by M. Martin St Ange, whose observations have since been confirmed by R. Wagner.[16] The testes are small, often leaden-coloured, either pear of finger-shaped, or branched like club-moss – these several forms sometimes occurring in the same individual; they coat the stomach, enter the pedicels, and even the basal segments of the rami of the cirri, and in some genera occupy certain / swellings on the thorax and prosoma, and in others the filamentary appendages: the testes seen in the apex in one of these appendages in Conchoderma (is represented in Pl. IX, fig. 5). The two vesiculae seminales are very large; they lie along the abdominal surface of the thorax, and generally (but not in some species of Scalpellum) enter the prosoma, where their broad ends are often reflexed; here the branched vessels leading from the testes enter. The membrane of the vesiculae seminales is formed of circular fibres; and is, I presume, contractile, for I have seen the spermatozoa expelled with force from the cut end of a living specimen. The two canals leading from the vesiculae generally unite in a single duct at the base of the penis; but in *Conchoderma aurita*, halfway up it. The probosciformed penis, except in certain species of Scalpellum, is very long; it is capable of the most varied movements; it is generally hairy, especially at the end; it is supported on a straight unarticulated basis, which in *Ibla quadrivalvis* alone (Pl. IV, fig. 9a), is of considerable length; in this species, the upper part is seen to be as plainly articulated as one of the cirri; in Alepas, the articulations are somewhat less plain, and in the other genera, the organ can be said only to be finely ringed, but these rings no doubt are in fact obscure articulations. In the females of *Ibla Cumingii* and *Scalpellum ornatum*, there is, of course, no penis.

Female organs. M. Martin St Ange has described how the peduncle[17]

---

[16] In *Müller's Archiv*, 1834, p. 467. I have already several times referred to M. Martin St Ange's excellent *Memoir*, read before the Academy of Sciences, and subsequently, in 1835, published separately.

[17] I may here mention, that in all sessile cirripedes, the ovarian branching tubes lie between the calcareous or membranous basis and the inner basal lining of the sack, and to a certain height upwards round the sack: the true ovaria and the two ducts occupy the same position as in the Lepadidae.

is gorged with an inextricable mass of branching ovarian tubes, filled with granular matter and immature ova. In Conchoderma and Alepas, the ovarian tubes run up in a single plane (Pl. IX, fig. 3), between the two folds of corium round the sack. Here the development of the ova can be well followed: a minute point first branches out from one of the tubes; its / head then enlarges, like the bud of a tulip on a footstalk; becomes globular; shows traces of dividing, and at last splits into three, four, or five egg-shaped balls, which finally separate as perfect ova. Within the peduncle, the ovarian tubes branch out in all directions, and within the footstalks of the branches (differently from what takes place round the sack), ova are developed, as well as at their ends. Close together, along the rostral (i.e. ventral) edge of the peduncle, two nearly straight, main ovarian tubes or ducts may be detected, which do not give out any branches till about halfway down the peduncle, where they subdivide into branches, which inosculate together, and give rise to the mass filling the peduncle, and sometimes, as we have just seen, sending up branches round the sack. These two main unbranched ovarian ducts, followed up the peduncle, are seen to enter the body of the Cirripede (close alongside the great double peduncular nerves), and then separating, they sweep in a large curve along each flank of the prosoma, under the superficial muscles, towards the bases of the first pair of cirri; and then rising up, they run into two glandular masses. These latter rest on the upper edge of the stomach, and touch the caeca where such exist; they were thought by Cuvier to be salivary glands. They are of an orange colour, and form two, parallel, gut-formed masses, having, in Conchoderma, a great flexure, and generally dividing at the end near the mouth into a few blunt branches. I was not able to ascertain whether the two main ducts, coming from the peduncle, expanded to envelope them, or what the precise connection was. The state of these two masses varied much; sometimes they were hollow, with only their walls spotted with a few cellular little masses; at other times they contained or rather were formed of, more or less globular or finger-shaped aggregations of pulpy matter; and lastly, the whole consisted of separate pointed little balls, each with a large inner cell, and this again with two or three included granules. These so closely / resembled, in general appearance and size, the ovigerms with their germinal vesicles and spots, which I have often seen at the first commencement of the formation of the ova in the ovarian tubes in the peduncle, that I cannot doubt that such is their nature. Hence I conclude, that these two gut-formed masses are the true ovaria. I may add, that several times I have seen in the two long, unbranched ducts, connecting the true ovaria and the ovarian tubes in the peduncle, pellets of orange-coloured cellular matter (i.e. ovigerms) forming at short intervals little enlargements in the ducts, and apparently travelling into the peduncle.

The structure here described is quite comfortable with that which we have seen in the larva; in the latter, two gut-formed masses of equal thickness extended from the caeca of the stomach to within the future peduncle, where the cement ducts entered them, and where, after a short period, they were seen to expand into a mass of ovarian tubes. In the mature cirripede, the cement ducts can still be found united to the ovarian tubes in the middle of peduncle; and the cause of the wide separation of the true ovaria and ovarian tubes, can be simply accounted for by the internal, almost complete intersection of the animal, which takes place during the last metamorphosis.

The ova, when excluded, remain in the sack of the animal until the larvae are hatched; they are very numerous, and generally form two concave, nearly circular, leaves, which I have called after Steenstrup and other authors, the *ovigerous lamellae* (Pl. IV, fig. 2b). These lamellae lie low down on each side of the sack: in *Conchoderma virgata*, however, there is often only a single lamella, forming a deeply concave cup: in *C. aurita* there are generally on each side four lamellae, one under the other. The ova lie in a layer from two to four deep; and all are held together by a most delicate transparent membrane, which separately enfolds each ovum: this membrane is often thicker and stronger round the margins of the lamellae, where they are united, in a peculiar manner, presently to be / described, to a fold of skin, on each side of the sack: these two folds, I have called the *ovigerous fraena* (Pl. IV, fig. 2f).

M. Martin St Ange, describes an orifice under the carina, by which he supposes the ova to enter the sack; this, after repeated and most careful examinations, I venture to affirm does not exist; on the contrary, I have every reason to believe that the ova enter the sack in the following curious manner. Immediately before one of the periods of exuviation, the ova burst forth from the ovarian tubes in the peduncle and round the sack, and, carried along the open circulatory channels, are collected (by means unknown to me) beneath the chitine tunic of the sack, in the corium, which is at this period remarkably spongy and full of cavities. The corium then forms or rather (as I believe) resolves itself into the very delicate membrane separately enveloping each ovum, and uniting them together into two lamellae; the corium having thus far retreated, then forms under the lamellae the chitine tunic of the sack, which will of course be of larger size than the last formed one, now immediately to be moulted with the other integuments of the body. As soon as this exuviation is effected, the tender ova, united into two lamellae, and adhering, as yet, to the bottom of the sack, are exposed: as the membranes harden, the lamellae become detached from the bottom of the sack, and are attached to the ovigerous fraena. To demonstrate this view, an individual should have been found, with both the old and new chitine tunic of the sack, and with the lamellae lying between them; this, I believe, I have seen, but it was before I understood the full importance of the fact: a great number of specimens would have to be examined in order to succeed again, for the changes connected with exuviation supervene very quickly. I have, however, several times found the ova so loose under the sack, as to be detached with a touch from the ovarian tubes; and I have twice carefully examined specimens, which had just moulted, as shown by even the mandibles being flexible, in which the lamellae had not become united to / the fraena, but still adhered to the newly formed chitine tunic of the sack; in these, the ova were so tender, that they broke into pieces rather than be separated from the membrane of the lamella, itself hardly perfectly developed, for pulpy cellular matter adhered outside some of the ova. These and other facts are quite inexplicable on any other view than that advanced.

As the lamellae are formed without organic union with the parent, they would be liable to be washed out of the widely open sack of the Lepadidae, if they had not been specially attached to the *fraena*. These fraena consist of a pair of more or less semicircular folds of skin, depending inside the sack, on each side of the point of attachment of the body. The fraena are often of considerable size, but in Ibla, they are very minute; they are formed of chitine tunic with underlying corium, like the rest of the sack; on their crests, there is a row, or a set of circular groups, or a broad surface, covered, either with minute, pointed, bead-like bodies mounted on long hair-like footstalks, or with staff formed bodies on very short footstalks. I measured some of the bead-like bodies, in *Lepas anserifera*, and they were $\frac{1}{2000}$th of an inch in diameter, and the footstalks three or four times as long as the elongated heads. These heads, of whatever shape they may be, have an opaque, and, I believe, glandular centre; I could not make out with certainty an aperture at their ends, but, I believe, such exists, and they seem to secrete a substance, which hardens into a strong membrane, serving to unite the crest of the fraenum to the edges of the lamellae. In one case, this bit of membrane seemed formed of a woven mass of threads. These little glandular bodies, with the membrane formed by them, are cast off at each exuviation, and new glands formed on the crest of the fraenum underneath. In some species of Pollicipes (viz. *P. cornucopia* and *elgans*), the fraena, though present and large, are functionless and destitute of the glands: I believe, they exist in this same functionless condition, and in rather a different position in the / sessile cirripedes, and that in this family they serve as Branchiae.

The above-described method by which Cirripedia lay their eggs, namely, united together in a common membrane, placed between their old outer and new inner integuments, and the manner in which the lamellae, when thus formed, are retained for a time fastened to the fraena, and are then cast off, appears to me very curious. In some of the lower Crustacea, it is known, that the ova escape by rupturing the ovisacs formed by the protruded ovarian tubes, and this is the nearest analogy with which I am acquainted. The ova are impregnated (as I infer from the state of the vesiculae seminales), when first brought into the sack, and whilst the membrane of the lamellae is very tender: the long prosbosciformed penis seems well adapted for this end. In the male of *Ibla Cumingii*, which has not a prosbosciformed penis, the whole flexible body, probably, performs the function of the penis: in *Scalpellum ornatum*, however, the spermatozoa must be brought in by the action of the cirri, or of the currents produced by them. That cross impregnation may and sometimes does take place, I infer from the singular case of an individual, in a group of Balani, in which the penis had been cut off, and had healed without any perforation: notwithstanding which fact, larvae were included in the ova.

Exuviation; rate of growth; size. I have had occasion repeatedly to allude to the exuviation of the Lepadidae: with the exception of the genus Lithotrya,[18] in which the calcareous scales on the peduncle, together with the membrane connecting them, is cast off, neither the

---

[18] The external integuments being moulted in Crustacea, but not in the Cirripedia may appear, at first, an important difference: but we here see that non-exuviation is not universal among the Lepadidae, and, on the other hand, according to M. Joly (*Annales des Sciences Naturelles*, 2nd series, Zoolog.), there is one true crustacean, the *Isaura cycladoides*, which has a persistent bivalve shell.

valves nor the membrane uniting them, nor that forming the peduncle with its scales and styles, are / moulted; but the surface gradually disintegrates and is removed, perhaps sometimes in flakes, whilst new and larger layers are formed beneath. In Scalpellum, I ascertained that the new membrane, connecting together the newly formed calcified rims under the valves of the capitulum, was formed as a fold, with the articulated spines which it bears, all adpressed in certain definite directions. This fold of new membrane, when the old membrane splits and yields, of course expands, and thus the size of the capitulum is increased. In the peduncle, lines of splitting can seldom be perceived, except, indeed, in the sub-globular, embedded, downward-growing peduncle of Anelasma, as described under that genus. I do not understand what determines the complicated lines of splitting of the old membrane between the several valves of the capitulum – without it be simply, that along these lines alone, the old membrane is not strengthened by the new membrane being closely applied under it, the new being formed, as we have just said, in a fold, in order to allow of increase in size. Although, as I believe, there is strictly no exuviation in the outer membranes of mature Lepadidae, it seems that narrow strips of membrane are cast off from between the valves, for the few first moults, after the final metamorphosis of the larva. I may here remark that, in most sessile cirripedes, the outside membrane connecting the operculum and shell, is regularly moulted.

The delicate tuning lining the sack (a mere duplicature of that thick one, forming the outside of the capitulum, and generally transformed into valves), and the integuments of the whole body, are regularly moulted. With these integuments, the membrane lining the oesophagus, the rectum, and the deep olfactory pouches, and the horny apodemes of the maxillae, are all cast together. I have seen a specimen of Lepas, in which, from some morbid adhesion, the old membrane lining one of the olfactory pouches had not been moulted, but remained projecting from the orifice as a brown shrivelled scroll. The new / spines on the cirri (and on the maxillae) are formed within the old ones; but as they have to be a little longer than the latter, and as they cannot enter these up to their very points, their basal portions are not thus included, but are formed, running obliquely across the segments of the cirri; and what is curious, these same basal portions are turned inside out, like the fingers of a glove when hastily drawn off. After the exuviation of the old spines, the new spines have their inverted basal portions drawn out from within the segments, and turned outside in, so as to assume their proper positions.

All Cirripedia grow rapidly: the yawl of H.M.S. *Beagle* was lowered into the water, at the Galapagos Archipelago, on the 15th of September, and, after an interval of exactly thirty-three days, was hauled in: I found on her bottom, a specimen of *Conchoderma virgata* with the capitulum and peduncle, each half an inch in length, and the former 7/20ths in width: this is half the size of the largest specimen I have seen of this species: several other individuals, not half the size of the above, contained numerous ova in their lamellae, ready to burst forth. Supposing the larva of the largest specimen became attached the first day the boat was put into the water, we have the metamorphosis, an increase of length from about 0·05, the size of the larva, to a whole inch, and the laying of probably several sets of eggs, all effected in thirty-three days. From this rapid growth, repeated exuviations must be requisite. Mr W. Thompson, of Belfast, kept twenty specimens of *Balanus balanoides*, a form of much slower growth, alive, and on the twelfth day he found the twenty-first integument, showing that all had moulted once, and one individual twice within this period. I may here add, that the pedunculated cirripedes never attain so large a bulk as the sessile; *Lepas anatifera* is sometimes sixteen inches in length, but of this, the far greater portion consists of the peduncle. *Pollicipes mitella* is the most massive kind; I have seen a specimen with a capitulum 2·3 of an inch in width. /

Affinities. Considering the close affinity between the several genera, there are, I conceive, no grounds for dividing the Lepadidae into subfamilies, as has been proposed by some authors, who have trusted exclusively to external characters. In establishing the eleven genera in the Lepadidae, no one part or set of organs affords sufficient diagnostic characters: the number of the valves is the most obvious, and one of the most useful characters, but it fails when the valves are nearly rudimentary, and when they are numerous: the direction of their lines of growth is more important, and fails to be characteristic only in Scalpellum: with the same exception, the presence or absence of calcified or horny scales on the peduncle is a good generic character. For this same end, the shape of the scuta and carina, but not of the other valves, comes into play. In three genera, the presence of filamentary appendages on the animal's body is generic; in Pollicipes, however, they are found only on three out of the six species. The number of teeth in the mandibles, and the shape of the maxillae, often prove serviceable for this end; as does more generally the presence of caudal appendages, and whether they be naked or spinose, uniarticulate or multiarticulate; in Pollicipes alone this part is variable, being uni- and multiarticulate; and in one species of Scalpellum they are absent, though present in all the others. The shape of the body, the absence or presence of teeth on the labrum, the inner edge of the outer maxillae being notched or straight, the prominence of the olfactory orifices, the arrangement of the spines on the cirri, and the number and form of their segments, are only of specific value.

Comparing the pedunculated and sessile cirripedes, it is, I think, impossible to assign them a higher rank than that of families. The chief difference between them consists, in the Lepadidae, in the presence of three layers of striaeless muscles, longitudinal, transverse and oblique, continuously surrounding the peduncle, but not specially attached to the scuta and terga; and on the other hand, in the Balanidae, / of five longitudinal bundles of voluntary muscles, with transverse striae, fixed to the scuta and terga, and giving them powers of independent movement. In the Lepadidae, the lower valves, or when such are absent, the membranous walls of the capitulum, move with the scuta and terga when opened or shut; and the lower part of the capitulum is separated by a movable peduncle from the surface of attachment; in the sessile cirripedes, the lower valves are firmly united together into an immovable ring, fixed immovably on the surface of attachment. I will not compare the softer parts, such as the cirri and trophi, of the Lepadidae with those of the Balanidae, as my examination of this latter family is not fully completed: I will only remark, that there is a very close general resemblance, more especially with the subfamily Chthamalinae.

Geographical range; habitats. The pedunculated cirripedes extend over the whole world; and most of the individual species have large ranges, more especially, as might have been expected, those attached to floating objects; excepting these latter, the greater number inhabit the warmer temperate, and tropical seas. Of those attached to fixed objects, or to littoral animals, it is rare to find more than three or four species in the same locality. On the shores of Europe I know of only three, viz. a Scalpellum, Pollicipes, and Alepas. At Madeira (owing to the admirable researches of the Rev. R. T. Lowe), two Paecilasmas, a Dichelaspis, and an Oxynaspis are known. In New Zealand, there are two Pollicipes and an Alepas, and, perhaps, a fourth form. From the Philippine Archipelago, in the great collection made by Mr Cuming, there are a Paecilasma, an Ibla, a Scalpellum, Pollicipes, and Lithotrya. Of all the Lepadidae, nearly half are attached to floating objects, or to animals which are able to change their positions; the other half are generally attached to fixed organic or inorganic bodies, and more frequently to the former than to the latter. Most of the species of Scalpellum are inhabitants / of deep water; on the other hand, most of Pollicipes,[19] of Ibla, and Lithotrya are littoral forms. The species of Lithotrya have the power of excavating burrows in calcareous rocks, shells, and corals; and the singular manner in which this is effected, is described under that genus. Anelasma has its subglobular peduncle deeply embedded in the flesh of Northern Sharks; and I have seen instances of the basal end of the peduncle of *Conchoderma aurita*, being sunk into the skin of Cetacea; in the same way the point of the peduncle in the male of Ibla, is generally deeply embedded in the sack of the female. I believe in all these cases, the cementing substance affects and injures the corium or true skin of the animal on which the creature is parasitic, whilst the surrounding parts, being not injured, continue to grow upwards, thus causing the partial embedment of the cirripede. In the case of Anelasma, we have growth at the end of the peduncle, and consequently downward pressure, and this may possibly cause absorption to take place in the skin of the shark at the spot pressed on.

Geological history. Having treated this subject at length, in the volume of the Palaeontographical Society for 1851,[a] I will not here enter on it: I will only remark, that the Lepadidae or pedunculated cirripedes are much more ancient, according to our present state of knowledge, than the Balanidae. The former seem to have been at their culminant point during the Cretaceous period, when many species of Scalpellum and Pollicipes, and a singular new genus, Loricula, existed; Pollicipes is the oldest genus, having been found in the Lower Oolite, and, perhaps, even in the Lias. The fossil species do not appear to have differed widely from existing forms. /

---

[19] I am informed by Mr L. Reeve that *Pollicipes mitella* is eaten on the coast of China; and Ellis states (*Phil. Trans.*, 1758) that this is the case with *P. cornucopia* on the shores of Brittany. It is well known that the gigantic *Balanus psittacus*, on the Chilian coast, is sought after as a delicacy; and I am assured, by Mr Cuming, that it deserves its reputation.

[a] *A Monograph of the fossil Lepadidae* (1851), reprinted in vol. 14, *Works of Darwin*.

GENUS: LEPAS

PLATE I

LEPAS. Linnaeus,[20] *Systema Naturae*, 1767
ANATIFA. Brugière,[21] *Encyclop. Method. (des Vers)*, 1789
ANATIFERA. Lister et plerumque auctorum Anglicorum
PENTALASMIS. Hill Leach, *Journal de Physique*, July, 1817
PENTALEPAS. De Blainville, *Dict. des Sci. Nat.*, 1824
DOSIMA. J. E. Gray, *Annals of Philosophy*, vol. x, 1825

Valvae 5, approximatae: carina sursùm inter terga extensa, deorsùm aut furcâ infossâ aut disco externo terminata: scuta subtriangula, umbonibus ad angulum rostralem positis.

Valves 5, approximate: carina extending up between the terga, terminating downwards in an embedded fork, or in an external disc: scuta subtriangular, with their umbones at the rostral angle.

Filaments seated beneath the basal articulation of the first cirri; mandibles with five teeth; maxillae stepformed; caudal appendages uniarticulate, smooth.

Distribution. Mundane; attached to floating objects.

Description. Capitulum flattened, subtriangular, composed of five approximate valves. The valves are / either moderately thick and translucent, or very thin and transparent; and hence, though themselves colourless, they are often coloured by the underlying corium. Their surfaces are either smooth and polished, or striated, or furrowed, and sometimes pectinated. They are not subject to disintegration; they are generally naked, except on the borders, where they

---

[20] Linnaeus, as is well known, included under this genus both the pedunculated and sessile cirripedes. According to the rules of the British Association, the name Lepas must be retained for part of the genus; and as the sessile division was named Balanus, by Lister and Hill, even before the invention of the binomial system, and subsequently, in 1778, by Da Costa, and again, in 1789, by Brugière, there can be no question that Lepas must be applied to the pedunculated section of the genus. In this instance it is particularly desirable to recur to the Linnean name, as no other name has been *generally* adopted. Had not Lister and Sir J. Hill published before the binomial system, their names of Anatifera and Pentalasmis would have had prior claims to Lepas.

[21] The date of this publication is almost universally given as 1792, apparently caused by an error in the title-page of the First Part, which has consequently been cancelled. The First Part contains Anatifa and Balanus, and was published in 1789. The Second Part was published in 1792, and has a corrected title-page for the whole *volume*.

are coated, and held together by membrane; in *L. fascicularis*, however, the valves are covered with thin membrane, bearing very minute spines. The manner of growth of the valves will be best described under each. All the valves, even in the same species, are subject to considerable variation in shape, more especially the terga.

Scuta. These valves are subtriangular in outline, with the basal margin straight and rather short; and with occludent and tergo–carinal margins more or less protuberant; in *L. fascicularis*, however, the basal (Pl. I, fig. 6), and occludent margins are slightly reflexed and prominent. A ridge generally runs from the umbo to the upper point. Internally, there is no conspicuous pit for the adductor muscle; under the umbones, there is generally either on both valves, or only on the right-hand side (Pl. I, fig. 1c), a small calcareous projection or tooth, of variable size and shape, even in the same species; it is generally largest on the right-hand valve; these teeth at first sight appear to form a hinge, uniting the opposite scuta at their umbones, but this is not really the case, and their use appears to be only to give attachment to the membrane uniting the valves together, and to the peduncle. The basal margin is internally strengthened by a calcified rim, more or less developed. The umbones (and primordial valves when distinguishable), are seated at the rostral angles; during growth the basal margin is not added to, and the occludent margin only to small extent; hence the main growth of the valve is at the upper end, and along the carina–tergal margin. In *L. fascicularis*, however, the basal reflexed margin is slightly added to beneath the umbo. /

Terga. Flat, small compared with the scuta, usually of an irregular quadrilateral figure, with the two upper or occludent margins very short, in proportion to the two (carinal and scutal) lower margins; all the margins are nearly straight. The two occludent margins, generally meet each other at about right angles, forming a small triangular projection; in *L. fascicularis*, however, the occludent margin is formed by a single, slightly curved line. The umbones (and primordial valves when distinguishable) are not seated at the uppermost point, but at the angle where the carinal margin unites to the upper of the two occludent margins: during growth the terga are added to, both on the occludent and on the scutal margins, and slightly along the carinal margin; hence their growth is unequally *quaquaversal*, except at one angle of the irregular quadrilateral figure.

Carina. This is always very narrow and curved, concave within, often carinated and barbed exteriorly; it extends upwards between the terga for one half or two-thirds of their length; at the lower extremity it ends (with the exception of *L. fascicularis*), in a small fork (Pl. I, fig. 1*a*, *b*) rectangularly inflected and embedded in the membrane, beneath the basal margin of the scuta. From comparing this lower part of the carina in *L. australis* (fig. 5*a*), with the same part in some of the species of the allied genus Paecilasma, it would appear that the fork is formed by an oblong disc, more and more notched at the end, and with the rim between the two points more or less folded backwards: conformably with this view, in very young specimens of *L. australis*, instead of a large and sharp fork, there is a small disc. The only use of the fork appears to be to give firm attachment to the membrane uniting the valves and peduncle. In *L. fascicularis*, instead of a fork, there is a broad, oblong disc (figs 6, 6*a*), rectangularly inflected; it is much longer than the fork, in proportion to the upper part of the carina; the disc is not more deeply embedded than the upper part. The umbo (and primordial valve when / distinguishable), of the carina is seated just above the embedded fork (or disc in *L. fascicularis*), at the point where the inflection takes place; hence the main growth of the carina is upwards – the fork, however, being of course, likewise added to at its point: in *L. fascicularis*, the growth is both upwards and downwards.

Peduncle and attachment. The peduncle is generally quite smooth: though with a high power its surface may be seen to be studded with minute beads, or larger discs, of yellowish and hard chitine; in the young of *L. australis*, and I suspect of some other species, it is covered with very minute spines. The peduncle in this genus attains its greatest development. The cement tissue debouches, I believe, only through the functionless larval antennae, except in one species, *L. fascicularis*, in which a ball of this substance is formed in a most peculiar manner round the peduncle (Pl. I, fig. 6), apparently for the purpose of serving as a float, as will be presently described.

Size and colour. The species of this genus are the largest of the Pedunculata, with the exception of some Pollicipes: even in the smallest species (*L. pectinata*), the capitulum sometimes attains a length of about half an inch. The peduncle varies much in length in the same species: in *L. anatifera*, it is occasionally above a foot long. The colours of *L. anatifera*, *L. Hillii*, and *L. anserifera*, are very bright and striking; the membrane bordering the valves and that round the top of the peduncle in two of the species, is of the brightest scarlet-orange; the valves, owing to the underlying corium, are pale blueish-grey, and the interspaces between them dark leaden-purple. The cirri and trophi are generally dark purple or lead-colour.

Filamentary appendages. These are attached to beneath the basal articulation of first pair of cirri; they vary in the several species, from one to five or six on each side, the lowest being always the longest. Several of them are occupied by testes. In *L. pectinata*, generally, not even one is developed. They are subject to great / variation in their proportional lengths, and in number, in the same species. These organs have generally been considered to serve as branchiae; I see no reason to believe that they are more especially designed for this end, than is the general surface of the body.

Mouth. The labrum is moderately bullate, the longitudinal diameter of this part equalling about one-third, or half of that of the rest of the mouth. The palpi are moderately developed. The mandibles (Pl. X, fig. 5) have five teeth with the inferior point either broad, or very narrow and tooth-like. The maxillae are step-formed (Pl. X, fig. 9); the first step is sometimes indistinct and curved; and in *L. pectinata*, all the steps vary much, and are more or less blended together. The outer maxillae (like those at Pl. X, fig. 16), are internally clothed continuously with spines. The olfactory orifices are not at all prominent.

Cirri. The first pair is placed near the second pair, and is of considerable length; the second has the anterior ramus thicker than the posterior ramus, and the segments brush-like; the segments (Pl. X, fig. 26) of the four posterior cirri bear from four to six pair of long spines, with a row of small intermediate spines: in the posterior cirri of *L. australis* the lateral rim spines are much developed; and in those of *L. fascicularis*, the usual pairs of large spines are lost in a broad triangular brush, formed by the increase of the lateral marginal, and intermediate spines.

Caudal appendages (Pl. X, fig. 18*b*), very small, either blunt or pointed, and quite destitute of spines.

The prosoma is well developed. The stomach is surrounded in the upper part by a circle of large branching caeca. The generative system is highly developed; the testes coating the whole of the stomach, entering the filamentary appendages and the pedicels of the cirri; the two ovigerous lamellae contain a vast number of ova; they are united to rather large fraena, of which the sinuous margin supports either a continuous row or separate tufts of glands. /

Distribution. The species abound over the arctic, temperate and tropical parts of the Atlantic, Indian and Pacific oceans, and are always, or nearly always, attached to floating objects, dead or alive. The same species have enormous ranges; in proof of which I may mention that of the six known species, five are found nearly all over the world, including the British coast; and the one not found on our shores, the *L. australis,* apparently inhabits the whole circumference of the southern ocean.

General remarks and affinities. The first five species form a most natural genus; they are often sufficiently difficult to be distinguished, owing to their great variability. The sixth species (*L. fascicularis*) differs to a slight extent in many respects from the other species, and has considerable claims to be generically separated, as has been proposed by Mr Gray, under the name of Dosima; but as it is identical in structure in all the more essential parts, I have not thought fit to separate it. As far as external characters go, some of the species of Paecilasma have not stronger claims, than has *L. fascicularis,* to be generically separated; and I at first retained them altogether, but in drawing up this generic description, I found scarcely a single observation applicable to both halves of the genus; hence I was led to separate Lepas and Paecilasma. If I had retained these two genera together, I should have had, also, to include the species of Dichelaspis and Oxynaspis; and even Scalpellum would have been separable only by the number of its valves; this would obviously have been highly inconvenient. Although some of the species of Paecilasma so closely resemble externally the species of Lepas, yet if we consider their entire structure, we shall find that they are sufficiently distinct; as indirect evidence of this, I may remark that Conchoderma (as defined in this volume), includes two genera of most authors, and yet certainly comes, if judged by its whole organization, nearer to Lepas than does Paecilasma. /

1. LEPAS ANATIFERA

PLATE I. FIG. 1 (var.)

L. ANATIFERA. Linnaeus, *Systema Naturae*, 1767
ANATIFA vel ANATIFERA vel PENTALASMIS laevis,[22] plerumque auctorum
ANATIFA ENGONATA [!].[23] Conrad, *Journal Acad. Nat. Sc. Philadelphia*, vol. vii, p. 262, Pl. xx, fig. 15, 1837
ANATIFA DENTATA (var.). Brugière, *Encyclop. Meth. (des Vers)*, 1789
PENTALASMIS DENTATUS (var.). Brown, *Illust. Conch.*, Pl. lii, fig. 5
ANATIFA . . . Martin St. Ange, *Mem. sur l'organisation des Cirripedes*, 1835

L. valvis aut laevibus aut delicate striatis: è duobus scutis, dextro solùm dentea interno umbonali instructo; pedunculi parte superiore fuscâ.
Valves smooth, or delicately straited. Right-hand scutum alone furnished with an internal umbonal tooth: uppermost part of peduncle dark-coloured.

Filaments, two on each side.

Var. (*a*). (Fig. 1.) Scuta and terga with one or more diagonal lines of dark greenish-brown, square, slightly depressed marks.

Var. (*b*). (Fig. 1*b*.) Carina strongly barbed.

Extremely common; attached to floating timber, vessels, seaweed, bottles, etc., and to each other, in the Atlantic Ocean, Mediterranean, West Indies, Indian Ocean, Philippine Archipelago, Sandwich Islands, Bass's Straits, Van Diemen's Land.

General appearance. Valves white, more or less translucent and thick, with a tinge of blueish-grey, from the underlying corium; sometimes brownish cream-coloured, rarely with a tint of purple. Surfaces smooth, / with traces of very fine lines radiating from the umbones, sometimes rather plain on the basal part of the scuta. Length in proportion to the breadth of the capitulum variable, owing to the varying degree to which the scuta and terga have their apices produced. *Scuta* with the occludent margin either considerably curved or nearly straight. The internal tooth of the right-hand scutum, close to the umbo, varies in size and form, being either pointed, square, or

---

[22] As this, though the commonest species, has never been defined, I give only a few synonyms and references, it being quite impossible to distinguish, in any published description, this species from *A. Hillii* of Leach; this latter species I recognize under this name only from having authentic specimens from the British Museum, as Leach overlooked every one of the real diagnostic characters.

[23] I have used, in conformity with botanists, the mark of interjection, to show that I have seen an authentic specimen.

obliquely truncated on either side, or it has a notch on the summit; internal basal rim of the scuta either plainly developed or nearly absent. In many specimens (Pl I, fig. 1), on the scuta, or on the scuta and terga (and sometimes more on one side of the individual than on the other), a nearly straight line, running diagonally across the capitulum, of slight, quadrilateral depressions, of a dirty greenish colour, with the edges blending away, is either conspicuously developed, or can only just be discerned. These marks increase in size from the umbones to the margins of the valves. There are sometimes two or even three rows on the scuta. They are formed by the retention of a portion of the chitine membrane, which is cast off the rest of the surface; the margins of the valves are occasionally notched slightly on the line of marks; there is no difference along this line in the underlying corium. Specimens both with and without a barbed carina are thus characterized. *Carina*; the interspace between the carina and the scuta and terga is not wide. The carina exteriorly, is either convex and smooth, or furnished with knobs or with extremely sharp, long teeth (Pl. I, fig. 1*b*); small specimens, with the capitulum under half an inch in length, are generally most strongly barbed.[24] Apex more or less acuminated; width and thickness variable; sides strongly furrowed. Fork (fig. 1*a*) generally less wide than the widest upper part of the valve, with the two prongs diverging from each / other at less than a right angle; their sharpness and precise form variable; rim between them reflexed (figs 1*a* and *b*), making a slight notch behind. *Peduncle* smooth, wrinkled, length in proportion to that of the capitulum varying, from barely equalling it, to six or seven times as long. I have noticed a specimen including mature ova, with a capitulum under half an inch long.

Filamentary appendages; never more than two on each side, with sometimes only one developed; of variable length; one seated on the flank of the prosoma, under the first cirrus; the second close under the basal articulation of this cirrus, on the posterior face of a slight swelling: these appendages correspond with *g* and *h* in Fig. 4, Pl. IX.

Mouth. Mandibles (Pl. IX, fig. 5), with, as usual, five teeth, all pointing downwards. Maxillae (Pl. IX, fig. 9), with the lower step of variable width compared to the two upper steps. *Cirri*; posterior cirri

---

[24] Mr W. Thompson found that 15 specimens, out of about 200, attached to a vessel which came from New Orleans into Belfast, had their carinas barbed.

with segments (fig. 26) bearing six pair of spines; intermediate fine spines rather long; first cirrus, anterior ramus longer by only about two segments than the posterior ramus; second cirrus with anterior ramus, with very broad transverse rows of bristles; spine-bearing surfaces considerably protuberant; caudal prominences smooth, rounded.

Size. The largest specimen which I have seen had a capitulum two inches in length; the longest, including the peduncle, was sixteen inches.

Colours. Calcareous valves already described. Edges of the orifice bright scarlet orange; basal edges of the scuta, and sometimes of all the valves, with a torn border of orange membrane. Interspaces between the valves dull orange-brown. Peduncle darkish purplish-brown, with the lower part sometimes pale; chitine membrane itself tinted orange; in young specimens, peduncle pale, the colour first appearing in the uppermost part, close under the capitulum; this upper part is often darker than the other parts, and never orange-coloured, as in *L. Hillii* and *L. anserifera*. Sack internally dark purplish lead colour, / sometimes with a tinge of orange, darkest under the growing edges of the valves; body of animal pale purplish lead colour. The four posterior cirri blackish purple; the second, and often the third cirrus, appear as if the colour had been laterally abraded off; these latter cirri have sometimes a tinge of orange. In very young specimens, the cirri are only barred with purple. The ova and the contents of the ovarian tubes are of a beautiful azure blue, becoming yellow in spirits.

In museums a vast amount of difference is seen in the colours of this species, caused by the method of preparation: if dried without having been in spirits, and subsequently kept dry, the orange tint round the orifice is preserved; if kept long in spirits, this is quite lost; but sometimes in specimens in spirits the colour of the membrane of peduncle is preserved and rendered pinker. The colours of the sack and animal are either quite discharged or rendered extremely dark. The valves themselves also often become more opaque. In some specimens well preserved in spirits, the sack and cirri were purplish-brown or lead colour, tinted with dirty green, or orange, or bright yellow, or brick-red.

General remarks. From the foregoing description it will be seen how extremely variable almost every part of this species is. I find, in the

British Museum, ten distinct specific names given by Dr Leach to different varieties, or rather to different specimens, for some of them are undistinguishable. A specimen from the Sandwich Islands, sent by Mr Conrad to Mr Cuming, is marked *A. engonata*.

.In looking over a large collection of specimens in a museum, the most distinctive characters appear at first to be the colours, the dentation or barbed condition of the carina, the row of square marks on the scuta and terga, and the more or less produced form of the whole capitulum: all these characters are absolutely worthless as distinctive characters, and blend into each other. In a fresh condition, the colours of this species, and of / *L. anserifera* and *L. Hillii* are surprisingly alike, though in *L. anatifera* alone, the uppermost part of the peduncle is dark. As far as I have seen, the smoothness of the valves, together with the presence of a tooth beneath the umbo, on the right-hand scutum, and its entire absence on the left side (in other species it is smaller on this, than on the right-hand side), is an unfailing diagnostic mark. I believe this species is always attached to floating objects, though there are some very young specimens in the British Museum, collected by Sir G. Grey, adhering to sandstone, but this may have been buoyed up by some large seaweed. Mr Peach has given me the particulars of two instances, in which, after gales of wind, this species, of nearly full size, adhering to *apparently* freshly broken-off Laminariae, has been cast upon the coast of England and Scotland.

2. LEPAS HILLII

PLATE I. *FIG. 2*

ANATIFA vel PENTALASMIS LAEVIS [!] plerumque auctorum
PENTALASMIS HILLII [!]. Leach, *Tuckey's Congo Expedit.*, p. 413, 1818
PENTALASMIS CHELONIAE [!]. Leach, *Tuckey's Congo Expedit.*, p. 413, 1818
ANATIFA TRICOLOR [?]. Quoy et Gaimard, *Ann. des Sc. Nat.*, 1st series, tom. x, Pl. vii, fig. 7, 1827, et *Voyage de l'Astrolabe*, Pl. xciii, fig. 4
ANATIFA SUBSTRIATA [!]. Conrad, *Journal Acad. Nat. Sc., Philadelphia*, vol. vii, p. 262, Pl. xx, fig. 14, 1837

L. valvis laevibus; scutorum dentibus internis umbonalibus nullis; carinâ à caeteris valvis, furcâ etiam a scutorum basali margine, paululum distante; pedunculi parte superiore aut pallidâ aut aurantiacâ.

Valves smooth; scuta destitute of internal umbonal teeth; carina standing a little separate from the other valves, with the fork not close to the basal margin of the scuta; uppermost part of peduncle either pale or orange-coloured.

Filaments three on each side. /

Extremely common; attached to ships' bottoms, from all parts of the world; on floating timber; associated with *L. anatifera* and *L. anserifera.* Mediterranean. Attached to turtles, in the Atlantic, lat. 30° north. West Indies. Falkland Islands. 'South Seas', collected by A. Menzies. Port Stephen, Australia.

General appearance. Capitulum laterally flat; length varies in proportion to the breadth; valves white, somewhat translucent, moderately thick, very smooth, but with faint traces of radiating lines; in some varieties, surface rather irregular along the zones of growth. *Scuta* without any internal teeth, and with scarcely any trace of the internal basal rim; upper angle little acuminated; the occludent margins of the two scuta stand rather separate from each other, showing a wide space of corium between them: these margins are arched and protuberant, but with the lower part a little hollowed out; basal margin a little curved. In one specimen alone, I saw a trace of a diagonal line of square coloured marks, like those common in *L. anatifera. Terga* rather broad, with the basal angle not much acuminated. The degree of prominence and outline of the double occludent margin varies very much. *Carina,* separated by a rather wide space from the scuta and terga; of very varying shape, the upper part not much acuminated, generally very flat, sometimes exteriorly marked by a central depressed line; never barbed; occasionally (in a specimen from Australia), middle part so wide as almost to become spoon-shaped; on the other hand occasionally of nearly the same width throughout; somewhat constricted above the fork. Fork deeply embedded as usual; situated, in fresh specimens, a little way beneath the basal margins of the scuta, instead of touching them, as in the other species; forks of varying width, not so abruptly inflected as in many species; sometimes much narrower than the upper widest part of the valve, sometimes nearly twice as wide; prongs of fork not very sharp, diverging at about a right angle, with the rim between them reflexed. The apex of the carina extends up between the terga for barely half their length, / instead of up fully three-fourths of their length, as in *L. anatifera.*

The chitine membrane at the base of the capitulum, especially at the anterior and posterior ends, is covered with beautiful, little, embedded, yellowish beads about 3/2000th of an inch in diameter; above this, on each side of the carina, there is a space with similar but smaller little spheres, and still higher up still minuter ones; others occur on different parts of the capitulum; these spaces are seen to be distinctly separated from each other, and present a beautiful appearance under a high power.

Peduncle, as long as, or rather longer than, the capitulum: in one set of specimens, however, it was thrice or four times as long as the capitulum. The peduncle, in some specimens, was conspicuously covered with transverse plates of yellowish hard chitine.

Filamentary appendages. Three on each side; one on the flank of the prosoma, with a pair beneath the basal articulation of the first cirrus; relative lengths various, but the posterior filament of the pair under the cirrus, is the shortest. *Mouth*; palpi not much acuminated; maxillae step-formed, but with the upper or first step in some specimens indistinct, or forming a curve. *Cirri*; the segments of the first cirrus and of the posterior arm of the second cirrus are highly protuberant, the protuberances sometimes equalling half the thickness of the segments themselves. Caudal appendages smooth, rounded.

Size. The largest specimen which I have seen in the collection of Mr Cuming, had a capitulum 1 1/10th of an inch long, and 1 1/4 wide; therefore not quite equalling in size the largest specimens of *L. anatifera*.

Colours. When fresh, valves blueish-grey from the underlying corium, edges of all the valves and round the orifice, and round the top of the peduncle, bright orange-yellow, passing into the finest scarlet, and varying slightly in tint in different specimens. Space between the carina and the other valves, and between the occludent margins of the scuta, rich purplish-brown; peduncle / either pale or purplish-brown, or only clouded on the sides with the same. In young specimens, peduncle nearly colourless; and in those under a quarter of an inch long in the capitulum, the top of the peduncle has not acquired its orange tint. Sack pale, leaden-purple, body the same, but paler and more reddish; cirri (but only the tips of first pair) tinted with fine golden orange. Immature ova in peduncle beautiful blue. After being long kept in spirits, the colours are changed, weakened, or discharged, as in *L. anatifera* and *L. anserifera*, and the valves become opaque. In some long-kept specimens the corium everywhere had become pale brown; more usually it assumes a dirty purplish lead colour.

Monstrous variety. Among a set of ordinary specimens from a ship from Genoa, sent me by Mr Stutchbury, there were three, one full-grown and two very young, with the whole capitulum (and likewise with the scuta and terga taken separately), not above half the usual length in proportion to the breadth. Neither the colours nor animal in this variety presented any difference.

General remarks. This species is almost universally confounded with *L. anatifera.* Quoy and Gaimard, however, appear to have distinguished it, under the name of *A. tricolor*, from its colours. Leach named it accidentaly, for he specifies not one distinctive character, and besides his two published names, he has appended two other names to specimens in the British Museum. A specimen, from the Sandwich Islands, sent by Mr Conrad to Mr Cuming, is marked *A. substriata.* In a dry state, from the shrinking of the membranes, and consequent approach of the carina to the other valves, and of the fork to the basal margin of the scuta, it is most difficult to distinguish this species, though so decidedly distinct, from *L. anatifera*; the absence, however, of a tooth on the underside of the right-hand scutum is at once characteristic. Even in specimens kept in spirits, in which there has been no shrinking, but in which the colours have changed, and taking into / account the variation in the carina and upper part of the terga, this species is not always readily distinguished from *L. anatifera*, without opening the valves and looking for the right-hand tooth of the latter. In fresh specimens, the orange ring at the top of the peduncle, and the broad purplish interspace between the carina and other valves, are characteristic. In all states, the filamentary appendages offer a good character.

3. LEPAS ANSERIFERA

PLATE I. FIG. 4

L. ANSERIFERA. Linnaeus, *Syst. Naturae*, 1767
ANATIFA STRIATA. Brug, *Encyclop. Meth. (des vers)*, Pl. clxvi, fig. 3
PENTALASMIS DILATATA [!] (young). Leach, *Tuckey's Congo Expedit.*, p. 413, 1818
ANATIFA SESSILIS [?]. Quoy et Gaimard, *Voyage de l'Astrolabe*, Pl. xciii, fig. 11
LEPAS NAUTA.[25] Macgillivray, *Edin. New Phil. Journ.*, vol. xxxviii, p. 300
PENTALASMIS ANSERIFERUS. Brown, *Illust. Conch.*, Pl. li, fig. 1, 1844

[25] Professor Macgillivray does not consider the species, which he has described

L. valvis approximatis leviter sulcatis (tergis praecipuè); scuto dextro dente forti interno umbonali, laevo aut dente exiguo, aut merâ cristâ instructo; margine occludente arcuato, prominente: pedunculi parte superiore aurantiacâ.

Valves approximate, slightly furrowed, especially the terga; right-hand scutum with a strong internal umbonal tooth; left-hand with a small tooth, or mere ridge; occludent margin arched, protuberant: uppermost part of peduncle orange-coloured. /

Filaments five or six on each side.

Var. (*dilatata*, young); valves rather thin, finely furrowed, often strongly pectinated; scuta broad, with the occludent margins much arched, making the space wide between this margin and the ridge connecting the umbo and the apex: carina often barbed.

Common on ships' bottoms from the Mediterranean, West Indies, South America, Mauritius, Coast of Africa and the East-Indian Archipelago. Central Pacific Ocean. China Sea. Chusan. Sydney. Attached to pumice, various species of fuci, Janthinae, Spirulae; often associated with *L. anatifera* and *L. Hillii*, and, in a young state, with *L. fascicularis*.

General appearance. Capitulum more or less elongated relatively to its breadth; in two specimens, with scuta of equal width, one was longer than the other by the whole of the occludent margin of the terga. Valves white, thick (in young specimens sometimes diaphanous and thin), closely approximate to each other; surfaces furrowed to a very variable amount. Terga generally more plainly furrowed than the scuta, of which the basal portion is generally less furrowed than the upper part; ridges, often rough, generally much narrower than the furrows: in half-grown specimens (var., *dilatata* of Leach), the ridges are frequently denticulated, and there is even sometimes a row of bead-like teeth along the basal margins of the scuta. The ridges vary much, sometimes alternately wide and narrow; in two specimens of equal size, there were, in one, thirty-two ridges, and in the other only eighteen, on the scutum.

Scuta, with the occludent margin rounded and protuberant to a

---

under *L. nauta*, and which I cannot doubt is the same with the present species, as the *L. anserifera* of Linnaeus; but I find it so named in all old collections, and it seems to agree very well with Linnaeus's description. There has been much groundless confusion about this species; I have no hesitation in giving *A. striata*, of Brugière, as a synonym, though I have received from Paris the *Lepas pectinata* of this volume, named as the *A. striata*; and on the other hand, Poli has incorrectly called a common variety of *L. pectinata* by the name of *L. anserifera*.

variable degree, but always leaving a rather wide space between the margin, and the ridge which runs from the umbo to the apex; apex pointed. Right-hand internal tooth considerably larger than that on the left, which is often reduced to a mere ridge; internal basal rim thick, sometimes furrowed along its upper edge, but of variable thickness, sometimes not extending as far as the baso–carinal angle. *Terga*, sometimes / equalling, sometimes only two-thirds of, the length of the scuta; in young specimens, the two occludent margins form a right-angle with each other; in older specimens they form less than a right-angle, and hence the portion of valve thus bounded is unusually protuberant. *Carina*, within deeply concave; exterior sides finely furrowed longitudinally, generally denticulated; valve only slightly narrowed in above the fork, of which the prongs diverge at an angle of 90°, or rather more, and are wider than the widest upper part of the valve; rim between the prongs reflexed; the heel or external angle, just above the fork, sometimes considerably prominent. I have seen only a single large specimen with its carina barbed. In half-grown specimens (var. *dilatata*, Leach), the carina is often strongly barbed, with the upper point much acuminated, the fork about twice as wide as the widest upper part, and the prongs diverging at rather more than a right-angle. In some specimens, especially very young ones, there are at the base of the carina, above the fork, some strong, downward-pointed, inwardly-hooked, calcareous teeth; such occur also in some specimens along the basal margins of the scuta, two of these hooked teeth under the umbones of the scuta being larger than the rest: specimens conspicuously thus characterized came from the Navigator Islands; in these, I may add, the acutely triangular primordial valves were quite plain.

Peduncle, generally about as long as the capitulum; in young specimens generally short.

Filamentary appendages, generally five, sometimes six, on each side; one is seated on the side of the prosoma, and the four others placed in pairs beneath the basal articulation of the first cirrus; the lowest posterior filament of the four generally is the largest. In young specimens, having a capitulum only half an inch long, the upper pair of the four often is not developed, or is represented by mere knobs. The mouth presents no distinctive characters. *Cirri*, with the longer ramus of the first pair almost equal to the shorter arms of the / second pair; spine-bearing surfaces only slightly protuberant. Caudal appendages smooth, curved, pointed.

Size. The largest specimen which I have seen, had a capitulum one inch and a half in length.

Colours. The white valves are edged with bright orange membrane; and are so close to each other that no interspaces, coloured from the underlying corium, are left. Peduncle, dark orange-brown, with the uppermost part under the capitulum bright orange all round; the chitine membrane itself being thus coloured. Sack, internally, dark purplish lead colour. Body and cirri, either nearly white or pale purplish-lead colour, with the arms of the second, third, and fourth cirri, and pedicels of the fifth and sixth, more or less tinted with orange. A specimen preserved during fourteen months in good spirits had only a tinge of orange left round the orifice and round the upper part of peduncle, and on the cirri. In some other specimens, badly preserved, the chitine membrane was quite colourless, and sack and cirri dirty lead colour. Fresh ova, peach-blossom-red; immature ova, in ovarian tubes, pale pink.

Monstrous variety. In Mr Stutchbury's collection, there was a specimen, with the scuta, broad, smooth, thin, and fragile, without any ridge running from the umbo to the apex, and with the occludent margin reflexed. This seemed caused by the shell having been attacked by some boring animal, and from having supported Balani. In the same specimen the first cirrus on one side was monstrously thick and curled; the second cirrus had its posterior ramus in a rudimentary condition. In Mr Cuming's collection, there are small specimens with the zones of growth overlapping each other, with thick irregular margins, and with the carina distorted.

This species has cost me much trouble: I have examined vast numbers of specimens, from a tenth to half an inch in length, attached to light floating objects, such as Janthinae and Spirulae from the tropical oceans, which all resembled each other, and slightly differed / from the common appearance of *L. anserifera*: this variety is the *Pentalasmis dilatata* of Leach; and for a long time I considered it as a distinct species. It differs from *L. anserifera*, in the less thickness of the valves, in their being more finely and yet plainly furrowed; in the greater width of the scuta; and more especially of that part of the valve lying between the occludent margin, and the ridge running from the umbo to the apex; in the less elongation of the area in the terga, bounded by the two occludent margins; and, lastly, in the less size of the whole individual. The trophi and cirri are absolutely identical. Lately, however, in carefully going over a great suite of specimens, all the above few distinctive characters broke down and insensibly graduated away; and I am convinced that this form is only a variety of *L. anserifera*; its different aspect being caused partly by youth, but chiefly, I suspect, from being attached to light objects floating close to the surface of the sea.

The *Lepas anserifera* can be distinguished by the slight furrows on its valves from all the other species, excepting *L. pectinata*: this latter species can be readily known, by the close proximity in the scuta of the occludent margin, and the ridge extending from the umbo to the apex; by its carina being very narrow above the fork; by the prongs of the fork diverging at an angle of from 135° to 180°; by the thinness of its valves; by the coarseness of the furrows on them; and lastly, by there being at most in *L. pectinata* only one filamentary appendage beneath the first cirrus.

4. LEPAS PECTINATA

PLATE I. FIG. 3

LEPAS PECTINATA. Spengler, *Skrifter Naturhist. Selbskabet*, 2, B, 2, H., Tab. X, fig. 2, 1793

LEPAS MURICATA (var.). Poli, *Test. Utriusque Scicil.*, vol. i, Pl. vi, figs 23, 29, 1795

LEPAS ANSERIFERA. Poli, *Test. Utriusque Scicil.*, vol. i, Pl. vi, figs 25–7

LEPAS SULCATA. Montagu, *Test. Brit.*, Pl. i, fig. 6, 1803

PENTALASMIS SULCATA. Leach, *Encyclop. Brit. Suppl.*, tom. iii, Pl. lvii, 1824

PENTALASMIS SPIRULAE [![(var). Leach, *Tuckey's Congo Expedit.* Appendix, 1818

PENTALASMIS RADULA (var.) et SULCATUS. Brown, *Illust. of Conchology*, Pl. li, figs 3–6, 1844

PENTALASMIS INVERSUSS. CHENU, *Illust. Conchy.*, Pl. i, fig. 14.

ANATIFA SULCATA. Quoy et Gaimard, *Voyage de l'Astrolabe*, Pl. xciii, figs 18, 20[26]

L. valvis tenuibus, crassè sulcatis, saepe pectinatis; scutorum cristâ prominente ab umbone ad apicem juxta marginem occludentem pertinente: furcae carinalis cruribus inter angulos 135° et 180° divergentibus.

Valves thin, coarsely furrowed, often pectinated. Scuta with a

---

[26] I may add, that I have received many specimens incorrectly labelled *A. striata*, which is properly a synonym of *L. anserifera*.

prominent ridge extending, from the umbo to the apex, close to the occludent margin; fork of the carina with the prongs diverging at an angle of from 135° to 180°.

Filaments absent, or only one on each side.

Var. (Pl. I, fig. 3a), upper part of the terga (bounded by the two occludent margins) produced and sharp; surface of all the valves often coarsely pectinated, and with the carina barbed.

Atlantic Ocean, from the North of Ireland to off Cape Horn; common, under the tropics; Mediterranean: attached to wood, cork, charcoal, seaweed, a reed-like leaf, spirulae, cuttlefish bones, to a bottle together with *L. anatifera*; to a ship's bottom, Belfast (W. Thompson). Often associated with *L. fascicularis*. Montagu states ('Test. Brit.', p. 18) that this species is sometimes attached to the fixed *Gorgonia flabellum*.

General appearance. The capitulum varies considerably in length compared to its breadth, caused chiefly by the greater or less production of the occludent portion of the terga; valves thin, brittle; the furrowed surface varies / much in character, narrow and broad ridges often alternating; frequently each ridge (but more especially the ridge running from the umbo to the apex of each scutum, and sometimes that alone), is covered with prominent, curled, flat calcareous spines, giving the shell an appearance like that of many mollusca. Other specimens show no trace of these calcified projections. From the thinness of the valves and the depth of the furrows, the margins of the valves are sinuous. *Scuta*: the ridge running from the umbo to the apex is unusually prominent and curved; it runs very close to the occludent margin, so that, differently from in all the other species, only a very narrow space is left between this margin and the ridge. Internal teeth, under the umbones, either sharp and prominent, or mere knobs; sometimes that on the right side is much larger than that on the left; sometimes they are nearly equal; sometimes that on the left is scarcely distinguishable. Internal basal rim absent, or barely developed.

Terga. These valves have a conspicuous notch to receive the apex of the scuta; the two occludent margins either meet each other at a rectangle, or at a much smaller angle, causing the portion thus bounded to vary much in outline, area, and degree of prominence. This at first led me to think that the *P. spirulae* of Leach, in which the point is very sharp and prominent, was a distinct species; but there are so many intermediate forms, that the idea must be given up. I may remark, that in all the species of Lepas, the upper part of the tergum seems particularly variable. The degree of acumination of the basal portion of the tergum also varies; the internal surface sometimes has small crests radiating from the umbo.

Carina, broad, within deeply concave; edges sinuous, externally sometimes strongly barbed; narrow above the fork, which latter is wider than the widest upper part of the valve; prongs sharp, thin, diverging at an angle of from 135° to 180°; the rim connecting the prongs not, or only slightly, reflexed.

Peduncle, narrow, shorter than the capitulum. /

Filamentary appendages, none, or only one, short, obtuse projection on each side, on the posterior face of the swelling under the first cirrus.

Mouth. Mandibles, with the inferior point produced into a single pectinated tooth, rarely into two pectinated teeth; on one side of one specimen, there were only four instead of five teeth. Palpi very narrow. Maxillae highly variable; they may be described as formed of five steps, of which the two lower ones are generally united into a single one, divided by a mere trace of a notch; or with the three lower steps blended into an irregular, projecting surface, and with even the fourth step indistinct. I have seen these two extreme forms on opposite sides of the mouth of the same individual – on one side the maxillae being regularly step-form, on the other the whole inferior part forming an almost straight edge, standing high up above the first notch or step which bears the two upper great spines.

Cirri. First pair rather far removed from the second pair, with the longer ramus about three-fourths of the length of shorter ramus of second cirrus; spine-bearing surfaces, hardly at all protuberant; lateral marginal spines on the posterior cirri rather long; caudal appendages smooth, rounded, extremely minute: penis very spinose.

Size. Capitulum in the largest specimen, six-tenths of an inch long; only a few arrive at this size.

Colours, after having been kept in spirits – sack and cirri, especially first cirrus, clouded with pale purple; peduncle brownish; valves appear blueish in specimens not long preserved, but in specimens kept longer they become perfectly and delicately white.

General remarks. Under the head of *L. anserifera,* I have made some remarks on the diagnostic characters of this species. In the thinness of the valves – form of the carina, with the rim connecting the prongs being not, or scarcely, reflexed – and in the shortness and narrowness of the peduncle, there is some approach to *L. australis,* and thence to *L. fascicularis.* In the form of the maxillae / – in one specimen having the mandible on one side bearing only four teeth – and in the frequent absence of filamentary appendages, there is some approach to the genus *Paecilasma*; but there is no such approach in the characters derived from the capitulum. We have seen that, as in so many other species of this genus, most of the parts are variable, and this is the case to a most unusual extent in the form of the maxillae. Dr Leach has attached eight specific names to the specimens preserved in the British Museum.

5. LEPAS AUSTRALIS

PLATE I. *FIG.* 5

L. Valvis glabris, tenuibus, fragilibus; scutorum dentibus umbonalibus utrinque internis; carinae parte superiore latâ, planâ, suprâ furcam valdè constrictâ; furcae cruribus latis, planis, tenuibus, acuminatis, intermedio margine non reflexo.

Valves smooth, thin, brittle; scuta with internal umbonal teeth on both sides. Carina with the upper part broad, flat; much constricted above the fork, which has wide, flat, thin, pointed prongs, with the intermediate rim not reflexed.

Filaments, two on each side.

Common on Laminariae in the whole Antarctic Ocean: Bass's Straits, Van Diemen's Land: Bay of Islands, New Zealand, lat. 35°S.: lat. 50°S., 172°W.: coast of Patagonia, lat. 45°S.: attached to bottom of H.M.S. *Beagle,* lat. 50°S., Patagonia: attached to a Nullipora (I presume a drift piece), British Museum.

General appearance. Capitulum rather obtuse and thick; valves thin, brittle, approximate, either white and transparent, or dirty-brown and opaque; or sometimes tinted internally with purple (perhaps the effects of being preserved in spirits); surface plainly marked by lines of growth, rarely marked with traces of lines radiating from the umbones. *Scuta* with teeth on both sides, / nearly equal; internal basal rim rather wide, sometimes furrowed; basal margin considerably curved inwards. *Terga* rather wide; basal angle blunt; angle formed by the two occludent margins blunt and rounded. *Carina* (fig. 5a) with the apex blunt, flat; the middle part generally very broad; much constricted above the fork, where it is internally deeply concave, and externally carinated; fork twice as broad as the broadest upper part of the valve; with the prongs flat, broad, thin, pointed, diverging at about an angle of 75°, with the intermediate rim not at all reflexed; the fork generally not deeply imbedded in the chitine membrane of the peduncle, so as to be quite easily visible externally; sometimes there is an internal, transverse, depressed line on the fork. In young specimens, with the capitulum about a quarter of an inch long, the fork of the carina is not developed, the lower slightly inflected portion consisting simply of an oval plate, twice as wide as the upper part. Until I had carefully examined a perfect series, showing the gradual changes in this part, I did not doubt that the young specimens formed a distinct species, and named it accordingly: the shortness of the penis first made me perceive that the specimens were immature. At this early age, I may add, the filamentary appendages were not developed. *Peduncle* either quite short, or as long as the capitulum, close under which it is considerably constricted all round.

Filamentary appendages. Two on each side; one long, tapering, placed on the prosoma (in one specimen represented by a mere knob), and the second shorter, situated on the posterior margin of the swelling beneath the first cirrus.

Mouth. Maxillae, with three large spines at the upper angle, and with the first step distinct, but narrow; mandibles with five teeth; in young specimens the inferior point ends in a single spine; sides of the supra-oral cavity very hairy; the membrane, forming the inner fold of the labrum, yellow and thickened in the form of a spoon.

Cirri. In the posterior cirri there are, at the upper / lateral edges of the segments on *both* sides, small spines; the segments in the first cirrus, and in the broad anterior ramus of the second cirrus, are hemispherically and considerably protuberant. Caudal appendages smooth.

Size. The largest specimen had a capitulum one inch long.

The *Colours* (after having been long in spirit) of the valves have already been given; sack and peduncle dirty yellowish-brown, with the parts corresponding to the margins of the valves much darker brown, or almost black; segments of the cirri clouded with dark brown; body and pedicels of the cirri dirty yellowish. I have reason to believe that the colours are totally different in living specimens.

Monstrous varieties. Most of the specimens from lat. 50°S., on the coast of Patagonia, were more or less deformed, with the successive zones of growth overlapping each other, and forming coarse concentric ridges. The carina in several specimens was laterally distorted.

I have already remarked that this species has some affinity to *L. pectinata*; but it is much more closely related to *L. fascicularis*, the affinity being clearly shown by the thinness and translucency of the valves, their convexity, by the width and little acumination of the upper part of the carina, by the width of the fork, and by its not being deeply imbedded. In young specimens, moreover, before the fork is fully developed, there is a remarkable similarity between the two species, in the form of this lower part of the carina. Again, the narrowness and inflection of the peduncle under the capitulum in *L. australis*, and lastly, the lateral marginal spines on both sides of the segments of the posterior cirri, all clearly indicate this same affinity to *L. fascicularis*.

I believe this species is confined to the southern ocean; and perhaps there represents *L. fascicularis* of the northern and tropical seas. It must, judging from the number of specimens brought home by Captain Sir J. Ross, and from those previously in the British Museum, and from those collected by myself, be a very common species. /

6. LEPAS FASCICULARIS

PLATE I. FIG. 6

LEPAS FASCICULARIS. Ellis and Solander, *Zoophytes*, Pl. xv, fig. 5, 1786
LEPAS FASCICULARIS. Montagu, *Test. Brit. Suppl.*, pp. 5, 164, 1808
LEPAS CYGNEA. Spengler, *Skrifter Naturhist. Selbskabet*, Bd. i, Pl. xi, fig. 8, 1790
LEPAS DILATA. Donovan, *British Shells*, 1804
PENTALASMIS FASCICULARIS. Brown, *Illust. Conch.*, Pl. li, fig. 2, 1844
PENTALASMIS SPIRULICOLA [!] et DONOVANI [!]. Leach, *Tuckey's Congo Expedit.*, p. 413, 1818
ANATIFA VITREA. Lamarck, *Animaux sans Vertebres*
DOSIMA FASCICULARIS [!]. J. E. Gray, *Annals of Philosophy*, vol. x, 1825
PENTALEPAS VITREA. Lesson, *Voyage de la Coquille, Mollusca*, Pl. xvi, fig. 7, 1830
ANATIFA OCEANICA [!]. Quoy et Gaimard, *Voyage de l'Astrolabe*, Pl. xciii

L. valvis glabris, tenuibus, pellucidis; carinâ rectangulè flexâ, parte inferiore in discum planum oblongum expansâ.

Valves smooth, thin, transparent; carina rectangularly bent, with the lower part expanded into a flat oblong disc.

Filaments, five on each side; segments of the three posterior cirri with triangular brushes of spines.

Var. (*Donovani*, of Leach). Carina with the upper part flat, spear-shaped, externally with a narrow central ridge.

Var. (*Villosa*, Pl. I, figs 6*b*, *c*). Valves placed rather distant from each other; carina extremely narrow, with the upper part of nearly the same width throughout; terga with the lower part much acuminated; body of animal finely villose.

Coasts of Great Britain and France; Baltic Sea, according to Montagu Southern United States (from Agassiz); tropical Atlantic Ocean; East-Indian / Archipelago, off Borneo and Celebes; Pacific Ocean, between the Sandwich and Mariana Archipelagos; New Zealand: attached to fuci, Spirulae Janthinae, Velellas, often to feathers and cork; often associated with the young of *L. anserifera* (var. *dilatata*), and *L. pectinata*.

General appearance. Capitulum highly variable in all its characters; thick and broad in proportion to its length, but the breadth is variable – in some specimens, the capitulum being longer by one-fifth of its total length than broad; in others, one-fifth broader than long. Valves generally approximate; in some varieties, however, from the narrowness of the carina and terga, the valves stand far apart, there being an interval between the carina and scuta of nearly half the breadth of the latter. Valves excessively thin, brittle, transparent, colourless, smooth, but generally sinuous along the zones of growth, which are conspicuous: valves generally covered throughout by thin chitine membrane, which is thickly clothed, especially in the interspaces between the valves, with minute spines, barely visible to the naked eye. *Scuta* with the lower part of the tergo–carinal margin extremely protuberant; occludent margin, more or less, but slightly reflexed, with a depressed line running from the umbo to the apex; basal margin much reflexed, but to a variable extent and at a varying angle, even up to a right angle – an external rim or collar being thus formed. There are no distinct *internal* teeth, but the basal margin under the umbones, is more or less distinctly produced into a rounded disc or projection, which is generally not so much outwardly reflexed as the rest of the basal margin; there is no distinct internal basal rim. The primordial valves are generally visible, but they do not lie, as in all other species, close to the basal margin, but a little above it – the lower reflexed portion having been subsequently developed. *Terga* flat, with the occludent margin slightly arched, and not, as in the foregoing species, formed of two sides; apex bent towards the carina; width of the lower half highly variable, owing to the varying extent to which the scutal margin is hollowed / out; in some specimens, the whole lower half beneath the apex of the scuta is of nearly the same width throughout; in other specimens this lower part is pear-shaped. The widest part of the tergum either equals in width, or is only two-thirds of the width of the widest part of the carina beneath its umbo. *Carina* (Pl. I, fig. 6a) highly variable in shape, with the part above the umbo either spear-shaped and slightly concave within, or nearly flat and furnished with a central external ridge; or the upper part (fig. 6c) is of equal and extreme narrowness throughout, and deeply concave within, appearing as if only the central ridge had been developed. The part below the umbo (answering to the fork in the foregoing species), is about one-third of the length of the whole valve, and generally twice as wide as the upper part, but in the variety with the upper part of the carina equally narrow throughout, the lower part is thrice as wide as the upper; the disc, or lower part, is generally slightly concave within, exteriorly either with or without a central ridge; basal margin rounded; lateral margin more or less curved, according to the form of the upper part. The disc is not more deeply imbedded in membrane than is the upper part of the valve. The heel or umbo is either angular and prominent, or rounded. In very young specimens the carina is simply bowed, instead of being rectangularly bent.

Peduncle – short, narrow, being abruptly inflected all round under the basal edges of the capitulum; lower part of very variable shape, being often suddenly contracted into a mere thread (fig. 6b), which sometimes widens again at the extreme end. The external membrane is very thin, and is penetrated by the usual fine tubuli leading to the corium; its surface is wrinkled and destitute of spines, or with extremely few. The peduncle is often completely surrounded by a yellowish ball (of which I have seen specimens from the coast of England, and from off Borneo), sometimes half as wide as the capitulum, composed of very tender, vesicular, structureless / membrane, and of a pulpy substance: perhaps the yellow colour may be owing to long immersion in spirits. Some authors have supposed that the ball was the ovisac of the animal; and for the first few minutes, deceived by the numerous included spores of, as I believe, Bacillariae, I thought that this was the case; others have supposed that it consisted of some encrusting algae or other foreign organism; but it is, in reality, a most singular development of the cement tissue, which ordinarily serves to attach cirripedes by their bases to some extraneous object, but here surrounding that object and the peduncle, gives buoyancy, by its vesicular structure, to the whole. The membrane of the ball falls to pieces in caustic potash, differently from the chitine membrane of the enclosed peduncle, and this shows that there is some difference in composition from ordinary cement. The ball, when cut in two, exhibits an obscure concentric structure. The whole is excreted by the two cement ducts, through two rows of orifices, one on each side of the surrounded portion of the peduncle; and I actually traced, in one case, the yellow pulpy substance coming out of the cement ducts. The upper apertures are in gradation larger than those below them, and they stand a little further apart from each other; these are figured as seen from the outside, much magnified (Pl. I, fig. 6d). I did not succeed in finding the cement glands, but I followed the ducts, of rather large size, running for a considerable distance as usual along and within the longitudinal muscles of the peduncle. Nearly opposite the uppermost aperture, on each side, the duct passes out through the corium, and becomes laterally attached to the outer membrane of the peduncle, at which point an aperture is formed (as in other cases, by some unknown process), thus giving exit to the contents of the duct. Beneath this upper aperture the duct runs down the peduncle, between the corium and the outer membrane, till it comes to the next aperture, to which it is also attached, and so on to all the lower ones; but I / believe no cement tissue continues to pass out through these lower apertures. Beneath the lowest aperture the two ducts run into the two prehensile antennae of the larva, which, as usual, terminate the peduncle. The antennae are attached to some small foreign body in the centre of the vesicular ball, by the usual tough, light brown, transparent cement. The two upper apertures are nearly on a level with the outside surface of the ball; and it was evident that as the animal grows, new apertures are formed higher and higher up on the sides of the peduncle, and that out of these, fresh vesicular membrane proceeds, and grows over the old ball in a continuous layer. It appears that the growth of the vesicular ball is not regular – that it is not always formed – and that when formed the whole, or the lower part sometimes disintegrates and is washed away. As that portion of the peduncle which is enclosed ceases to grow, and has its muscles absorbed, retaining only the underlying corium, whereas the upper unenclosed portion, and likewise (as it appears), lower portions once enclosed but since denuded, continue to increase in diameter, the peduncle, when the vesicular ball is removed, often has the most irregular outline, contracting suddenly into a mere thread, and then occasionally expanding again at the basal point.

Frequently two or three specimens have their peduncles imbedded in one common ball, of which there is a fine specimen in the College of Surgeons (Pl. I, fig. 6), the ball being about one inch and a quarter in diameter, with a slice cut off. In this specimen, it is seen that the vesicular membrane proceeding from several individuals, unites to form one more or less symmetrical whole, and that the original common object of attachment is entirely hidden. Dr Coates[27] gives a curious account of the infinite number of specimens, through which he sailed during several days, in the Southern Atlantic Ocean: the balls appeared like bird's eggs, and were mistaken for some / fucus, which was supposed to have encrusted the scales of the Velellae, to which the cirripede had originally become attached. Several individuals had their peduncles imbedded in the same ball, 'which floated like a cork on the water'. As this species grows into an unusually bulky animal, we here see a beautiful and unique contrivance, in the cement forming a vesicular membranous mass, serving as a buoy to float the individuals, which, when young and light, were supported on the small objects to which they originally had been cemented in the usual manner.

Filamentary appendages. Five on each side, of which four lie in pairs at the base of the first cirrus (of these, only three are sometimes developed), and one on the flank of the prosoma.

Mouth. Palpi much acuminated. Mandibles with five teeth; the first not far remote from the second; inferior point rather broad and finely pectinated. Maxillae with two large, unequal, upper spines, and four regular steps.

Cirri. Posterior cirri, with the upper parts of the segments slightly protuberant; in young specimens, the spines can be seen to consist of five pairs, placed in two converging lines in the upper half of each segment, with numerous minute, latero-marginal, and intermediate little bristles: in large specimens, all these latter have so increased in

---

[27] *Journal of the Acad. Nat. Sc.*, Philadelphia, vol. vi, p. 138, 1829.

number, that the normal five pair cannot be distinguished, and the front of each segment is covered by a triangular thick brush of bristles, all pointing in the same direction, thus giving a very unusual character to the posterior cirri: the dorsal tuft on each segment consists of six or seven large spines, with from one to three dozen fine ones. First cirrus and anterior ramus of second cirrus with broad brushes of bristles. The pedicels of all the cirri are thickly covered with bristles. *Caudal* appendages smooth, with rounded summits.

Penis very hairy: vesiculae seminales purple, much convoluted, lying within the prosoma; testes dendritic, scarcely enlarged at their terminal points, purplish; / ovigerous fraena large with sinuous margins, the glandular beads being arranged in groups.

Size. The largest specimen (from the coast of Devonshire) had a capitulum 1·6 of an inch long, and 1·2 broad, and of unusual thickness.

Colours, after having been in spirits: front surfaces of the segments of the cirri and of the pedicels purple. In some specimens from off Borneo, parts of the sack and the interspaces between the two scuta, were of a fine purple. Montagu states, that the whole shell and body of animal, when fresh, are pale blue, with the cirri spotted with brown.

General remarks. The extreme variability of this species is remarkable. In the College of Surgeons, there is a group of specimens collected by Mr Bennett, I believe, in the Atlantic, in which the extreme narrowness of the carina and of the terga (Pl. I, fig. 6*b*, *c*) (with consequent wide spaces of membrane left between these valves), led me, at first, to entertain no doubt, that it was quite a distinct species, which was strengthened by finding that the whole surface of the cirri were villose, with very minute spines; hence I called this variety, *villosa*. On the closest examination, however, I could detect no other differences, and the narrowness of the carina and terga varied very considerably: moreover, in one of the specimens, which was about intermediate in the form of its valves between this variety and the common form, the surfaces of the cirri were not in the least degree villose. Again, in some other specimens, the terga were as narrow as in Mr Bennett's, whilst the carina had its usual outline.

In a var. (called by Leach, *P. Donovani*), from the Atlantic, under the Equator, the carina is remarkable from the extreme flatness of the upper part, and from the presence of an exterior, narrow, central ridge. In one specimen from Jersey, in the British Museum, the carina made an extremely near approach to this same form.

Affinities. This species is certainly much the most / distinct of any in the genus, and Mr Gray has proposed to separate it under the name of Dosima; but considering the close similarity of the whole organization of the internal parts, together with the transitional characters afforded by *L. australis*, I think the grounds for this separation are not quite sufficient. I have remarked, under *L. australis*, on the affinity between that and the present species. In the carina terminating in a disc (though here not imbedded), there is some slight affinity to *Paecilasma eburnea* and *crassa*, and markedly so in the arrangement of the bristles on the posterior cirri. In the valves being covered with villose membrane, and to a certain extent in the form of the carina and of the occludent margin of the terga, and especially in the two rows of cement orifices in the peduncle, there is some affinity to Scalpellum.

POECILASMA. *Nov. Genus*[28]

PLATE II

ANATIFA. J. E. Gray, *Proc. Zoolog. Soc.*, p. 44, 1848
TRILASMIS. Hinds, *Voyage of the Sulphur, Mollusca*, 1844

Valvae, 3, 5, aut 7, approximatae: carina solùm ad basales apices tergorum extensa, termino basali aut truncato aut in discum profunde infossum producto: scuta paenè ovalia, umbonibus ad angulum rostralem positis.

Valves, 3, 5, or 7, approximate: carina extending only to the basal points of the terga; with its lower end either truncated or produced into a deeply imbedded disc. Scuta nearly oval, with their umbones at the rostral angle.

Mandibles with four teeth; maxillae notched, with the lower part of edge prominent; anterior ramus of the / second cirrus not thicker than the posterior ramus; caudal appendages uniarticulate, spinose.

Generally attached to Crustacea.

[28] Ποκιλοσ, various, and ελασμα, plate or valve. I have not been able to adopt Mr Hinds' name for this genus, as it would be too glaringly incorrect to call a five-valved species, a *Trilasmis*.

I have already given my reasons for instituting and separating this genus from Lepas; as far as the capitulum is concerned, the differences between these genera certainly appear trivial; they consist in the carina not extending up between the terga, and in the lower end being either truncated, or produced into an imbedded disc: the terga have a single occludent margin. The included animal's body differs in more important respects; for both mandibles and maxillae are very distinct; the cirri of some of the species also differ; and the caudal appendages are here always spinose: there are no filamentary appendages: and lastly, the habits are different.

The genus may be divided into two sections, firstly, *P. Kaempferi* and *P. aurantia*, which have their carinae basally truncated, the basal angles of their terga cut off, and the anterior rami of their second cirri shorter than the posterior rami; and, secondly, *P. crassa*, *P. fissa*, and *P. eburnea*, which in these several respects are otherwise characterized. The *P. eburnea*, however, differs rather more from *P. crassa* and *P. fissa*, than these two do from each other; but certainly not enough to allow of the retention of Mr Hinds' genus of Trilasmis. *P. crassa*, in an especial degree, connects together all the forms.

General appearance. Capitulum oval, more or less produced, flat or gibbous; formed of three, five, or seven approximate valves; the lesser number arising from the abortion of the terga, and the greater number from the scuta being divided into two segments. Valves moderately thick, either white or reddish, smooth or striated, and sometimes partly covered by membrane, bearing minute spines. *Scuta* oval, of varying proportions; the basal margin is generally narrow, and blends into the / carina–tergal margin; the internal basal rim generally is well developed, sometimes with, and sometimes without internal teeth beneath the umbones. In *P. eburnea*, and sometimes in *P. crassa*, there is a line of apparent fissure, and in *P. fissa* of actual disseverment, running from the umbo to the apex of each scutum, nearly in the line in which a ridge extends in Lepas: the primordial valves of the scuta in these three species, are seated at the basal angles of the lateral and larger segments. The positions of the primordial valves, and the direction of growth in the calcified valves, are, in all the species, the same as in Lepas. In several of the species attached to Crustacea, the two scuta are unequally convex, which is caused, as was pointed out to me by Mr Gray, by that valve which lies close and nearly parallel to the body of the crab, being least developed. The *Terga* are either quite absent, or rudimentary as in *P. crassa*, or pretty well developed as in the other species: the occludent margin is single, and not double as generally in Lepas; the basal angle is either pointed or truncated. The *carina* varies considerably in shape, but never extends up between the terga, nor ends downwards in a fork; in the first two species it is truncated; in the others, it terminates in a deeply imbedded oblong disc, which in *P. eburnea* seems almost entirely (but of course not quite) to separate the inside of the capitulum from the peduncle; a similar separation is effected in *P. fissa*, where the imbedded disc is small, by two large teeth on the internal basal rims of the two scuta. The carina is always narrow, and either solid internally or very slightly concave.

Peduncle, is very short and narrow; the membrane is generally ringed with thicker, yellower portions, and often bears very minute spines.

Size. All the species are small, with a capitulum not exceeding half an inch in length.

Filamentary appendages. None.

Mouth. Labrum generally considerably bullate in / the upper part, with a row of teeth on the crest. The *mandibles* have four teeth, with the inferior point narrow and spine-like, or rudimentary and absent. The *maxillae* have, under the two or three upper great spines a deep notch itself bearing spines; beneath this, the lower part is straight and considerably prominent (Pl. X, fig. 15). Outer maxillae are covered on their inner sides continuously with spines.

Cirri. The first pair is sometimes seated very distant from the second. The arrangement of the spines on the posterior cirri varies, to an unusual degree within the limits of the same genus. We have either the ordinary structure of anterior pairs, with single fine intermediate spines (as in *P. Kaempferi* and *aurantia*), or we have the pairs increased by one or two additional longitudinal lateral rows, as in *P. eburnea*; or we have the front spines forming a single transverse row, as in *P. crassa* and *P. fissa* (Pl. X, fig. 29a). The segments in none of the species are protuberant; the anterior ramus of the second cirrus does not seem to be thicker than the posterior ramus, as is usually the case. The rami of the second, and of most of the other cirri, are unequal in length – the anterior ramus, contrary to the ordinary rule, being longer in *P. eburnea*, *P. fissa*, and *P. crassa*, than the posterior ramus by several segments; I have hitherto observed this inequality only in the sessile genus Chthamalus.

The *caudal appendages* are small, uniarticulate, and always furnished with bristles.

Distribution. Four out of the five species live attached to Crustacea in the European and Eastern warmer temperate and tropical oceans; the fifth species was found attached to the dead spines of an Echinus, off New Guinea. It is probable that several more species will be hereafter discovered.

1. POECILASMA KAEMPFERI

PLATE II. FIG. I

P. valvis 5; carinae basi truncatâ et cristatâ: scutorum dentibus internis umbonalibus fortibus: tergorum / acumine basali truncato, margini occludenti paene parallelo.

Valves 5; carina with a truncated and crested base; scuta with strong internal umbonal teeth; terga with the basal point truncated, almost parallel to the occludent margin.

Maxillae with short thick spines in the notch under the two upper great spines; caudal appendages with scattered bristles on their summits, and along their whole outer margins.

Japan; attached, in great numbers, to the upper and undersides of the *Inachus Kaempferi* of De Haan, a slow-moving brachyourous crab, probably from deep water. British Museum.

General appearance. Capitulum rather compressed, narrow, and produced. Valves white, tinged with orange, smooth, moderately thin, occasionally with faint traces of striae radiating from the umbones. *Scuta,* apex pointed, with a very slight ridge running to the umbo; basal margin equalling two-thirds of the length of the terga, with an internal basal rim; on the under side of each valve, beneath the umbo, there is a strong tooth. Out of the numerous specimens, all excepting one had their scuta unequally convex, with their occludent margins unequally curved, that of the more convex valve at the umbo, curling beyond the medial line. The basal end of the carina is, likewise, slightly curved laterally, and always turns towards the more convex valve. This inequality, as Mr Gray pointed out to me, depends on the position of the specimens; the flatter side lying close to the carapace of the crab.

Terga, flat, oblong, nearly rectangular; occludent margin straight; basal angle, truncated, almost parallel to the occludent margin; in width, three or four times as wide as the carina. *Carina* (fig. 1a), short, narrow, slightly curved, upper part broadest, with the apex rounded, only just passing up between the basal broad ends of the terga; externally carinated, internally very slightly concave; basal end / abruptly truncated, crested, not deeply imbedded in the membrane of the peduncle.

Peduncle, barely as long as the capitulum, apparently (for specimens dry and much shrunk) narrow, surrounded by rings or folds of thicker yellowish membrane, of which the upper ones retain moderately long spines; low down these rings become confluent; whole surface finely dotted, dots largest on the rings.

Mouth. Labrum highly bullate in the upper part, with a row of teeth on the crest; mandibles with four teeth, the fourth close to the inferior apex, which is very little developed, sometimes making the fourth tooth appear simply bifid. Maxillae with two large spines on the upper angle, beneath which there is a large depression, bearing one rather long and thick, and four short and thick, spines; inferior upraised part with a double row of longer and thinner spines.

Cirri. Posterior cirri with segments bearing five pairs of spines, of which the lowest pair is very minute; intermediate spines minute; spines of the dorsal tuft thin, of nearly equal size; segments not at all protuberant, elongated. First cirrus, standing far separated from the second (as in Scalpellum), with its nearly equal rami rather above half as long as those of the second cirrus. Second cirrus with anterior ramus not thicker, and scarcely more thickly clothed with spines, than the posterior ramus, but shorter than it by three or four segments; the spines not forming a very thick brush on the anterior ramus. Both rami of third cirrus with a longitudinal row of minute spines, parallel to the main pairs. Between the bases of the pedicels of the first pair of cirri, there are two closely approximate, conical flattened protuberances, like the single one to be described in Ibla.

Caudal appendages, about one-third of the length of the pedicel of the sixth cirrus, with some moderately long and strong spines at the end, and down the whole outer sides.

Ova, much pointed. *Penis*, hairy. /

Size. Capitulum in largest specimens half an inch long.

2. POECILASMA AURANTIA

PLATE II. FIG. 2

P. valvis 5; carinae basi truncatâ: scutis ovatis, margine basali perbrevi, dentibus parvis, internis, umbonalibus instructo: tergorum acumine basali peroblique truncato.

Valves 5; carina with a truncated base; scuta oval, with the basal margin very short, furnished with small internal umbonal teeth; terga, with the basal point very obliquely truncated.

Maxillae with fine spines in the notch under the three great upper spines; caudal appendages with scattered bristles on their summits, and along only the upper part of their outer margins.

Madeira; found by the Rev. R. T. Lowe, attached to the rare *Homola Cuvierii*, probably a deep-water crab. British Museum.

General appearance. This species so closely resembles *P. Kaempferi*, that it is superfluous to describe it in detail; and I will indicate only the points of difference. When the valves have been well preserved, they are of fine pale orange colour, and hence the name above given, which was proposed by the Rev. R. T. Lowe.

Scuta, with the internal umbonal teeth small; basal internal marginal rim very prominent, furrowed within; basal margin short (only equalling half the length of terga), owing to the great curvature of the lower part of the carino–tergal margin; hence, the outline of the scuta is almost pointed oval. I saw no appearance of inequality in the two sides.

Terga, rather smaller in proportion to the scuta, than in *P. Kaempferi*, with the basal end very obliquely truncated, so as to appear at first simply pointed, not parallel to the occludent margin; apex considerably more pointed and produced than in the foregoing species. /

Carina, almost of equal narrowness throughout, barely concave within; lower end triangular, abruptly truncated, and not crested.

Primordial valves very plain, with the usual hexagonal structure: those of the terga, rounded at both ends, instead of being square, as in the mature calcified valves.

Peduncle short, narrow, not half as long as the capitulum; paved with minute equal beads, as in the genus Dichelaspis.

Mouth. Mandibles with the fourth tooth very small; inferior angle rudimentary. Maxillae, with three great upper spines, beneath which there is a deep notch bearing some delicate spines; inferior upraised part, as in *P. Kaempferi.*

Cirri. Rami of the first cirrus hardly more than one third as long as the rami of the second cirrus, which latter rami are unequal in length by only two segments; the posterior ramus being the longer one.

Caudal appendages, with only two or three lateral bristles, besides those on the summit.

Size. Capitulum, three- to four-tenths of an inch long.

General remarks. This species has the closest general resemblance to *P. Kaempferi*, and is evidently a representative of it. On close examination, however, almost every part differs slightly; the chief points being the narrowness of the basal margin of the scuta; the obliqueness of the truncated basal end of the terga and the sharpness of the upper end; the rudimentary state of the inferior angle of the mandibles; the character of the spines on the maxillae; the proportional lengths of the cirri, and the fewness of the spines on the outer sides of the caudal appendages. The fact of Madeira having this Paecilasma, a representative both in structure and habits of a Japan species, is interesting, inasmuch, as I am informed by Mr Lowe, that some of the Madeira fishes are analogues of those of Japan. /

3. POECILASMA CRASSA

PLATE II. *FIG. 3*

ANATIFA CRASSA. J. E. Gray, *Proc. Zoolog. Soc.*, p. 44, Annulosa, Pl. iii, figs 5, 6, 1848

P. valvis 5; carinae termino basali in discum parvum infossum producto: scutis convexis, dentibus internis umbonalibus nullis: tergis paene rudimentalibus, vix carinâ latioribus.

Valves 5; carina with the basal end produced into a small imbedded disc; scuta convex, without internal umbonal teeth; terga almost rudimentary, scarcely broader than the carina.

Spines on the segments of the posterior cirri arranged in single transverse rows.

Madeira; attached to the *Homola Cuvierii*, Rev. R. T. Lowe. British Museum.[29]

General appearance. Capitulum highly bullate, or thick. Valves rather thick, opaque, either pale or dark flesh-red, smooth, yet rather plainly striated from the umbones. There are a few very minute spines on the membranous borders of the valves.

Scuta highly convex, broadly oval, apex broad rounded; basal margin narrow, much curved; no internal, umbonal teeth; basal internal rim strong, running up part of the occludent margin. A slightly prominent ridge, either rounded or angular, but in one specimen a narrow depressed fissure-like line, runs parallel to the occludent margin and ends near the apex in a slight notch; this fact is of interest in relation to the structure of the scuta in *P. eburnea* and *P. fissa*. The scuta are either equally or very unequally convex; in the latter case, the occludent margin of one valve is curled, so that its umbo is not quite medial. /

Terga, minute, almost rudimentary, scarcely broader than the carina, and half as long as the chord of its arc; carinal margin slightly curved; scutal margin straight, with a slight prominence fitting into a notch in the scuta; basal end bluntly pointed.

Carina (fig. 3, *a*), rather shorter than the scuta, extending up only to the basal ends of the terga; moderately curved; apex moderately sharp; middle part broadest, externally carinated; internally not concave, with the inner lamina of shell, at the basal end, produced into a very small oblong disc or tooth, which is only as wide as the narrowest upper part of the valve. The exterior keel does not extend on to this disc, which is slightly constricted at its origin.

Peduncle very short, narrow, ringed, and apparently without spines.

Size. Capitulum four-tenths of an inch long.

The following parts of the animal are described from some small and not well preserved specimens from Madeira, which I owe to the kindness of Mr Lowe.

Mouth. Labrum highly bullate in the upper part, with large, inwardly pointed, unequal teeth. Mandibles, with four large, pointed, equal-

---

[29] It is stated in *Zoolog. Proc.* (1848, p. 44), that this species was attached to a gorgonia, from Madeira; I cannot but suspect that there has been some confusion with the *Oxynaspis celata* from Madeira, which is thus attached.

sized teeth, with the inferior angle very narrow, acuminated like a single spine. Maxillae, with three [?] large upper spines, of which the middle one is extremely strong and long, beneath which, there is a deep notch with a single strong spine, and with the whole inferior part square and much upraised, so as to stand on a level almost with the tips of the great upper spines.

Cirri in a miserable state of preservation; first cirrus short, second cirrus with rami unequal, and I suspect the anterior one the longest; some of the other cirri also have unequal rami. The segments of the posterior cirri are not protuberant, they have on their anterior faces a single transverse row of bristles: in the upper segments, some of the spines in each dorsal tuft (which is much spread out), are *much* thicker, though rather / shorter than those on the anterior face. This peculiar structure is common to all five posterior cirri.

Caudal appendages. I can only say that they are spinose on their summits.

Affinities. This species is allied to *P. eburnea* in the rudimentary condition of its terga; in the disc-shaped basal end of its carina; and in the presence in some specimens, of a fissure-like line on the scuta parallel to their occludent margins. Its affinity, however, is closer to *P. fissa*, as is more especially shown by the remarkable arrangement of the spines on the five posterior cirri.

4. POECILASMA FISSA

PLATE II. FIG. 4

P. valvis 7; scuto utroque è duobus juxtapositis segmentis formato; segmento altero intus dentato: tergis brevibus, ter aut quater carinâ latioribus: carinae termino basali in discum parvum angustum infossum producto.

Valves 7; each scutum being formed of two closely approximate segments; of which one is internally toothed: terga short, three or four times as wide as the carina: carina with the basal end produced into a small, narrow, imbedded disc.

Spines on the segments of the posterior cirri arranged in single transverse rows.

Philippine Archipelago; Island of Bohol; parasitic on a spinose crab, found under a stone at low water; single specimen, in *Mus.*, Cuming.

General appearance. Capitulum gibbous, broadly oval, nearly a quarter of an inch long. Valves white, smooth, moderately thick, marked by the lines of growth. The occludent segments of the scuta, and nearly the whole of the terga, and the whole of the carina, enveloped in lemon-yellow membrane, tinged with orange, but the specimen had long been kept dry.

Scuta formed of two, apparently always separate, segments, closely united, so that externally their separation / is hardly visible, and does not allow of movement; the fissure thus formed runs almost in the line connecting the umbo and apex (where in most species a ridge extends), but a little on the carinal side of it. The occludent segment is narrowly bow-shaped, pointed at both ends, with the upper end projecting slightly beyond the apex of the lateral segment, and with the occludent margin regularly curved from end to end. The lateral segment is large, of an oval shape, with a narrow strip cut off on one side. Primordial valves very plain at the umbones of the lateral segments, but none are visible on the occludent segments; and this makes me believe that these two pieces are normally parts of a single valve; having only one specimen of *P. fissa*, I was not able to make out quite certainly whether the two segments are continuously united at their umbones by a non-calcified portion of valve, as is certainly the case with Dichelaspis. The basal margin of the lateral segment is narrow, inflected, and blends with the carino–tergal margin; it has an internal, prominent, basal rim, and towards the occludent margin a large, prominent, internal tooth. This internal basal rim is not parallel to the outer basal margin, but rises to a point a little way up the occludent margin, in the same manner as in *P. eburnea*, but in a lesser degree; in this latter species the peduncle is internally almost cut off by the large disc of its carina; here, on the other hand, it is internally almost cut off by these rims and the two large teeth of the lateral segments of the scuta.

Terga subtriangular, short, nearly half as broad as long; three or four times as wide as the carina, and rather wider than the occludent segment of the scuta; occludent margin single, arched; carinal margin slightly arched; basal angle bluntly pointed.

Carina very narrow, much arched, running up just between the basal ends of the terga; exterior ridge enveloped in membrane; heel blunt, prominent; internally, not concave, even slightly convex, produced at the lower / end into a very narrow, short, imbedded disc (or rather tooth), which is itself a little curved downwards and blunt at the end.

Peduncle very narrow, about half as long as the capitulum; yellow, finely beaded, plainly ringed, without spines.

Mouth. Labrum, with a row of minute teeth; palpi narrow. Mandibles with all the lower part narrow; of the four teeth, the second and third are narrow, the fourth is pectinated and placed very close to the inferior angle, which is produced into a long thin tooth. Maxillae unknown.

Cirri. First pair lost. The arrangement of the spines on all is most abnormal (Pl. X, fig. 29): dorsal tuft long, arranged in a transverse line and seated in a deep notch; in the sixth cirrus, the spines on the lower segments are fine, those on the upper segments are thick and claw-like, mingled with some fine spines; in the four anterior cirri the spines of the dorsal tufts are even thicker and more claw-like. On the anterior faces, also, of all the segments the spines form a single row; they are shorter than those composing the dorsal tuft; hence the spines on each segment are arranged in a circle, interrupted widely on the two sides: this arrangement is common to all five posterior cirri. Second cirrus, with the *anterior* ramus one-third longer and thinner than the posterior ramus (this is the reverse of the usual arrangement); this longer ramus equals in length the sixth cirrus. Third cirrus, with the anterior ramus considerably longer than the posterior ramus; in the three posterior pair of cirri, also, the anterior rami are a little longer than the posterior: except in length, there is little difference of any kind between the five posterior pair of cirri. Pedicels of the cirri long; rami rather short; segments elongated, not protuberant.

Caudal appendages nearly as long as the pedicels of the sixth cirrus, thickly clothed with very fine bristles, like a camel's-hair pencil brush. /

Affinities. In the structure of the carina, and more especially of the scuta, there is a strong affinity between the present and following species; for we shall immediately see that in *P. eburnea* there is evidence of the scuta being composed of two segments fused together; and the larger segment is furnished with an internal oblique, strong, basal rim. To this same species there is an evident affinity in the form of the mandibles and of the caudal appendages, and in the anterior rami of the cirri being longer than the posterior rami. Notwithstanding these points of affinity, I consider that *P. fissa* is more closely related in its whole organization, as more particularly shown in the arrangement of the spines on the cirri and in the presence of terga, to *P. crassa* than to *P. eburnea*. Although in Dichelaspis, the scuta are invariably composed of two almost separate segments, yet *P. fissa* shows no special affinity to this genus.

5. POECILASMA EBURNEA

PLATE II. FIG. 5

TRILASMIS EBURNEA. Hinds, *Voyage of Sulphur*, vol. i, Mollusca, Pl. xxi, fig. 5, 1844

P. valvis 3; scutis acuminatis, ovatis; ad pedunculum paene transversè spectantibus; dentibus internis umbonalibus fortibus: tergis nullis: carinae termino basali in discum amplum oblongum infossum producto.

Valves 3; scuta pointed, oval, placed almost transversely to the peduncle; internal umbonal teeth strong: terga absent: carina with the basal end produced into a large, oblong, imbedded disc.

Spines on the upper segments of the posterior cirri, arranged in three or four approximate longitudinal rows, making small brushes.

Habitat. New Guinea, attached to the spines of a dead Echinus. Brit. Mus., and Cuming.

General appearance. Capitulum flat, pear-shaped, / placed almost transversely to the peduncle. Valves white, smooth, moderately thick.

Scuta: the basal margin, as seen externally, is narrow, and can hardly be separated from the carinal margin; but an internal basal rim (fig. 5, *b*) (along which the imbedded disc of the carina runs), shows where, in the other species, the basal and carinal margins are separated. This basal internal rim is not parallel to the external basal margin, but runs upwards to the occludent margin, leaving beneath it a large triangular space, to which the membrane of the peduncle is attached; and this makes it appear as if the rostral umbones of these valves had grown downwards; but, judging from the allied species, *P. fissa*, I have no doubt that the primordial valves really lie on the umbones, and that the growth has been in the usual direction, that is, exclusively upwards.

The occludent margin is curved, and blends by a regular sweep into the carinal margin, so that there is no acute upper angle. A distinct line can be seen, as if two calcareous valves had been united, running from the umbo to the upper end of the valve, thus in appearance separating a slip of the occludent margin; internally this appearance is more conspicuous; this structure is important in relation to that of *P. fissa*. The pointed umbones are divergent, and internally under each, there is a large tooth. The two valves are equally convex.

Terga, entirely absent.

The *carina* (Pl. II, fig. 5, *a*, *c*), including the disc, is three-fourths as long as the scuta; it is placed almost transversely to the longitudinal axis of the peduncle; it is narrow and internally convex; the imbedded disc is very large, forming a continuous curve with the upper part of the carina; this disc runs along the internal basal rim of the scuta, and hence almost separates, internally, the peduncle from the capitulum; it equals one-fourth of the total length of the valve, and is thrice as wide as the upper part; it is oval, externally marked by a central / line, and with a slight notch at the end, giving a divided appearance to the whole, and indicating how easily a fork might be formed from it. The carina is thick, measured from the inner convex to the exterior surface, which is carinated; heel prominent.

Peduncle, narrow, very short, not nearly so long as the capitulum.

Mouth. Labrum considerably bullate, with the lower part much produced towards the adductor muscle; crest with small bead-like teeth; palpi small, pointed; mandibles, with the first tooth standing rather distant from the second; inferior angle spine-like and bifid; maxillae (Pl. X, fig. 15), with two considerable spines (only one is shown in the Plate) beneath the upper large pair; the inferior upraised part bears seven or eight pair of spines, and its edge is not quite straight; close to the main notch, lying under the four upper spines, there are two minute notches, with the interspace bearing a tuft of fine spines and a pair of larger ones.

Cirri. The rami in all are rather unequal in length, the anterior rami being rather the longest; the anterior rami of the second and third cirri are not thicker than the posterior rami. The segments in the three posterior cirri are not protuberant; the upper segments bear three or four pair of spines, with some minute intermediate ones, and with the lateral marginal spines unusually large and long, so as to form, with the ordinary pairs, a third or fourth longitudinal row; hence a small brush is formed on each segment. The dorsal tuft is large and wide, so as to contain even fourteen spines, of which some are as long as those in front. In the lower segments of these same posterior cirri, the lateral marginal spines are not so much developed (nor is the dorsal tuft), and hence the segments can hardly be said to be brush-like. The first cirrus is placed rather distant from the second pair. The second and third cirri differ from the three posterior pair, only in the bristles being slightly more numerous, and in the dorsal tufts being more spread out. /

Caudal appendages about half the length of the lower segments of the pedicels of the sixth cirrus; truncated and rounded at their ends; thickly clothed with long excessively fine bristles, so as to resemble camel-hair pencils.

The *Stomach*, I believe, is destitute of caeca; in it was a small crustacean.

General remarks. I was at first unwilling to sacrifice Mr Hind's genus, Trilasmis, which is so neatly characterized by its three valves; moreover, the present species does differ, in some slight respects, from the other species of Paecilasma; but under the head of *P. fissa* I have shown how that species, *P. crassa* and *P. eburnea* are tied together. The absence of terga, which are rudimentary in *P. crassa* (and we shall hereafter see, in *Conchoderma*, how worthless a character their entire absence is), and the arrangement of the spines in the upper segments of the posterior cirri, are the only characters which could be used for a generic separation.

GENUS: DICHELASPIS

PLATE II

OCTOLASMIS.[30] J. E. Gray, *Annals of Philosophy*, vol. x, new series, p. 100, August, 1825
HEPTALASMIS. Agassiz, *Nomenclator Zoologicus*

[30] From διχηλοσ, bifid, and ασπισ, a shield, or scutum. The name Octolasmis was given by Mr Gray under the belief that there were eight valves. Leach (as stated in

Valvae 5, quae ferè pro septem haberi possent, scuto in segmenta planè duo, ad angulum autem rostralem conjuncta, diviso: carina plerumque sursum inter terga extensa, deorsum aut disco infosso aut furcâ aut calyce terminata. /
Valves 5, generally appearing like 7, from each scutum being divided into two distinct segments, united at the rostral angle; carina generally extending up between the terga, terminating downwards in an imbedded disc, or fork, or cup.

Mandibles, with three or four teeth; maxillae notched, with the lower part of edge generally not prominent; anterior ramus of the second cirrus not thicker than the posterior ramus, not very thickly clothed with spines; caudal appendages uniarticulate, spinose.

Distribution. Eastern and Western warmer oceans in the Northern hemisphere, attached to crustacea, sea-snakes, etc.

Description. The capitulum appears to contain seven valves; but, on examination, it is found that two of the valves on each side, are merely segments of the scutum; these are united at the umbo, in three of the species, by a narrow, non-calcified portion of valve, where the primordial valve is situated; in *D. orthogonia*, however, the junction of the two segments is perfectly calcified, and of the same width as the whole of the basal segment. The capitulum is much compressed, broad at the base, and extends a little beneath the basal segments of the scuta. The valves are very thin, often imperfectly calcified, and generally covered with membrane. They are not placed very close together, and in all the species a considerable interspace is left between the carina and the two other valves: in the *D. Grayii* the valves are so narrow that they form merely a calcified border round the capitulum. The membrane between the valves and over them, is very thin, and is thickly studded, in some of the species, with minute blunt conical points apparently representing spines. The valves in the same species present considerable variations in shape; in their manner or direction of growth, and in the position of their primordial valves, they agree with Lepas and Paecilasma.

---

the *Annals of Philosophy*), had proposed, in MS., the name Heptalasmis, and this is now used in the British Museum by Mr Gray, and thus appears in Agassiz's *Nomenclator Zoologicus*. Although, strictly, there are only five valves, I continued to use, in my MS., the term Heptalasmis, until I examined the *D. orthogonia*, where it was so apparent to the naked eye that there were only five valves, the scuta in this species being less deeply bifid, that I was compelled to give up a name so manifestly conveying a wrong impression, and hence adopted the one here used.

Scuta. In three of the species the two segments, / named the occludent and basal, appear like separate valves, but these, by dissection, can be most distinctly seen to be united at the rostral angle. The primordial valve, formed of the usual hexagonal tissue, is elliptic, elongated, and placed in the direction of the occludent segment; calcification commences at its upper point, so as to form the occludent segment, and afterwards at its lower point, but rectangularly outwards, to form the basal segment; in the minute space between these two points of the primordial valve, there is, in four of the species, no calcification; so that the two segments are united by what may be called a flexible hinge; in *D. orthogonia* the two calcareous segments are absolutely continuous. The occludent segment is longer than the basal segment; it either runs close along the orifice, or in the upper part bends inwards; both segments are narrow, except in *D. Warwickii*, in which the basal segment is moderately broad; the two segments are placed at an angle, varying from 45° to 90°, to each other. The capitulum generally extends for a little space beneath the basal segments of the scuta, where it contracts to form the peduncle.

The *terga* present singular differences in shape, and are described under the head of each species; scarcely any point can be predicated of them in common, except that they are flat and thin.

The *carina* is much bowed, narrow, and internally either slightly concave or convex and solid; the upper end extends far up between the terga; the lower end is formed by a rectangularly inflected, imbedded, triangular or oblong disc, deeply notched at the end, or as in *H. Lowei*, of a fork, the base, however, of which is wider than the rest of the carina, so as to present some traces of the disc-like structure of the other two species; or lastly, as in *D. orthogonia*, it terminates in a crescent-formed cup.

Peduncle. This is narrow, compressed, and about as long, or twice as long, as the capitulum; in *D. Warwickii* it is studded with minute beads of yellowish chitine. /

Size. Small, with a capitulum scarcely exceeding a quarter of an inch in length.

Filamentary appendages. None. There are two small ovigerous fraena, which, in *D. Warwickii*, had the glands collected in seven or eight little groups on their margins.

Mouth. Labrum highly bullate, with small teeth on the crest; palpi small, not thickly covered with spines. *Mandibles* narrow, with three or four teeth. Maxillae small, with a notch beneath the two or three great upper spines; lower part bearing only a few pair of spines, generally not projecting, but in *D. orthogonia* largely projecting. Outer maxillae, with their inner edges continuously covered with bristles.

Cirri. First pair short, situated rather far from the second pair; second pair with the anterior ramus not thicker than the posterior ramus, and hardly more thickly clothed with spines than it, excepting sometimes the few basal segments. All the five posterior pair of cirri resemble each other more closely than is usual. In *D. Lowei*, the segments of the posterior cirri bear the unusual number of eight pair of main spines.

Caudal appendages. Uniarticulate, spinose; in *D. pellucida* they are twice as long as the pedicels of the sixth cirrus, but I could not perceive in them any distinct articulations.

Distribution. Attached to crabs at Madeira, and off Borneo; to sea-snakes in the Indian Ocean. The individuals of all the species appear to be rare.

General remarks. Four of the five species, forming this genus, though certainly distinct, are closely allied. I have already shown, that although the characters separating Lepas, Paecilasma, and Dichelaspis are not very important, yet if they be neglected these three natural little groups must be confounded together. Dichelaspis is much more closely united to Paecilasma than to Lepas, and, as far as the more important characters of the animal's body are concerned, there is no important difference / between them. Consequently, I at first united Paecilasma and Dichelaspis, but the latter forms so natural a genus, and is so easily distinguished externally, that I have thought it a pity to sacrifice it. The carina (which seems to afford better characters than the other valves in Dichelaspis), from generally running up between the terga and in ending downwards, in three of the species, in a deeply notched disc or fork, more resembles that in Lepas than in Paecilasma; in the manner, however, in which the imbedded disc, in *D. Warwickii* and *D. Grayii*, nearly cuts off the inside of the capitulum from the peduncle, there is a resemblance to *Paecilasma eburnea*. In the extent to which the valves are separated from each other, in the bilobed form of the scuta (the two segments in Dichelaspis, perhaps, answering to the upper and lateral projections in the scuta of *Conchoderma virgata*), and in the basal half of the scuta not descending to the base of the capitulum, there is a considerable resemblance to Conchoderma; in both genera the adductor muscle is attached under the umbones of the scuta; but the structure of the mouth and cirri and caudal appendages shows that the affinity is not stronger to Conchoderma than to Lepas. It appears at first probable, that Dichelaspis would present a much closer affinity to *Paecilasma fissa*, in which, owing to the scuta being formed of two segments, there are seven valves, than to any other species of that genus; but in *P. fissa* the primordial valve is triangular and is situated on the basal segment, whereas, in Dichelaspis, it is elliptic and is seated between the two segments, and is more in connection with the occludent than with the basal segment; and this I cannot but think is an important difference: in other respects, *P. fissa* shows no more affinity to Dichelaspis than do the other species of the genus. Finally, I may add that Dichelaspis bears nearly the same relation to Paecilasma, as Conchoderma does to Lepas. /

1. DICHELASPIS WARWICKII

PLATE II. FIGS. 6, 6a, b

OCTOLASMIS WARWICKII. J. E. Gray, *Annals of Philosophy*, vol. x, p. 100, 1825; *Spicilegiao Zologica, t. vi, fig. 16, 1830*

D. scutorum segmento basali duplo latiore quam segmentum occludens: tergorum parte inferiore paulò latiore quam occludens scutorum segmentum.
Scuta, with the basal segment twice as wide as the occludent segment; terga, with the lower part slightly wider than the occludent segment of the scuta.

Mandibles, generally with four teeth.

Off Borneo, attached to a crab (Belcher): China Sea. British Museum.

General appearance. Capitulum much compressed, elongated with the valves not very close together, the carina being separated by a rather wide space from the scuta and terga. Valves variable in shape, very thin and translucent, covered by thin membrane, which, over the whole capitulum, is studded with minute blunt points.

Scuta. Segments without internal teeth or an internal basal rim; the occludent segment long, narrow, pointed, not quite flat, sometimes slightly wider in the upper part; about one-third of its own length longer than the basal segment; occludent margin slightly arched; basal segment about twice as wide as the occludent segment, triangular, slightly convex; in young specimens (Pl. II, fig. 6*b*), the carinal margin of the basal segment is protuberant, and the occludent margin hollowed out; in old specimens the occludent margin of the basal segment is straight, and the carinal margin much hollowed out. In very young specimens the basal segment is very small compared to the occludent.

Terga, variable in shape; flat, lower part wider than the occludent segment of the scuta; occludent margin / double, forming a considerable rectangular projection, as in the terga of Lepas; scutal margin deeply excised at a point corresponding with the apex of the scuta, a flat tooth or projection being thus formed; there is sometimes a second tooth (fig. 6*b*) a little above the basal point. The terga, in the first variety, somewhat resemble in shape the scuta of *Conchoderma aurita*.

Carina, much bowed, narrow, slightly concave within (in the Borneo specimen, rather wider and more concave), extending up between the terga for half their length, terminating downwards in a rectangularly inflected, deeply imbedded, oblong, rather wide, flat disc, at its extremity more or less deeply notched. This disc is externally smooth; internally it sometimes has two divergent ridges on it; it extends across about two-thirds of the base of the capitulum (fig. 6*a*, as seen from beneath, when the peduncle is cut off), to under the middle of the basal segments of the scuta.

Peduncle, narrow, flattened; united to the capitulum some little way below the scuta; about as long as the capitulum; the membrane of which it is composed is thin, externally studded with bluntly conical beads of yellowish chitine, of which the largest were $\frac{1}{2000}$ of an inch in diameter; on their internal surfaces these are furnished with a small central, circular depression, apparently for a tubulus; the arrangement of the beads varied in concentric zones. Similar conical points on the capitulum have an internal concave surface about $\frac{1}{3000}$ in diameter, with a central circle $\frac{1}{12000}$ in diameter, for the insertion, as I believe, of a tubulus.

Size. The largest specimen had a capitulum a quarter of an inch long.

Mouth. Labrum highly bullate; crest with not very minute, blunt teeth, which towards the middle lie closer and closer to each other, so as to touch. Palpi rather small, with a few very long bristles at the apex.

Mandibles, narrow, produced, with four teeth, and the inferior angle tooth-like and acuminated; in one specimen, / on one side of the mouth, the mandible had only three teeth.

Maxillae, small; at the upper angle there are two large spines and a single small one, beneath which there is a deep notch, and beneath this a straight but projecting edge, bearing a few moderately large and some smaller spines. Outer maxillae sparingly covered with bristles along the inner margin.

Cirri. First pair far removed from the second pair, and not above half their length; segments rather broad, with transverse rows of bristles not very thickly crowded together; terminal segments very obtuse, and furnished with thick spines. The segments of the three posterior pair have each three or four pair of spines, with a few minute spines scattered in an exterior, parallel, longitudinal row; dorsal tufts, with four or five long spines. The second cirrus has its anterior ramus not thicker, but rather shorter than the posterior ramus; the former is only a little more thickly clothed with spines, owing to those in the longitudinal lateral row being longer and more numerous, than is the sixth pair of cirri. Bristles not serrated.

Caudal appendages, narrow, thin, slightly curved, about half as long as the pedicels of the sixth cirrus; in young specimens, the appendage bore seven or eight pair of long bristles rectangularly projecting; in some older specimens, there was a tuft of bristles on the summit, and two other tufts on the sides.

I at first thought that the Borneo specimen was a distinct species, but after careful comparison of the external and internal parts, the only difference which I can detect is, that the terga are slightly larger, and that the carina, to a more evident degree, is wider, more especially in the middle and lower portions. /

2. DICHELASPIS GRAYII

PLATE II. FIG. 9

D. scutorum segmento basali angustiore quam segmentum occludens; longitudine paene dimidiâ: tergis bipenniformibus, margine crenato, spinâ posticâ, manubrio angustiore quam occludens scutorum segmentum.

Scuta, with the basal segment narrower than the occludent segment, and about half as long as it. Terga like a battleaxe, with the edge crenated and a spike behind; the handle narrower than the occludent segment of the scuta.

Mandibles with three teeth; cirri unknown.

Attached to the skin of a sea-snake, believed to have been the *Hydeus* or *Pelamis bicolor*, and therefore from the Tropical, Indian or Pacific Oceans; associated with the *Conchoderma Hunteri*; single specimen, in a very bad condition, in the Royal College of Surgeons.

General appearance. Capitulum much compressed, elongated, formed of very thin membrane, with the valves forming round it a mere border. Valves thin, imperfectly calcified, covered with membrane.

Scuta formed of two narrow plates at very nearly right-angles to each other, one extending along the occludent, and the other along the basal margin; both become very narrow at the point of junction, and are there not calcified, but are evidently continuous and form part of the same valve; the basal segment is about half as long and narrower than the occludent segment, flat and bluntly pointed at the end; occludent segment slightly curled, and therefore the whole does not lie quite in the same plane; narrow close to the umbo, with a very minute tooth on the underside; apex rounded. In the upper part, the occludent segments leave the membranous margin of the orifice, and run in near to the terga, bending towards them at an angle of 45° with their lower part. I was unable to distinguish the primordial valves. /

Terga. These valves are of the most singular shape, resembling a battleaxe, with a flat and rather broad handle; the upper part consists of an axe, with a broad cutting crenated edge, behind which is a short blunt spike. The spike and cutting edge together answer to the double occludent margin of the tergum in Lepas. The whole valve is flat, thin, and lies in the same plane; the carinal margin is nearly straight; the scutal margin bulges out a little, and at a short distance above the blunt basal point is suddenly narrowed in, making the lower-most portion very narrow; the widest part of the handle of the battleaxe, is narrower than the occludent segment of the scuta. The two spikes behind the cutting and crenated edges of the two terga, are blunt and almost touch each other; above their point of juncture, the membrane of the orifice forms a slight central protuberance.

Carina, very narrow throughout, concave within, much bowed; upper point broken and lost, but it must have run up between the terga for more than half their length; basal portion inflected at nearly right angles, and running in between, and close below, the linear basal segments of the scuta, so as almost entirely to cut off internally the peduncle and capitulum. This lower inflected and imbedded portion, or disc, gradually widens towards its further end, which is, at least, four times as wide as the upper part of the carina, and is deeply excised, but to what exact extent I cannot state, as the specimen was much broken. On each side of this elongated triangular disc, there is a slight shoulder corresponding to the ends of the basal segments of the scuta; and on the upper surface of each shoulder, there is a small tooth or projection. The middle part of the disc is barely calcified, and is transparent.

Peduncle, rather longer than, and not above half as wide as, the capitulum; the latter being nearly 2/10ths of an inch in length: the membrane of the peduncle is thin, naked and structureless.

Mouth. Labrum highly protuberant in the upper part, / with a row of beads on the crest. Palpi small, with few bristles. *Mandibles*, with the whole inferior part, very narrow; three teeth very sharp, with a slight projection, perhaps, marking the place of a fourth tooth; inferior angle ending in the minutest point; first tooth as far from the second, as the latter from the inferior angle. *Maxillae* with a *broad* shallow notch; inferior angle much rounded, bearing only four or five pair of spines.

Cirri. First pair apparently remote from the second pair; all five posterior pair lost; first pair short, with the rami unequal by about two segments; segments clothed with several transverse rows of bristles; terminal segments blunt.

3. DICHELASPIS PELLUCIDA

PLATE II. FIG. 7

D. valvarum singularum acuminibus superioribus et inferioribus vix intersecantibus: scutorum segmento basali multo angustiore quam segmentum occludens; longitudine ferè dimidiâ: tergis bipenniformibus, margine integro, manubrii acumine ad carinam flexo.

Valves with the upper and lower points of the several valves only just crossing each other. Scuta with the basal segment much narrower than the occludent segment, and about half as long as it. Terga like a battle-axe, with the edge smooth, and the point of the handle bent towards the carina.

Mandibles with four teeth; caudal appendages twice as long as the pedicels of the sixth cirrus.

Indian Ocean; attached to a sea-snake.

This species comes very close to the *D. Grayii*, which likewise was attached to a snake; but I cannot persuade myself, without seeing a graduated series, that the differences immediately to be pointed out can be due to ordinary variation. I am much indebted for specimens to the kindness of Mr Busk. /

General appearance. The membrane of the capitulum and peduncle is surprisingly thin and pellucid, so that the ovarian tubes within the peduncle can be traced with the greatest ease. The valves are small, the apices only just crossing each other, and are composed of yellow chitine, with mere traces of calcification. The capitulum is pointed, oval, 0·15 of an inch long; the peduncle is narrow, and fully twice as long as the capitulum.

Scuta. The two segments stand at right-angles to each other; the basal segment is linear and pointed, fully half as long, but only one-third as wide, as the occludent segment. The point of junction of the two segments is wider than the rest of the basal segment. This latter segment lies some little way above the top of the peduncle. The occludent segment is bluntly pointed; it is directed a little inwards from the edge of the orifice towards the terga; the apex reaches up just above the slightly reflexed lower point of the terga. The adductor muscle is fixed under the point of junction of the two segments.

The *terga* are battleaxe shaped, with the blade part very prominent, smooth-edged; behind the blade there is a short upwardly-turned prominence. The lower point of the handle of the axe is bent towards the carina. The tergum, measured in a straight line, equals in length two-thirds of the occludent segment of the scutum, the handle being rather narrower than this same segment.

The *carina* is extremely narrow and much bowed; the apex reaches up only to just above the lower bent points of the terga. The basal end is rectangularly inflected, and stretches internally nearly across the peduncle; it consists (fig. 7a) of a triangular disc of yellow thin membrane, four or five times as wide as the upper part of the valve; the end of this disc is hollowed out; its edges are thickened and calcified, and hence, at first, instead of a disc, this lower part of the carina appears like a wide fork; the tips of the prongs stretch just under the tips of the basal segments of the scuta. /

Peduncle. Its narrowness and transparency are its only two remarkable characters.

Mouth. All the parts closely resemble those of *D. Grayii*, but being in a better state of preservation I will describe them. The labrum is highly bullate, with a row of minute teeth on the crest, placed very close together in the middle. Palpi small, thinly clothed with spines; mandibles extremely narrow, hairy, with four teeth, but the lower tooth is so close to the inferior angle, as only to make the latter look double. Maxillae, with a very deep broad notch, dividing the whole into two almost equal halves; in the upper part there are three main spines.

Cirri. The first pair are placed at a considerable distance from the second pair; they are short with equal rami, and rather broad segments furnished with a few transverse rows of bristles. The five posterior cirri have singularly few, but much elongated segments, bearing four pair of spines: the two rami of the second pair are alike, and differ only from the posterior cirri in a few of the basal segments having a few more spines.

The *caudal appendages* are twice as long as the pedicels, and nearly half as long as the whole of the sixth cirrus; they have a small tuft of long thin spines at their ends, and a few in pairs, or single, along their whole length; at first I thought that they were multi-articulate, but after careful examination I can perceive no distinct articulations; I have seen no other instance of so long an appendage without articulations.

Diagnosis. This species differs from *D. Grayii* in all the valves being shorter, so that their points only just cross each other; but this, I conceive, is an unimportant character. In the scuta, the basal segment is here narrower, but the point of junction of the two segments wider than in that species; in the terga, the edge of the axe is smooth instead of being crenated, and the handle and the point behind are of a rather different shape; in the carina the imbedded basal disc has no shoulders and / small teeth, as in *D. Grayii*. Notwithstanding these differences, I should not be much surprised if the present form were to turn out to be a mere variety.

4. DICHELASPIS LOWEI

PLATE II. FIG. 8

D. scutorum segmento basali angustiore quam occludens segmentum, longitudine ferè ⅘: tergorum parte inferiori duplo latiore quam occludens scutorum segmentum.

Scuta with the basal segment narrower than the occludent segment, and about four-fifths as long as it. Terga with the lower part twice as wide as the occludent segment of the scuta.

Mandibles with four teeth; segments of the three posterior cirri with eight pair of main spines.

Habitat. Madeira; attached to a rare Brachyourous Crab, discovered by the Rev. R. T. Lowe. Very rare.

General appearance. Capitulum much compressed, subtriangular, formed of very thin membrane; valves imperfectly calcified, and thin.

Scuta formed of two narrow plates placed at about an angle of 50° to each other, and united at the umbo by a non-calcified flexible portion. The primordial valve is situated at this point, but chiefly on the occludent segment. The occludent segment is about twice as wide and about one-fifth longer than the basal segment, which latter is rather sharply pointed at its end. The occludent segment is slightly arched, a little narrowed in on the occludent margin close to the umbo; its upper end is broad and blunt; it runs throughout close to the edge of the orifice of the sack, and its longer axis is in the same line with that of the terga. Close to the umbones, on the underside of the basal segment, there is, on each valve, a longitudinal calcified fold, serving as a tooth. /

Terga broad, with a deep notch corresponding to the apex of the occludent segment of the scuta; the part beneath the notch is of nearly the same width throughout, and is twice as broad as the occludent segment of the scuta; it has its basal angle very broad and blunt. The entire length of the terga equals two-thirds of that of the occludent segment of the scuta; occludent margin simply and slightly curved.

The *carina* is of nearly the same width throughout, with the upper part rather the widest, and the apex blunt; within *convex*; it extends up between three-fourths of the length of the terga, terminating downwards in a fork with very sharp prongs, standing at right-angles to each other (fig. 8a). The fork, measured from point to point, is thrice as wide as, and measured across at the bottom of the prongs it is wider than, the widest upper part of the valve – a resemblance being thus shown with the triangular notched disc in *D. Grayii*. The points of the prong extend under about one-fourth of the length of the basal segments of the scuta.

Peduncle rather longer than the capitulum, which, in the largest specimen, was 9/10ths of an inch in length; peduncle narrow, close under the capitulum; membrane thin and structureless. The larger specimen had almost mature ova in the lamellae.

Mouth. Labrum with a few bead-like teeth on the crest, distant from each other even in the central part; palpi rather small, moderately clothed with bristles.

Mandibles, with four teeth; the inferior angle blunt and broad, showing, apparently, a rudiment of a fifth tooth; the first tooth is as far from the second, as is this from the inferior angle; second, third, and fourth teeth very blunt, whole inferior part of mandible not much narrowed. Maxillae small, with a small notch under the three upper spines, which are followed by five or six pair, nearly as large as the upper spines.

Cirri. First pair remote from the second; their rami nearly equal, and about one-third of the length of the / rami of the second cirrus; thickly clothed with bristles: rami of the second cirrus of equal thickness, but little shorter than those of the sixth cirrus; the three or four basal segments of the anterior ramus are thickly clothed with spines; the other segments, and all the segments on the third pair, resemble the segments of the three posterior pair. These latter are elongated, not protuberant, and support eight pairs of spines with very minute intermediate spines; those in the dorsal tufts are numerous and long.

Caudal appendages nearly as long as the pedicels of the sixth cirrus; oval, moderately pointed, with their sides, for one-fourth of their length, thickly clothed with long very thin spines.

Affinities. In the form of the scuta and of the carina this species is most nearly allied to D. *Grayii* or D. *pellucida*, in the form of the terga to D. *Warwickii*.

5. DICHELASPIS ORTHOGONIA

PLATE II. FIG. 10

D. scutorum basali segmento angustiore quam occludens segmentum; longitudine ferè dimidiâ; duorum segmentorum junctione calcareâ: tergorum prominentiis marginalibus inaequalibus quinque: carinâ deorsum in parvo calyce lunato terminatâ.

Scuta with the basal segment narrower than the occludent segment, and about half as long as it; junction of the two segments calcified. Terga with five unequal marginal projections. Carina terminating downwards in a small crescent-formed cup.

Maxillae with the inferior part of edge much upraised.

Habitat unknown; associated with *Scalpellum rutilum*, apparently attached to a horny coralline. British Museum.

The specimens are in a bad condition, not one with all the valves in their proper positions, and most of them broken; animal's body much decayed and fragile. /

General appearance. Capitulum apparently much flattened; valves naked, coloured reddish, separated from each other by thin structureless membrane.

The *Scuta* consist of two bars placed at right-angles to each other, with the point of junction fully as wide as any part of the basal segment, and perfectly calcified; the primordial valve lies at the bottom of the occludent segment. The basal segment is equally narrow throughout, and very slightly concave within; the occludent segment widens a little above the junction or umbo, and then keeps of the same width to the apex, which is obliquely truncated; internally this segment is concave; externally it has a central ridge running along it; the occludent segment is twice as long and twice as broad as the basal segment. Both segments are a little bowed from their junction to their apices.

Terga. These are of a singular shape; they are about three-fourths as long as the occludent segment of the scuta, and in their widest part, of greater width than it. They consist of four prominent ridges proceeding from the umbo, and united together for part only of their length, and, therefore, ending in four prominences; one of these, the longest, has the same width throughout, and forms the basal point; a second, very small one, is seated high up on the carinal margin just above the apex of the carina; the third and fourth, are nearly equal in length, and project one above the other on the scutal margin. There are two occludent margins, meeting each other at right angles, and forming a prominence, as in Lepas; and this gives to the margin of the valve the five prominences. The whole valve internally is flat; externally, it is ridged as described.

Carina (fig. 10*a*, *b*), much bowed, narrow, long; externally, the central ridge is quite flattened; internally, slightly concave, but scarcely so towards the lower part, which is narrow; the upper part widens gradually, and the apex is rounded. The basal embedded portion is as wide as the uppermost part, and forms a cup, unlike anything / else known: the outline of this cup is semi-oval and crescent-formed; it is moderately deep; it is formed by the external lamina of the carina bending rectangularly downwards and a little outwards, whereas the inner lamina of the lower part (which is slightly concave), is continued with the same curve as just above, and forms the concave chord to the semi-oval rim of the cup. This cup, I believe, lies under the points of the basal segments of the scuta.

Peduncle unknown, probably short.

Length of capitulum, above $\frac{2}{10}$ths of an inch.

Mouth. Labrum with the upper part highly bullate, and produced into a large overhanging projection; crest with a row of rather large bead-like teeth; *palpi* small, their two sides parallel, very sparingly covered with long bristles.

Mandibles, narrow, produced, with four teeth, and the inferior angle produced into a single strong spine: the distance between the tips of the first and second teeth almost equals that between the tip of the second tooth and of the inferior angle.

Maxillae with three large upper unequal spines, beneath which, there is a deep and wide notch (bearing one spine), and the inferior part projects highly, bearing three or four pairs of spines, and is, itself, obscurely divided into two steps.

Outer maxillae, very sparingly covered with bristles; outline, hemispherical.

Cirri. The rami of the five posterior pair are extremely long, as are the pedicels; the segments are much elongated, with their anterior faces not at all protuberant; each bears five pair of very long and thin spines, with an excessively minute one between each pair; the dorsal tuft consists of very fine and thin spines. The second cirrus has its anterior ramus not at all thicker than the posterior ramus; but has an exterior third longitudinal row of small bristles. First cirrus, separated by a wide interval from the second pair; very short with the two rami / slightly unequal in length; the segments are broad, and are paved moderately thickly with spines; the terminal spines not particularly thick.

Caudal appendages consist of very small and narrow plates, about half the length of the pedicels of the sixth cirrus, with a few long spines at their ends.

This well-marked species, I think, has not more affinity to one than to another of the previous species: it differs from all, in the junction between the two segments of the scuta being perfectly calcified; in the peculiar cup, forming the base of the carina; and lastly, in the inferior part of the maxillae projecting.

OXYNASPIS.[31] Gen. Nov.

PLATE III

Valvae 5, approximatae: scutorum umbones in medio marginis occludentis positi: carina rectangulè flexa, sursùm inter terga extensa, termino basali simpliciter concavo.

Valves 5, approximate; scuta with their umbones in the middle of the occludent margin; carina rectangularly bent, extending up between the terga, with the basel end simply concave.

Mandibles with four teeth; maxillae notched, with the lower part of edge nearly straight, prominent; anterior ramus of the second cirrus thicker than the posterior ramus; caudal appendages, uniarticulate, spinose.

Attached to horny corallines.

I have most unwillingly instituted this genus; but it will be seen by the following description, that the one known species could not have been introduced into Lepas or Paecilasma, without destroying these genera, although it has a close general resemblance with both. As far as the valves are concerned, it is more nearly related to Lepas than to Paecilasma; but taking the entire animal, its / relation is much closer to the latter genus than to Lepas: it differs from both these genera in the manner of growth of the scuta, which is both upwards and downwards, the primordial valve being situated in nearly the middle of the occludent margin. In this respect, and in the shape of the carina and terga, there is an almost absolute identity with Scalpellum; I may, however, remark that in Scalpellum, the scuta first grow downwards, and afterwards in most of the species upwards, whereas here from the beginning, the growth is both upwards and downwards. In the mouth and cirri, there is rather more resemblance to Scalpellum than to Paecilasma and Lepas: in habits, also, this genus agrees with Scalpellum, and if it had possessed a lower whorl of valves, it would have quite naturally entered that genus. It is unfortunate, that so insignificant and poorly characterized a form should require a generic appellation. In natural position, it appears to lead from Scalpellum through Paecilasma to Lepas.

---

[31] From ὀξυνω, to sharpen, and ασπιζ, a shield or scutum.

1. OXYNASPIS CELATA

PLATE III. FIG. 1

Madeira; attached in numbers to an Antipathes; Rev. R. T. Lowe. Mus., Hancock.

General appearance. The capitulum is rather thin, and broad in proportion to its length; it seems always entirely covered by the horny muricated bark of the Antipathes, and hence externally is coloured rich brown and covered with little horny spines. The membrane over the valves is very thin, and is with difficulty separated from the Antipathes; it has, I believe, no spines of its own. The corium lining the peduncle is a fine purple. All the individuals are attached to the coralline, with their capitulums upwards in the direction of the branches, and in this respect fig. 1 is erroneous.

The valves, when cleared of the bark, are white, or are strongly tinged with pinkish-orange. The upper parts of the scuta and terga are plainly furrowed in lines / radiating from their umbones; hence their margins are serrated with blunt teeth; their surfaces, moreover, are sparingly studded with small calcareous points.

Scuta (fig. 1a), subtriangular, with the lower part rounded and protuberant, the upper produced and pointed. The umbo is situated in the middle of the occludent margin, instead of at the rostral angle, as in the foregoing genera. The occludent margin is straight, and is bordered by a narrow step or ledge, formed of transverse growth-ridges, and therefore has its edge serrated: the rostral angle is often slightly produced into a small projection. The basal margin is short, and forms an angle above a rectangle with the occludent margin: the tergal margin is straight; the carinal margin is rounded, protuberant, and of unusual length compared to the basal margin. The surface of the valve is convex near the umbo; and beneath there is a large deep hollow for the adductor muscle.

Terga (fig. 1b) large, flat, triangular, as long as the scuta or the carina, all three valves being nearly equal in length; occludent margin straight, or slightly arched, basal angle broad, not very sharp.

Carina short (fig. 1c, drawn rather too long), deeply concave, rectangularly bent, with the lower part not quite as long as the upper, and a little wider: the basal margin is truncated, rounded and slightly sinuous. The umbo is situated at the angle, and therefore nearly central. The umbo of the terga, I may add, is in the same place, as in Lepas.

The *peduncle* is very short and narrow, and is, I believe, without spines; it is enveloped by the bark of the Antipathes. The capitulum in the largest specimens was 0·2 of an inch in length.

Filamentary appendages, apparently none.

Mouth, with the orifice rather inclined abdominally.

Labrum, with the upper part extremely protuberant, forming a projecting horn; no teeth on the crest. Palpi rather small, with only a few bristles at the end.

Mandibles, with four teeth and the inferior angle / pointed: first tooth as far from the second, as in the latter from the inferior angle; in one specimen, on one side, there were five teeth.

Maxillae with three great spines at the upper angle, beneath which a deep notch, and with the inferior part much upraised; this lower part rather rounded at both corners, with the upper spines longer than the lower.

Outer maxillae with the bristles continuous in front; externally, slightly protuberant, with a tuft of bristles longer than those on the front side. Olfactory orifices apparently not protuberant; but all the specimens were in a bad state.

Cirri. Prosoma very little developed. First cirrus very far removed from the second. The three posterior cirri are straight and long; the segments are elongated and bear four or five pairs of very long spines, with a single minute intermediate spine between each pair; dorsal tufts, with long spines. First cirrus, rami unequal by two or three segments, and thickly covered with spines; the first cirrus is short compared to the second, owing to the length of the pedicel of the latter, though the longer ramus of the first, nearly equals the shorter ramus of the second pair. Second cirrus, with its anterior ramus shorter by two or three segments than the posterior ramus, and thicker than it, with the segments covered like brushes with bristles; posterior ramus, and both rami of the third cirrus, a little more thickly clothed with bristles than are the three posterior cirri.

Caudal appendages, minute, broadly oval, with six or seven long bristles on their summits.

GENUS: CONCHODERMA

PLATE III

CONCHODERMA. Olfers, *Magaz. der Gesellsch. Natuforsch. Freunde zu Berlin*, Drittes Quartel, 1814[32] /
LEPAS. Linnaeus, *Systema Naturae*, 1767
BRANTA. Oken, *Lehrbuch der Naturgeschichte*, Th. 2, p. 362, 1815
MALACOTTA et SENOCLITA. Schumacher, *Essai d'un Nouveau Syst. des Habitations des Vers.*, 1817
OTION et CINERAS. Leach, *Journal de Phys.*, vol. lxxxv, p. 67, July, 1817
GYMNOLEPAS. De Blainville, *Dict. des Sci. Nat.*, Art. Mollusca, 1824
PAMINA. J. E. Gray, *Annals of Philosophy*, vol. x (Second Series), August, 1825[33]

Valvae 2 ad 5, minutae, inter se remotae: scuta bi- aut tri-lobata, umbonibus in medio marginis occludentis positis: carina arcuata, terminis utrinque paene similibus.

Valves 2 to 5, minute, remote from each other: scuta with two or three lobes, with their umbones in the middle of the occludent margin: carina arched, upper and lower ends nearly alike.

Filaments seated beneath the basal articulations of the first pair of cirri, and on the pedicels of four or five anterior pairs; mandibles, with five teeth, finely pectinated; maxillae step-formed; caudal appendages, none.

Distribution. Mundane, throughout the equatorial, temperate, and cold seas; attached to floating objects, living or inorganic.

The *capitulum* is formed of smooth membrane, including five small valves, of which the terga and carina are often quite rudimentary or absent. Valves minute, thin, generally more or less linear, placed far distant from each other; sometimes imperfectly calcified and covered by chitine membrane, or imbedded in it. The umbones / of the valves (together with the primordial valves) are nearly central, so that they

---

[32] The general title to the volume, containing four quarterly parts, is dated 1818; but as in the *Journal de Physique*, for July, 1817, the editor / refers to Conchoderma, the quarterly part containing this genus must have appeared before 1818: Lamarck gives the year 1814 as the date of the paper in question, and I have accordingly followed him. From a similar reference by the editor, it appears that Schumacher's volume appeared before the number of the *Journal de Physique* containing Leach's Paper.

[33] Under these nine generic names, the two common species of Conchoderma have received thirty-three different specific denominations, caused partly by changes of nomenclature, and partly from varieties having been ranked as species.

are added to at their upper and lower ends; hence their manner of growth is considerably different from that of the valves in Lepas. The adductor muscle is attached to a slight concavity on the underside of each scutum, at the point whence the lobes diverge.

The *Terga* are placed almost transversely to the scuta; at their lower ends, there is either a very slight prominence in the capitulum, or there is a large tubular, folded appendage, opening into the sack, and apparently serving for respiratory purposes.

Peduncle, smooth, moderately long; attachment effected by the cement stuff being poured out exclusively, as it appears, from the larval antennae. These antennae in *C. aurita* and *C. virgata*, resemble, in the form of the disc and in the long feathered spines on the ultimate segment, those in Lepas.

The *filamentary appendages* are highly developed; there are six or seven on each side; two are attached beneath the basal articulation of the first cirrus (as is usual in Lepas), and near them there are one or two small pap-formed projections of apparently similar nature; the rest of the filaments are attached to the posterior edges low down, on the lower segments of the pedicels of the cirri. I believe, in all cases, these appendages are occupied by testes.

Prosoma, moderately developed.

Mouth, situated not far from the adductor muscle; labrum considerably bullate, with the crest hairy and pectinated with inwardly pointing, approximate, flattened teeth: inner fold of the supra-oesophageal cavity slightly thickened and yellowish, villose on the sides.

Palpi of the usual shape, not meeting, moderately broad.

Mandibles, with five teeth, graduated in size, nearly equidistant, finely pectinated either on one or both sides towards their bases; inferior angle narrow, either produced into a fine tooth, or almost rudimentary. /

Maxillae, about ¾ths of the size of the mandibles, step-formed, with five steps generally distinct; at the upper angle there are two large unequal spines, of which the lower one is the largest, with a third long thin one on the first step; lower spines doubly serrated. Apodeme directed inwards and backwards.

Outer maxillae (Pl. X, fig. 16) simply arched; the membrane of the supra-oesophageal cavity under these maxillae is highly bullate and villose. Olfactory orifices not prominent.

Cirri. First pair not seated far distant from the second pair. The three posterior pair have the anterior faces of their segments considerably protuberant, supporting four or five pair of long bristles; between which, there is a row of minute, fine, upwardly pointing bristles: on the lateral upper margins of each segment, there are a few very minute spines; dorsal tuft short, with thick and thin spines intermingled. In the first cirrus (of which the rami are nearly equal in length), and in the anterior ramus of the second cirrus, the faces of the segments are highly protuberant, and clothed with thick transverse rows of finely and doubly serrated spines: the anterior ramus of the second cirrus is considerably thicker than the posterior ramus, which latter, together with both rami of the third cirrus, differ from the three posterior cirri only in the intermediate and in the lateral marginal spines being slightly more developed.

Caudal appendages, absent.

Alimentary canal. The upper part of the stomach has four large caeca, of which the posterior one is the largest; the whole surface, also, is covered with minute pits, arranged in transverse rows.

Generative system, developed to an extraordinary degree. The testes run into all the filamentary appendages, as well as more or less, into the pedicels of the cirri: the two vesiculae seminales unite *within* the penis, either just beyond its basal constriction, or up one-third of its length. Penis short, hairy. The ovarian tubes not / only fill the peduncle, but extend in a thin sheet between the two folds of corium all round the sack, close up to the terga. The two ovigerous fraena are present in the usual position; the ovigerous lamellae either form several layers, in pairs, one under the other, or are united in a single large cup-formed sheet enclosing the whole animal.

Colours. The prevailing tint is a dark purplish-brown, which forms, or tends to form, broad longitudinal bands on the peduncle and capitulum.

General remarks. This genus is intimately related, as has been remarked by Professor Macgillivray,[34] to Lepas; if we look to the body

---

[34] Remarks on the Cirripedia, etc.; *Edin. New Phil. Journal*, vol. xxxix, p. 171.

of the animal, which from being less exposed to external influences must, in the Cirripedia, offer the most trustworthy characters, we find that in Conchoderma there are additional filamentary appendages attached to the cirri, that there are no caudal appendages, that the teeth of the mandibles are finely pectinated, and that the ovarian tubes run higher up round the sack; in every other respect, there is the closest similarity, even to the arrangement of the bristles on the cirri. In the capitulum, the difference consists chiefly, though not exclusively, in the less development of the valves, and their consequent wide separation: the scuta, however, in Conchoderma, are added to beneath their umbones, or original centres of growth, which is never the case, or only to a very slight degree, in Lepas. Conchoderma has no very close affinity to any other genus. As the majority of authors have ranked the two common species under two distinct genera (Otion and Cineras), I may observe, that there is no good ground for this separation; in the above few specified points in which Conchoderma differs from the genus most closely allied to it, the two species essentially agree together. If we take the nearest varieties of *C. virgata* and *C. aurita*, there is but a very slight difference even in the form of their valves, and these hold the same relative positions to each other; the / carina, however, is always less developed in *C. aurita*; even the colouring in both tends to follow the same arrangement. The only obvious distinction between the two species, are the ear-like appendages of *C. aurita*, which, however, are not developed in its early age, are subject to considerable variation, are of no high functional signification, and are indicated in *C. virgata* by two prominences on the same exact spots. On these grounds I conclude, that the generic separation of the two species is quite inadmissible.

1. CONCHODERMA AURITA

PLATE III. FIG. 4

LEPAS AURITA. Linn.,[35] *Systema Naturae*, 1767
OTION CUVIERANUS [!] BLAINVILLIANUS [!] BELLIANUS [!] DUMERILLIANUS [!] RISSOANUS. Leach, *Encyclop. Brit.*, vol. iii, Supp., 1824, and *Zoological Journal*, vol. ii, p. 208, July, 1825
OTION DEPRESSA et SACCUTIFERA. Coates, *Journal Acad. Nat. Sci. of Philadelphia*, vol. vi, p. 132, 1829
OTION AURITUS. Macgillivray, *Edinburgh New Phil. Journal*, vol. xxxviii, 1845
LEPAS LEPORINA. Poli, *Test. utriusq. Sicil.*, pl. vi, fig. 21, 1795
LEPAS CORNUTA. Montagu, *Linn. Trans.*, vol. xi, p. 179, 1815
CONCHODERMA AURITUM et LEPORINUM. Olfers, *Magaz. der Gesell. Freunde zu Berlin*, 3d Quartel., p. 177, 1814
BRANTA AURITA. Oken, *Lehrbuch der Naturgesch.*, Th. 11, p. 362, 1815
MALACOTTA BIVALVIS. Schumacher, *Essai d'un Nouveau Syst., etc.*, 1817
GYMNOLEPAS CUVIERII. De Blainville, *Dict. des Sc. Nat.*, Art. Mollusc., Plate, fig. 1, 1824 /

C. capitulo duobus tubularibus quasi-auribus instructo, pone terga rudimentalia (saepe nulla) positis: scutis bilobatis: carinâ nullâ, aut omnino rudimentali: pedunculo longo, a capitulo distincte separato.

Capitulum with two tubular ear-like appendages, seated behind the rudimentary and often absent terga; scuta bilobed; carina absent, or quite rudimentary; peduncle long, distinctly separated from the capitulum.

Filaments attached to the pedicels of the second cirrus; two upper spines of the maxillae pectinated.

Habitat. Mundane; extremely common. On ships' bottoms from all parts of the world. Arctic Sea. Greenland. Pacific Ocean. Often attached to Coronulae on whales. On slow-moving fish, according to Dr A. Gould. Often associated with *C. virgata*, and *Lepas anatifera, L. Hillii*, and *L. ansifera*.

General appearance. The capitulum (seen from above in Pl. III, fig. 4a) is slightly compressed, almost globular, composed of thick membrane, with two large, ear-like, flexible, tubular, folded appendages, at the upper end, opening into the sack. These appendages are seated behind the rudimentary terga when such are present, or behind the

---

[35] Many authors (Poli, Montagu, etc.), have doubted from the strangely mistaken description, viz., 'ore octovalvi dentato', whether this species could be the *Lepas aurita* of Linnaeus. But in the Linnean Society, there is a proof plate from Ellis's 'Account of several rare Species of Barnacles', in *Phil. Trans.*, 1758, with an excellent figure of the *C. aurita*, and on the margin in Linnaeus's handwriting is the name *Lepas aurita*.

spots which they would have held if not aborted. In a young condition they are tubular, but not folded; and often, according to Professor Macgillivray, either one or both are at first imperforate. They are formed externally of the outer membrane of the capitulum (rendered thin where folded), and internally of a prolongation of the inner tunic of the sack; between the two, there is, as around the whole sack, a double layer of corium. A section across both appendages, near their bases (is given in Pl. III, fig. 4*b*), showing how they are folded – the chief fold being directed from below upwards, with a smaller fold, not always present, from between the two, outwards. The folds sometimes do not exactly correspond on opposite sides of the same individual; they are almost confined to the lower part, the orifice itself being often simply tubular. These appendages are sometimes very nearly as long as the whole capitulum: a section / near their bases is subtriangular. I shall presently make some remarks on their functions and manner of formation.

The *scuta*, as well as the other valves, are imperfectly calcified: shape, variable. They usually consist of two lobes or plates, placed at above a right angle to each other, and rarely (fig. 4*c*) almost in a straight line; the lower lobe is more pointed and narrower than the upper; the two correspond to the lower and middle lobes in the scuta of *C. virgata*, the upper one being here absent.

The *terga* are developed in an extremely variable degree; they are often entirely cast off and absent. In very young specimens, they are of the same length with the carina, but after the carina has ceased to grow, the terga always increase a little, and sometimes to such a degree as to be even thirty or forty times as long as the carina. When most developed (fig. 4*a*) they are not above one-third as long as the scuta, to which they lie at nearly right angles; they consist of imperfectly calcified plates, square at both ends, slightly broader and thinner at the end towards the carina, where they are a little curled inwards, than at the opposite end; they are not quite flat in any one plane; internally they are slightly concave; finally, I may add, they nearly resemble in miniature the terga of *C. virgata*. In full-grown specimens, the terga almost invariably drop out and are lost; but even in this case, a long brownish cleft in the membrane of the capitulum, marks their former position. The orifice of the capitulum is usually notched between the terga, or between the clefts left by them; on each side of the notch there is a slight prominence. In some few cases, however, there is no trace of this notch. Behind the terga or the clefts, the great ear-like appendages, as we have seen, are situated.

Carina, rudimentary (fig. 4) and often absent; it is pointed-elliptical, and is rarely above the 1/40th of an inch long. After arriving at this full size, calcareous matter is added to the under surface over a less and less area, so that it becomes internally pointed, and finally, in place of calcareous matter, continuous sheets of chitine are spread out beneath it; hence, during the disintegration of the outer surface, the carina comes to project more and more, and at last drops out; subsequently, even the little hole in which it was imbedded, disintegrates and disappears.

Peduncle, cylindrical, distinctly separated from the capitulum, and generally twice or thrice as long as it: the thickness of the outer membrane generally great, but variable: surface of attachment variable, either pointed, or widely expanded, or formed into divergent projections.

Filamentary appendages, seven on each side, highly developed, long and tapering; there are two beneath the basal articulation of the first cirrus, and one on the posterior margin of the pedicel of each cirrus, excepting the sixth pair; the filaments on the pedicels are nearly twice as long as the cirri themselves.

Mouth, mandibles, with the five teeth nearly equidistant, and towards their bases finely pectinated on both sides; inferior angle rudimentary, often represented by a single minute spine: in one specimen, there were only four teeth on one side. Maxillae, with five steps, not very distinct from each other, with the first step much curved. The larger of the two upper great unequal spines is pectinated, like the teeth of the mandibles; there is a third long finer spine beneath the upper large pair.

Cirri rather short, broad, with the anterior faces of the segments protuberant, especially those of the first cirrus and of the anterior ramus of the second pair: spines on the anterior cirri doubly serrated. Posterior cirri, with the intermediate spines between the pairs, long; dorsal tufts, minute. On the lower segment of the pedicels of the four posterior cirri, there are two separate tufts of bristles.

Colours extremely variable; sometimes five longitudinal bands of dark purple can be distinctly seen (as in *C. virgata*) on the peduncle, these bands becoming more or less confluent on the capitulum; at other times, the capitulum is more or less spotted, or often nearly uniformly purple: the sack, cirri and trophi are, also, purple. /

Size. The largest specimen which I have seen was, including the peduncle and ears, five inches in length, the capitulum itself being rather above one inch in length, and 7/10ths of an inch in breadth.

General remarks. I have come to the same conclusion with Professor Macgillivray, concerning the variability of this form, and I believe there is only one true species. With respect to Dr Coates's species, viz., *Otion depressa* and *O. saccutifera*, though I have not seen specimens, I can hardly doubt, from the insufficient characters given, that they are mere varieties.

With respect to the ear-like appendages, we shall presently see in *C. virgata*, that at corresponding points on the capitulum (Pl. III, fig. 2b), there are two slight, closed prominences. According to Professor Macgillivray, in *C. aurita*, every gradation can be followed by which the appendages, at first closed, become tubular and open. The opening would ensue, if the corium became absorbed at the bottom of the appendages whilst still imperforate, for then the inner tunic would be cast off at the next moult and would not be re-formed, whilst the outer membrane would gradually disintegrate together with the other external parts of the capitulum, and not being re-formed at this point, an aperture would at last be left. These appendages have no relation to the generative system: the ovarian tubes, which surround the sack do not extend into them; nor do the ovigerous lamellae. I believe, that their function is respiratory: the corium lining them is traversed by river-like circulatory channels, and their much-folded, tubular and open structure must freely expose a large surface to the circumambient water. Why this species should require larger respiratory organs than any other, I know not. In this species, moreover, the filamentary appendages are developed to a greater extent than in any other cirripede; in most genera, the surface of the body and of the sack suffices for respiration. /

2. CONCHODERMA VIRGATA
PLATE III. FIG. 2. PLATE IX. FIG. 4

LEPAS VIRGATA. Spengler, *Skrifter Naturhist. Selbskabet.*, B. i, Pl. vi, fig. 9, 1790
LEPAS CORIACEA. Poli, *Test. utriusque Sicil.*, Pl. vi, fig. 20, 1795
LEPAS MEMBRANACEA. Montagu, *Test. Brit. Supp.*, p. 164, et Linn. *Trans.*, vol. xi, Pl. xii, fig. 2, 1808
CONCHODERMA VIRGATUM. Olfers, *Magaz. Gesells. Naturfor. Freunde, Berlin*, p. 177 (3d Quartel), 1814[36]
BRANTA VIRGATA. Oken, *Lehrbuch der Gesell.*, Th. ii, p. 362, 1815
SENOCLITA FASCIATA. Schumacher, *Essai d'un Nouveau Syst.*, 1817
CINERAS VITTATA. Leach, *Encyclop. Brit. Supp.*, Tom. iii, Plate, 1824
CINERAS CRANCHII[!] CHELONOPHILUS[!] OLFERSII[!]. Leach, *Tuckey's Congo Expedition*, p. 412, 1818
CINERAS MAGALEPIS[!] MONTAGUI[!] RISSOANUS. Leach, *Zool. Journal*, vol. ii, p. 208, 1825
CINERAS MEMBRANACEA. Macgillivray, *Edin. New Phil. Journal*, vol. xxxix, p. 171, 1845
CINERAS BICOLOR. Risso, *Hist. Nat. des Productions, etc.*, Tom. iv, p. 383, 1826
CINERAS VITTATUS. Brown, *Illust. of Conch.*, Pl. li, figs 16–8, 1844
GYMNOLEPAS CRANCHII. De Blainville, *Dict. des Sci. Nat. Hist.*, 1824
PAMINA TRILINEATA[!] (Var. Monstr.). J. E. Gray, *Annals of Phil.*, vol. x, 1825

C. scutis trilobatis: tergis intùs concavis, apicibus introrsùm leviter curvatis: carinâ modicâ, leviter curvatâ: pedunculo in capitulum coalescente.
Scuta three-lobed: terga concave internally, with their apices slightly curved inwards: carina moderately developed, slightly curved: peduncle blending into the capitulum.

No filament attached to the pedicel of the second cirrus.

Var. chelonophilus (Pl. III, fig. 2c). Terga, minute, nearly straight, solid, acuminated at both ends, placed far distant from the other valves: carina, either minute / and acuminated at both ends, or moderately developed and slightly arched and blunt at both ends: lateral lobes of the scuta broad: valves imperfectly calcified.

Habitat. Mundane: extremely common on ships' bottoms from all parts of the world. Falkland Islands. Galapagos Islands, Pacific Ocean. Attached to seaweed, turtle and other objects. Often associated with *Conchoderma aurita, Lepas anatifera, L. Hillii,* and *L. anserifera.*

General appearance. Capitulum, flattened, gradually blending into the peduncle; summit square, rarely obtusely pointed. Membrane, thin.

---

[36] See page 113 [136] respecting this date.

Valves, thin, small, sometimes imperfectly calcified, very variable in shape and in proportional length, and therefore, situated at variable distances from each other, but always remote and imbedded in membrane.

Scuta, trilobed, consisting of an upper and lower lobe (the latter generally the broadest), united into a straight flat disc, with a third lobe standing out from the middle of the exterior margin, generally at an angle of from 50° to 70° (rarely at right angles) to the upper part, and generally (but not always) bending a little inwards. The shape of the lateral lobe varies from rounded oblong to an equilateral triangle; as it approaches this latter form, it becomes much wider than the upper or lower lobes. In one specimen, and only on one side, the scutum (fig. 2*d*) presented five points or projections. In some specimens, the scuta are very imperfectly calcified, and consist of several quite separate beads of calcareous matter of irregular shape, held together by tough brown membrane.

Terga, extremely variable in shape, placed at nearly right angles to the scuta: beyond their carinal ends (fig. 2*b*), the capitulum presents two small prominences, which are important as indicating the position of the homologous, ear-like appendages in *C. aurita*.[37] The upper ends of the terga are imbedded in membrane, and project freely like little horns for about one-third of their length: this free portion exactly answers to the projecting / portion, bounded by the two occludent margins, in the terga of Lepas. The freely projecting portion is generally curled inwards, and the carinal portion more or less outwards – the form of the letter S being thus approached; but the curvatures are not exactly in the same plane. The whole valve is generally of nearly equal width throughout, the carinal part being a very little (but in some specimens considerably) wider; internally, it is deeply concave; both points generally are blunt and rounded. In some rare varieties (*Cineras chelonophilus* of Leach, fig. 2*c*), the terga are much smaller and flat, with both points sharp, the whole upper portion being much and abruptly attenuated, and internally, without a trace of a concavity. Generally, the terga are about two-thirds of the length of the scuta, rarely only half their length; generally, they are separated from the apices of the scuta by about their own length,

---

[37] These have also been observed by Dr Coates; see *Journal of Acad. Nat. Sci. Philadelphia*, vol. vi, p. 134, 1829.

rarely by twice their own length. Generally, the terga are shorter than the carina, but sometimes a very little longer than it: generally they are distant by one-third or one-fourth of their own length from the apex of the carina, rarely by their entire length.

Carina (fig. 2a), lying nearly parallel to the scuta, concave within, very slightly bowed, of nearly the same width throughout, but with the lower third beneath the umbo, generally a trace wider than the upper part. Length, variable, generally rather longer (sometimes by even one-third of its own length) than the scuta, but sometimes equalling only three-fourths of the length of the scuta; generally longer than the terga. Upper and lower points rounded; in rare varieties, both ends are sharply acuminated. The carina and terga are generally most acuminated where they are smallest and least perfectly calcified; and consequently, in this same state, the valves stand furthest apart.

Peduncle, flattened, gradually widening as it joins the capitulum, to which it is generally about equal in length, or a little longer.

Filamentary appendages. Six on each side (Pl. IX, fig. 4), / of which one (*h*) is seated on the posterior margin of a swelling, beneath the basal articulation of the first cirrus, and this is the longest; the second (*g*) is short and thick, and is seated a little lower on the side of the prosoma (near to this, there are also two little pap-like eminences); the third (*i*) is seated on the posterior margin of the pedicel of the first cirrus, above the basal articulation; the fourth, fifth, and sixth (*j, k, l*) in similar positions on the pedicels of the third, fourth, and fifth cirri. These three latter filaments are shorter and smaller than the first three. At the base of the second cirrus, which has no proper filament, there is a swelling as if one had been united to it.

Mouth. Mandibles, with the basal edges of the five teeth pectinated by minute, short, strong spines on one side; inferior angle extremely short. In one specimen, there was a minute pectinated tooth between the first and second; in another, the second tooth was bifid on its summit; in another, the fourth was rudimentary.

Maxillae, with five steps: sometimes each step commences with a spine rather larger than the others; at the upper angle, there are two large unequal spines (neither pectinated), with a third longer and thinner, seated a little below. *Outer maxillae* (Pl. X, fig. 16), simple.

Cirri, with twice as many segments in the sixth cirrus as in first; spines on the first and second cirri doubly serrated.

Colours (when alive). Capitulum and peduncle grey, with a tinge of blue, with six black bands, tinged with purplish-brown. The two bands near the carina become confluent on the peduncle, and sometimes disappear; the carina is edged, and the interspace between the two scuta, coloured with the same dark tint. The whole body and the pedicels of the cirri are dark lead colour, with the segments of the cirri almost black: in some specimens, the colour seems laterally abraded from the cirri. Ova white, becoming in spirits pinkish, and then yellow. The dark bands on the capitulum and peduncle become in spirits purple; but are sometimes discharged; the general grey tint disappears. Professor Macgillivray / states that many individuals are light-brown or yellowish-grey, with irregular brown streaks, or crowded dots: he states that in very young specimens the colours are paler, and the valves spicular.

Size. The largest specimen which I have seen, had a capitulum rather above one inch long and three-fourths of an inch wide: growth very rapid.

Monstrous variety. In the British Museum, there is a dried and somewhat injured specimen of a monstrous variety, the *Pamina trilineata* of J. E. Gray: it differs from the common form only in having a tubular projection, just behind the notch separating the upper points of the terga; this tube springs from over the terga, and is, therefore, in a different position from the ear-like appendages in *Conchoderma aurita*. It does not open into the sack: the membrane composing it appears to have been double in the upper part, and to have been lined with corium: in short, this tube seems to have been an excrescence or tumour, of a cup or tubular form.

General remarks. It will have been seen how much subject to variation the valves of this species are. When I first examined the *Cineras chelonophilus* of Leach, from 36°N. lat., Atlantic Ocean, and found in many specimens, both old and young, that the terga were very small, flat, acuminated at both ends, with a projecting shoulder on the carinal margin, and situated at about their own length from the apex of the carina, and at twice their own length from the scuta; and when I found the carina acuminated at both ends, and the scuta very imperfectly calcified, with the lateral lobe broad, flat, and standing out at right angles; and lastly, when I found the whole capitulum bluntly pointed, instead of being square on the summit, I had not the least doubt, that it was a quite distinct species. Afterwards, I found in the *Cineras Olfersii* of Leach, from the South Atlantic, the same form of terga; but within slightly more concave or furrowed, and not nearly so small, and therefore not placed at above half so great a distance from the other valves; and here, / the carina had its usual outline, as had nearly the scutum on one side, whereas, on the other side, it presented a new and peculiar form, having five ridges or points, and was imperfectly calcified; seeing this, it was impossible to place much weight in the precise form or size (and therefore, relative separation), of the calcified valves; and on close examination, I found every part of the mouth and cirri identical in Leach's *Cineras chelonophilus* and *C. Olfersii*, and in the common form. Therefore, I conclude, that *C. chelonophilus*, and still more *C. Olfersii*, are only varieties; the terga presenting the greatest, yet variable, amount of difference, namely, in their acumination and flatness. We know, also, that in the species of the closely allied genus of Lepas, the terga are very variable in shape, and this is the case, even in a still more marked degree, in *Conchoderma aurita*. Professor Macgillivray, I may add, has come to a similar conclusion regarding the extreme variability of the valves of this species.

As the varieties here mentioned are very remarkable, and may perhaps turn out to be true species, I think they are worth describing in some detail: I will only further add, that we must either make several new species, or consider, as I have done, several forms as mere varieties.

C. VIRGATA, var. CHELONOPHILUS of *Leach*

PLATE III. FIG. 2C

Atlantic Ocean, 35° 15′ N., 16° 32′ W. On the Testudo caretta.

Capitulum not above half an inch long, composed of very thin membrane, with six bands (as stated by Leach) of faint colour; summit bluntly pointed; valves very small, far distant from each other; the scuta are imperfectly calcified, the central part of the umbo consisting of thick, brown chitine, with imbedded shelly beads; terga and carina perfectly calcified.

Scuta trilobed, flat, within slightly concave, upper lobe / rather more acuminated than the lower; lateral lobe triangular in outline, twice as wide as either the upper or lower lobes; lying in the same plane with them, and standing out at almost exactly right angles.

Terga, flat; placed obliquely to the scuta, and barely half as long; separated from them by nearly twice their own length; upper and lower points acuminated; the umbo on the carinal margin forms a projecting shoulder; the scutal margin is straight, they are separated by nearly their own length from the apex of the carina.

Carina narrow, very slightly arched, within slightly concave, both points acuminated; lower third rather wider than the upper part; in length equalling three-fourths of the scuta, and longer by one-third than the terga; about as wide as the latter.

Filaments, cirri, and mouth exactly as before.

In some specimens sent to me by the Rev. R. T. Lowe from off the *Testudo caretta*, taken near Madeira, the scuta have their lateral lobes broad and nearly rectangular: the carina extends nearly to between the terga: the terga are nearly straight, somewhat pointed at both ends, distant from the scuta, almost solid within, with their upper points bowed outwards: the whole capitulum is bluntly pointed, as in the *var. chenophilus*, to which form this makes a rather near approach.

C. VIRGATA, var. OLFERSII

CINERAS OLFERSII. Leach, *Tuckey's Congo Expedition*.

Habitat. South Atlantic Ocean.

Scuta, unlike on the opposite sides of the same individual, on one side with a single lateral lobe as usual, but this very narrow, on the other (fig. 2*d*), with five lobes or projections.

Terga slightly concave within, separated by a little more than their own length from the tips of the scuta, and by one-third of their own length from the tip of the carina. /

Carina longer than the scuta by about one-fifth or one-sixth of its own length, blunt at both ends, considerably bowed.

Again, I possess a group of remarkably fine specimens given me by Mr L. Reeve, from the southern ocean (as I infer from a young *Lepas australis* adhering to them), in which all the individuals, young and old, are characterized as follows: Scuta, with the lateral lobe generally broad, but to a very varying extent, with the upper and lower lobes extremely sharp. Terga separated from the scuta, by one and a fourth of their own length, and by their own length from the carina; somewhat acuminated at both ends, nearly straight, with a very slight shoulder near the umbo. Carina equalling the terga in length, and about three-fourths of the length of the scuta; neither the upper nor lower point much acuminated. All the valves most imperfectly calcified: in one specimen, the scutum on one side was simply horny, without a particle of calcareous matter. The summit of the capitulum nearly intermediate in outline between the common square, and bluntly pointed form of *var. chelonophilus*. I compared the cirri and trophi with those of a common variety, and could detect not the smallest difference. This variety differs from *var. Olfersii*, in the less development of its carina, and from *chelonophilus*, in the greater development of its carina, and especially of its terga. It would appear as if the great variability of the valves was connected with the absence of calcareous matter.

3. CONCHODERMA HUNTERI

PLATE III. FIG. 3

Cineras Hunteri. R. Owen, *Cat. Mus. Coll. of Surgeons*, Invert. Part I., p. 71, 1830

C. valvis angustis: scutis trilobatis, prominentiâ laterali non latiore quam inferior: tergorum parte superiore paene rectangulè secundùm aperturae marginem flexâ: carinâ valde arcuatâ: pedunculo brevi, in capitulum coalescente. /

Valves, narrow: scuta, trilobed, with the lateral lobe not wider than the lower one: terga, with the upper part bent almost rectangularly along the margin of the orifice: carina considerably arched: peduncle short, blending into the capitulum.

No filament attached to the pedicel of the second cirrus.

Var. Carina absent; scuta, with the upper lobe absent; terga, with the rectangular projection little developed.

Attached to the skin of a snake, probably the Hydeus or Pelamis bicolor, and therefore from the tropical Indian or Pacific Oceans. Mus. Coll. of Surgeons.[38]

Capitulum, with the membrane very thin; summit obtusely pointed. Valves linear and thin.

Scuta, elongated, flat, with the upper projecting lobe rather more acuminated than the lower, and equalling it in length; lateral lobe not wider than the lower, and about as long as it, forming an angle of about 55° with the upper one.

Terga, of somewhat variable length, generally about half as long as the carina, narrow, and of nearly equal width throughout; lower point sharp; externally convex; internally solid, with a trace of a central depressed line; the upper fourth part generally a little bowed out of the plane of the lower part, and abruptly bent at rather above a right angle along the occludent margin of the orifice. These valves are situated at about half their own length from the upper points of the scuta.

Carina considerably arched, extending to the lower points of the terga, or running up between them for even half their length; equally narrow throughout; scarcely broader than the terga; both points rounded; internally concave; the lower point does not extend as far down as that of the lower lobe of the scuta. /

Peduncle, narrow, shorter than the capitulum, which, in the largest specimen was 4/10ths of an inch long. Longitudinal purple bands appear to have originally existed on the peduncle.

Filamentary appendages, trophi and cirri all similar to the same parts in *C. virgata*; but perhaps the anterior faces of the segments in the posterior cirri are rather less protuberant; perhaps also the first cirrus is rather shorter in proportion to the sixth cirrus.

Variety (monstrous). Among the specimens, I found one very young one, in which the scuta had not upper lobes, so that in outline they exactly resembled the scuta in the quite distinct *C. aurita*: there was not even a rudiment of a carina: the tergum, *on one side*, was externally

---

[38] I owe to the kindness of Professor Owen, an examination of these specimens, and information regarding them.

bordered by a projecting, semicircular, calcified disc; and the upper points of both terga showed only traces of the rectangular projection, which is the chief characteristic of *C. Hunteri*. From these traces alone, and from the specimen being mingled with the others, do I here include this variety.

General remarks. I have very great doubts whether I have acted rightly in considering this as a species; but as there were many specimens, old and young, all differing remarkably from the common species, this form anyhow deserves description. The points by which it can be distinguished from *C. virgata*, are – the almost rectangular manner in which the upper portion of the tergum is bent outwards and along the orifice of the sack – the narrowness of all the valves, and especially of the lateral lobes of the scuta – and lastly, the greater curvature of the carina, which in some specimens runs up far between the terga; had this last character been constant, it would have been an important one, but such is far from being the case. Great as are these differences in the valves, and though common to many specimens, they are not sufficient to convince me that it is a true species, and I should not be at all surprised at varieties, intermediate between it and the common form, being hereafter found; / had a name not been already attached to it, I should not have given one. In the monstrous variety described, we see to what an extent the valves may vary. The *C. Hunteri* approaches nearest to the var. of *C. virgata*, called by Leach *Cineras chelonophilus*, for in both, the top of the capitulum is bluntly pointed and the terga are solid within; in the *Var. chelonophilus*, the terga and carina are minute, whereas here, though very narrow, they are much elongated. Certainly *C. chelonophilus* has almost as strong a claim to rank as a species as *C. Hunteri*; but in the former, by the aid of other varieties, the differences were almost reduced to the peculiarities in the terga – the valves, the most subject to variation. In *C. Hunteri* we have other differences, and the form of the terga is even still more peculiar. I have, therefore, provisionally attached to it the specific name by which it is designated in the Museum of the College of Surgeons. From having been long kept in spirits, all aid from colour is lost.

GENUS: ALEPAS

PLATE III

ALEPAS. Sander Rang, *Manuel des Mollusques*, 1829
ANATIFA. Quoy et Gaimard, *Voyage de l'Astrolabe*, 1834
TRITON. Lesson, *Voyage de la Coquille*, 1830
CINERAS. Lesson, *Secundum Sander Rang*

Capitulum aut sine valvis, aut scutis corneis, paene abditis.
Capitulum without valves,[39] or with horny, almost hidden, scuta.

Filaments seated beneath the basal articulations of the first pair of cirri; mandibles, with two or three teeth; / maxillae notched, with the lower part irregular, projecting; caudal appendages multi-articulate.

Attached to various living objects, fixed or floating.

Capitulum either entirely destitute of valves, or with transparent horny scuta, not containing any calcareous matter, and almost hidden in membrane. These scuta are formed of a lower and a lateral lobe, placed at above right angles to each other; they are added to by successive layers, and closely resemble in shape the scuta of the *Conchoderma aurita*. The orifice in *A. tubulosa* projects so much as to be almost tubular. In *A. parasita* and *A. minuta* it does not project, and is either moderately large, or very small in proportion to the length of the capitulum; from contraction it is much wrinkled. The membrane forming the capitulum is smooth and very transparent; it contains very few tubuli, except under certain irregular projections in *A. cornuta*.

The *peduncle* is rather short and narrow; it blends into the capitulum, and is not, in some of the species, separated from it by any distinct line; the surface of attachment is rather wide. Within the peduncle we have the three usual layers of striae-less muscles; namely, the innermost and longitudinal, which run lower down than the others; the middle and transverse; and, lastly, the exterior, oblique muscles, which cross each other (becoming transparent) on the rostral central line. These several muscles run up from the peduncle and

---

[39] Any one not attending to the characters derived from the softer parts of the Balanidae and Lepadidae, might easily confound with Alepas the genus Siphonicella (genus nov.), which, undoubtedly, though having the external appearance of a pedunculated cirripede, belongs to the Balaninae, and is closely related to Coronula.

surround the capitulum; from the transparency of the membranes they can be seen from the outside: they are particularly conspicuous round the orifice, which they probably serve to close. There is, in all cases, the usual adductor scutorum muscle (with transverse striae), which is attached under the horny scuta, where such exist. The fact of the striae-less muscles of the peduncle surrounding the whole capitulum, has been observed only in one other genus, namely Anelasma. In consequence of this structure, / the capitulum must possess considerable powers of contraction.

The antennae of the larva in the *Alepas cornuta* and *A. minuta* have the sucking disc nearly circular, with the spines unusually plain on the distal as well as proximal margin. Basal segment broad, much constricted where united to the disc. The ultimate segment has on the middle of the outer margin, in *A. cornuta*, two minute spines, which I have not observed in any other cirripede: on the summit there are the usual spines.

Size. Three of the species are small.

Filamentary appendages. These are rather small; there is only one on each side, situated on the posterior margin of a slight swelling, beneath the basal articulation of the first cirrus; and therefore in the position in which the filaments are most constant in Lepas, and where they likewise occur in Conchoderma.

Body. The prosoma is either pretty well developed or is small, according as the first cirrus is placed near to, or far from the second cirrus.

Mouth. Labrum moderately bullate, with the lower part more or less produced; crest with blunt, bead-like teeth, and short hairs.

Palpi (Pl. X, fig. 8), acuminated and narrow to an unusual degree.

Mandibles, with two or three teeth, and the inferior angle acuminated; the lateral bristles unusually strong, so as to give the main teeth the *appearance* of being pectinated.

Maxillae, widely notched, with three great upper spines; the part beneath the notch projecting, and either straight or irregular.

Outer Maxillae, with the inner bristles either continuous or divided into two groups: exteriorly there is a smaller or larger prominence, with long bristles. The olfactory orifices are either slightly, or not at all protuberant.

Cirri. In the three posterior pair, the segments have their bristles arranged in a transverse row, either in the / form of a narrow brush, or consisting only of a single pair with two or three minute, intermediate, and lateral marginal spines. The anterior ramus of the second cirrus is thicker, and more thickly clothed with spines than is the posterior ramus: this latter ramus, however, and both rami of the third cirrus, are rather more thickly clothed with spines than are the three posterior pair. The unique case in *A. cornuta* of the inner rami of the fifth and sixth cirri being rudimentary (Pl. X, fig. 28) will be minutely described under that species.

Caudal appendages, thin, tapering, multi-articulate, about as long as the pedicels of the sixth cirrus.

Stomach. The oesophagus runs in a somewhat sinuous course, and enters the top of the stomach obliquely. There are no caeca. The biliary envelope presents a reticulated structure, instead of the usual longitudinal folds.

Generative system. The penis is hairy, not very long, and ringed or articulated in an unusually plain manner; the space between each ring being about one-fourth of the diameter of the penis: the unarticulated basal portion or support is here remarkably long. The vesiculae seminales are long, tortuous, and enter the prosoma. The ovarian tubes are of wide diameter: in *A. cornuta* they surround the whole capitulum. The ovigerous fraena are small, constricted at the base, and square on the free margin, which is studded with minute glandular beads, borne on the finest footstalks.

Range. Southern shores of England, Mediterranean, Atlantic, West Indies, New Zealand, attached to various objects. *A. parasita* has been always taken on Medusae.[40]

Affinities. This genus differs from all, except Anelasma, in the manner in which the striaeless muscles of the peduncle run up and surround the capitulum, and likewise / in the reticulated character of the biliary envelope of the stomach. To Conchoderma, especially to *C. aurita,* there is a manifest affinity in the form of the horny scuta: there

---

[40] It appears that Solander (*Dillwyn Des. Cat.*, vol. i, p. 34) observed a species of this genus adhering to a Medusa on the coast of Brazil. Mr Cocks informs me that an Alepas, apparently *A. parasita*, has been cast on shore near Falmouth, attached to a Cyanaea; and that two other specimens adhered to the bottom of a vessel arriving at that port from Odessa.

is also some affinity to this same genus in the presence of filamentary appendages though here little developed, and in the circular form of the disc of the larval antennae, and, lastly, in the ovarian tubes in *A. cornuta* surrounding the capitulum. There is quite as close, if not closer affinity to Ibla, in the following peculiarities – in the curved oesophagus – in the general character of the cirri and trophi, with the olfactory orifices in one species in some degree prominent – in the multi-articulated caudal appendages – and in the plainly articulated penis, with its elongated unarticulated support, though both these characters are exaggerated in Ibla. Lastly, the scuta in Ibla, though not at all resembling in shape those of *A. cornuta*, are formed without calcareous matter; and again, in Ibla, the muscles of the peduncle run up to the bases of the valves, and so almost surround the space in which the animal's body is lodged.

The four species of Alepas appear to form two little groups; viz. *A. parasita* and *A. minuta* on the one hand, and *A. cornuta* and *A. tubulosa* on the other.

1. ALEPAS MINUTA

PLATE III. *FIG.* 5

ALEPAS MINUTA. Philippi, *Enumeratio Mollusc. Siciliae*, Pl. xii, fig. 23, 1836
ALEPAS MINUTA. A. Costa, *Esercitazione Accadem.*, vol. ii, part I, Naples, Pl. iii, fig. 5, 1840 (secundum Guerin in *Revue Zoolog.*, p. 250, 1841)
ALEPAS MINUTA. Chenu, *Illust. Conch.*, Pl. iii, figs 8–10

A. aperturâ non prominente, capituli longitudinis vix tertiam partem aequante: scutis corneis, paene absconditis: longitudine totâ ad quartam unciae partem.
Orifice not protuberant, one-third of the length of the / capitulum: scuta horny, almost hidden. Total length quarter of an inch.

Outer maxillae, with the spines in front continuous; posterior cirri, with several long spines arranged in a transverse row on each segment; caudal appendages longer than the pedicels of the sixth cirrus.

Sicily; attached to a Cidaris:[41] island of Capri (*A. Costa*).

Capitulum oval, blending insensibly into the peduncle; moderately flattened; composed of thin structureless membrane, with the exception of two horny, almost quite hidden scuta. Orifice situated near the

[41] I am greatly indebted to Professor J. Müller, of Berlin, for kindly lending me specimens.

summit, and in a line, which is oblique to the longitudinal axis of the peduncle; much wrinkled; barely one-third of the length of the whole capitulum.

The *scuta*, consist of yellowish, transparent, horny, laminated chitine without any calcareous matter; externally covered by the common integument of the capitulum; these valves are placed very near to each other, close under the orifice, and therefore high up on the capitulum; the membrane between them is smooth and unwrinkled; they are formed of two rather acuminated lobes, joining each other at above a right angle; one lobe (the longer one) stretching nearly transversely across the capitulum, the other running down parallel to its rostral margin: in shape and position they resemble the scuta of *Conchoderma aurita*; and if another lobe had been developed it would have run along the orifice, and then these valves would have resembled the scuta of *Conchoderma virgata*. In a specimen with a capitulum ²⁄₁₀ths of an inch long, the scuta from point to point were ½₀th of an inch in length.

Peduncle, much wrinkled, about one-third in diameter of the capitulum, and shorter than it; at the base it is generally expanded into two or three finger-like projections. / *Length* of the largest specimen, about one-fourth of an inch. *Colour*, according to A. Costa in the work above cited, 'rufo-flava vittatâ;' but after spirits the whole becomes uniformly yellowish.

Filamentary appendages, situated beneath the basal articulation of the first cirrus, on the posterior edge of the usual enlargement; acuminated, about two-thirds of the length of the shorter ramus of the first cirrus.

Prosoma well developed.

Mouth. On each side there are two slight prominences; one under the mandibles, the other transverse nearer to the adductor muscle.

Labrum, placed near the adductor muscle, with the upper part not more bullate than the lower part; crest with a row of blunt teeth, and many fine bristles growing chiefly outside the teeth; there are many fine bristles on the inner or supra-oesophageal fold of the labrum.

Palpi not nearly touching each other, pointing towards the adductor: much hollowed out on their inner sides, hence narrow and acuminated, with doubly serrated bristles.

Mandibles, with three teeth and the inferior angle ending in a single sharp spine; whole inferior portion narrow; first tooth as far from the second, as the latter from the inferior angle; owing to the presence of short thick spines projecting from the sides of the jaw, the lower edges of the second and third teeth appear pectinated.

Maxillae, nearly two-thirds of the width of the mandibles; beneath the three larger upper spines there is a considerable notch, and the whole lower part is very slightly upraised; edge irregular, with obscure traces of either two projections, or perhaps of four steps.

Outer Maxillae, with bristles in front continuous; exteriorly there is a slight prominence near each olfactory orifice, with a tuft of long bristles.

Cirri not much elongated; first pair placed not quite close to the second; five posterior cirri nearly equal in length; pedicels long, with irregularly scattered spines – those on the pedicel of the first cirrus beautifully and / conspicuously feathered. The segments of the three posterior pair are *not* very short or broad; very slightly protuberant, each with a long transverse, crescentic, narrow brush of bristles, which stand two or three deep in the middle, but on the sides are single: dorsal tufts long, and in the upper segments the spines are thick and claw-like. This structure is common to all the cirri. First cirrus with the rami unequal in length by two segments; from the shortness of the pedicel, this cirrus is much shorter than the second, but its rami are about two-thirds of the length of those of the second cirrus. Second cirrus (and in a less degree the third cirrus), with the anterior ramus a shade broader than the posterior ramus, and rather more thickly covered with spines than are the three posterior cirri. Fifteen segments in the sixth cirrus; nine in the longer ramus of the first cirrus.

Caudal appendages, rather longer than the pedicels of the sixth cirrus, composed of seven cylindrical, tapering segments, each with a circle of very fine bristles on its summit.

The acoustic [?] sacks are situated some way below the basal articulations of the first cirrus.

2. ALEPAS PARASITA

ALEPAS PARASITA. Sander Rang, *Man. des Mollusq.*, p. 364, Pl. viii, fig. 5, 1829[42]
ANATIFA UNIVALVIS. Quoy et Gaimard, *Annales des Sciences, Nat.*, tom. x, p. 234, Pl. vii, fig. 8, 1827
ANATIFA PARASITA. Quoy et Gaimard, *Voyage de l'Astrolabe*, Pl. xciii, 1834
TRITON (ALEPAS) FASCICULATUS. Lesson, *Voyage de la Coquille. Mollusc.* Pl. xvi, fig. 6, tom. ii, part I, p. 442, 1830 /

A. aperturâ non prominente, capituli longitudinis ⅔ aequante: scutis corneis: longitudine totâ ad 2 uncias.
Orifice not protuberant, equalling two-thirds of the length of the capitulum: scuta horny. Total length two inches.

Animal unknown.

Parasitic on Medusae, Mediterranean and Atlantic Oceans: south shore of England [?].[43]

I have not seen this species, and have drawn up the above specific character from the plates and brief descriptions in the voyages of the *Coquille* and *Astrolabe*. M. Lesson thinks that his species differs from that of M. Quoy and Gaimard; but as the peculiar yellow colour of the capitulum, general shape, short cirri, habits and range, are all common to both, I believe that they are identical. There is, however, one singular difference, namely, that the cirri are coloured bright blue in the plate in the *Voyage of the Astrolabe*, and yellowish in that in the *Voyage of the Coquille*: this possibly may have resulted from the drawing in the latter case having been made from a specimen long kept in spirits.

M. Lesson says that there are seven pair of cirri, from which I infer that this species has a pair of long, articulated, caudal appendages: he asserts that each cirrus has ten segments; the cirri are short and little curled. M. Lesson remarks, that 'deux languettes bifurques occupent le bas de l'ouverture ovale': I can hardly doubt but that these are horny

---

[42] M. Sander Rang rejects the specific name '*univalvis*', as signifying a generic character, and he has been followed in this by MM. Quoy and Gaimard themselves. This, according to the Rules of the British Association, would hardly have been a sufficient reason, but it appears that *A. parasita*, like *A. minuta*, has a pair of horny scuta or valves; and, therefore, the / name *univalvis* is too obviously false to be retained. With respect to the generic name Triton, I fully believe that it was applied by Linnaeus to the cast-off exuviae of sessile cirripedes.

[43] See footnote, p. 132 [159].

scuta of nearly the same shape as in *A. minuta*. The whole animal seems to be extremely transparent, and of a 'jaune-citron clair'. MM. Quoy and Gaimard, however, remark, that different specimens vary from white to yellow. Entire length two inches, of which the capitulum is fourteen French lines. The peduncle is narrow and short. /

3. ALEPAS CORNUTA

PLATE III. FIG. 6

A. aperturâ parvâ, leviter prominente: scutis nullis: capitulo plerumque tribus, parvis, compressis eminentiis secundum carinalem marginem instructo.
Orifice small, slightly protuberant; capitulum without horny scuta; generally with three small flattened projections along the carinal margin.

Outer maxillae with the inner bristles divided into two groups; segments of the posterior cirri extremely numerous, each with one pair of main spines; inner rami of the fifth and sixth cirri rudimentary.

St Vincent's, West Indies, attached to an Antipathes, collected by the Rev. L. Guilding.

Capitulum globular, slightly flattened, smooth, translucent, entirely destitute of valves; orifice slightly projecting or tubular, parallel to the longitudinal axis of the peduncle, with the edges sinuous; it appears more tubular than it really is, from the convexity of the part of the capitulum immediately beneath the orifice. Three small, flexible, horny, irregular prominences project from the carinal margin; one at the bottom of the capitulum; a second about halfway up it; and a third generally close to the orifice; but their positions vary a little, and the prominences vary still more in shape and size, being either rounded and very small, or much flattened and considerably prominent; they are imperforate; in the membrane under them a few tubuli may be seen, which are not elsewhere visible; their summits are roughened with very minute points and beads of chitine; others, still minuter, are scattered over the whole capitulum.

Peduncle short, narrower than the capitulum, into which it insensibly blends; strongly wrinkled; surface of attachment wide; position with respect to the branches of the coralline, various. /

Size and colour. The largest specimen, including the peduncle, was half an inch in length, and ³⁄₁₀ths of an inch across the capitulum; colour, after having been long in spirits, brownish-yellow.

Filamentary appendages, one on each side, short, tapering and pointed; seated on the posterior margin of a slight swelling beneath the basal articulation of the first cirrus; they are about equal in length to the pedicels of this cirrus.

The *mouth* is directed abdominally; labrum much produced downwards, so as to be far separated from the adductor muscle; moderately bullate, forming about one-third of the longitudinal axis of the entire mouth; upper part forming a slightly overhanging prominence; crest with a row of blunt, bead-like teeth, and externally to them there are numerous curved short bristles.

Palpi (Pl. X, fig. 8), unusually narrow, a little hollowed out along their inner margins; pointing towards the adductor muscle; thickly covered with doubly serrated bristles.

Mandibles, with either two or three teeth; inferior angle narrow and tooth-like; both sides covered with strong bristles or spines, projecting beyond the toothed edge.

Maxillae, with two large upper spines, and a third rather distant from them; beneath these, there is a wide notch or hollow; inferior part square, projecting, bearing six pair of moderately long spines (of which the central one is the longest), mingled with finer ones.

Outer maxillae, with a semicircular outline; the serrated bristles in front are divided into two groups; externally there is a rounded and very considerable projection covered with long bristles. Olfactory orifices slightly prominent, approximate, seated within and just beneath the rounded projections at the base of the maxillae.

Body. Prosoma little developed; thorax small.

Cirri, extremely long, but slightly curled, capable of being protruded so as almost to touch the base of the peduncle or the surface of attachment; segments short, / extraordinarily numerous. In the three posterior cirri (excepting the rudimentary rami), each segment supports two long, slightly serrated spines, with two or three minute intermediate ones, and with one or two very short, thick spines on the inner and upper lateral margins: dorsal tufts with only two or three long, fine, unequal spines. All the segments are extremely flat, broad, short, with their anterior faces not protuberant; the greater number of the segments, especially the lower ones, have very obscure articulations, to be seen only with a high power, and these can be capable of little or no movement.

First cirrus placed far from the second, with the top of its pedicel on a level with the top of the lower segment of the pedicel of the second cirrus; rami short, barely half the length of those of the second cirrus; unequal, the anterior ramus being only two-thirds of the length of the posterior one; the shorter ramus contains thirteen inverted conical segments, with one side rather protuberant; the longer ramus contains twenty-three thinner segments; the segments on both rami are clothed with bristles, arranged in two or three rows, forming narrow transverse brushes.

Second cirrus, with its pedicel long, and its rami nearly equalling in length those of the sixth pair; the two rami of nearly equal length; the anterior one rather thicker than the posterior one; this posterior ramus has fifty-five segments! The bristles on the second and third cirri are arranged on the same principle as on the three posterior pair; but from an increase in size and number of the little intermediate bristles between the main pairs, and of those on the lateral rims, the segments, especially the basal ones, of the anterior ramus of the second cirrus, are clothed with thin brushes of bristles; these same bristles, on the posterior ramus of the second, and on both rami of the third cirrus, can hardly be said to form brushes, though longer and more numerous than those on the three posterior pair of cirri.

Fifth and sixth cirri. These resemble each other, / and have their inner or posterior rami in an almost rudimentary condition. In the sixth cirrus (Pl. X, fig. 28) the outer ramus *a* has actually sixty-three segments, whereas the rudimentary ramus *k* has only eleven, nearly cylindrical segments. These are furnished with extremely minute spines, of which those on the dorsal face are longer than those on the anterior face; the spines on the summit of the terminal segment are the longest; the segments are not half as thick as the normal ones in the outer ramus. The rudimentary ramus is only one-seventh part longer than the pedicel which supports both it and the normal ramus. in the fifth cirrus, the rudimentary ramus rather longer, and has thirteen segments, resembling those in the rudimentary ramus of the sixth. In the fourth cirrus there is no trace of this peculiar structure, the rami being equal in length and strength. The two rudimentary rami on each side are nearly straight, and seem incapable of movement; they project out behind the normal rami, and closely resemble in general appearance, the two caudal appendages; hence this cirripede, at first sight, appears to be six-tailed.

Pedicels of cirri. The pedicel of the first pair is very short; that of the second is the longest; those of the posterior cirri decreasing in length. Upper segments short; lower segments in the second, third and fourth cirri, irregularly and rather thickly clothed with bristles, but in the fifth and sixth cirri, there is a regular double row of main spines, with some minute intermediate ones: hence there is a difference, both in the rami and in the pedicels, between the fourth cirrus and the fifth and sixth, and this is a unique case. On the dorsal surface of the pedicel of the second cirris, there is a tuft of much feathered fine spines.

Caudal appendages. Each consists of eight much tapering, very thin segments, furnished with a few short simple spines round their upper margins, and with a longer tuft on the terminal short segment; basal segments twice as thick as the middle ones. In length, these caudal appendages equal to the pedicels of the sixth / pair of cirri, and are a very little shorter than the rudimentary rami of these same cirri.

General remarks. Having examined this species first in the genus, I fully anticipated that the very remarkable character of the inner rami of the fifth and sixth cirri being rudimentary, and serving the same function (if any) with the caudal appendages, would have been generic; but this is not the case, for *Alepas cornuta* cannot be separated from *A. minuta* without violating a clear natural affinity.

4. ALEPAS TUBULOSA

Quoy et Gaimard, *Voyage de l'Astrolabe*, Pl. xciii, fig. 5, 1834

A. aperturâ parvâ prominente et tubulosâ: scutis et prominentiis secundùm marginem carinalem, nullis.
Orifice small, tubular, protuberant; capitulum without horny scuta or projections along the carinal margin.

Animal unknown.

New Zealand, Tolaga Bay. Attached to a living Palinurus.

I have given the above brief character from the plate, and imperfect description in the voyage of the *Astrolabe*. The small and distinctly tubular orifice, and the smooth carinated edge of the globose capitulum, appear sufficiently to distinguish this species from *A. cornuta*. The colour is stated to have been white with violent tints. Length, two (French) lines.

ANELASMA *Gen. Nov.*

PLATE IV

ALEPAS. Lovén, *Ofversigt of Kongl. Vetenskaps-Akad. Fördhandlinger: Forsta Argangen.* Stockholm, p. 192, Pl. 3, 1844

Capitulum sine valvis: aperturâ amplâ: pedunculus fimbriatus, sub-globosus, infossus. /
Capitulum without valves; aperture large; peduncle fimbriated, sub-globular, imbedded.

Cirri without spines; outer maxillae and palpi rudimentary, spineless; mandibles minute, with several small teeth irregularly placed; maxillae minute, with very minute irregularly scattered spines. No caudal appendages.

I owe to the great kindness of Professor Steenstrup, an examination of this very curious cirripede, well described and figured by Lovén, who considered it an Alepas. It lives parasitic, with its peduncle imbedded in the skin of sharks, in the North Sea. According to the principles of classification which I have followed, this cirripede cannot possibly remain in Alepas, and must form a new genus; for some time, indeed, I thought that a new family or subfamily ought to have been instituted for its reception; but when I considered that its highly peculiar characters are all negative, as the non-articular, non-spinose structure of the cirri, and that no new or greatly modified functional organ is present, I concluded that it might properly remain among the Lepadidae. We shall, moreover, hereafter see that the male of Ibla, which, of course, must remain in the same family with the female, is, in some analogous respects, even more abnormal than Anelasma.

1. ANELASMA SQUALICOLA

PLATE IV. FIGS 1–7

ALEPAS SQUALICOLA. Lovén, *ut supra*

North Sea. Parasitic on Squalus.

Capitulum, destitute of valves; oval, much flattened; the double membrane composing it, thin, highly flexible, coloured externally and internally, by the underlying corium, of a blackish purple; aperture, extremely large, extending from the upper end of the capitulum, to close above the peduncle, gaping, and not protecting (in the dead condition) the cirri and mouth. /

The *peduncle* is about half as long as the capitulum, but, according to Lovén, this part varies in length; it is a little narrower than the capitulum; colourless, from being imbedded in the shark's skin; subglobular; basal end almost hemispherical. Total length of animal 1·3; diameter of peduncle 0·4 of an inch.

The external membrane of the capitulum is not nearly so thick as is usual in other cirripedes, and is, therefore, unusually flexible. The internal membrane, on the other hand, is very much thicker than is usual, being only a little thinner than the outside coat; this circumstance, as well as the similarity in colour on both sides, is evidently due to the remarkable openness of the sack, and consequent exposure of its inside. The inner membrane, when viewed under a high power, is seen to be covered with the minutest spines; the external membrane is structureless, except that there are a few rows of very minute beads of hard chitine, like those which occur on the capitulum of *Conchoderma aurita*. Lovén, however, states that there are imbedded in the outer membrane, scattered, minute, dendritic, calcareous particles. Of these, I could see no trace. There is a very thin muscular layer between the two coats, all round the capitulum, and this layer becomes rather thicker round the base, near the peduncle. The adductor muscle occupying its usual place close below the mouth, is thinner than in any other cirripede of the same size seen by me; nor does it end so abruptly at each extremity, as is usual: where attached to the outer coat, no impression is left. It is a singular fact, that in this cirripede alone, the fibres of the adductor, and of the muscles of the cirri, and of the trophi of the mouth, are destitute of transverse striae; but it is not singular, that the muscles surrounding the capitulum should, also, be destitute of striae, for this is the case with the muscles which, running up from the peduncle, surround the capitulum in Alepas, and partly surround it in Conchoderma. It must not be inferred from the absence of transverse striae in the muscular fibres of the / adductor and of the cirri and trophi, that they are involuntary, but only that they are in an embryonic condition, for I find in the natatory larva, that all the muscles, with the exception of some connected with the eyes, are similarly destitute, and yet perform voluntary movements.[44]

Although in the dead state, the aperture of the capitulum seems to be always gaping, yet I have little doubt that the living animal can fold the flexible membrane like a mantle, round its thorax and cirri, and thus protect, though feebly compared with most cirripedes, these organs. I suspect that the mouth is always exposed.

Peduncle. The membrane of the peduncle is thin; the whole surface is sparingly and quite irregularly studded with minute, much-branched filaments (Pl. IV, fig. 3, highly magnified); these are occasionally as much as $\frac{1}{5}$th of an inch in length; the degree of branching varies much, but is generally highly complex; the ordinary diameter of the branches is about $\frac{1}{200}$th of an inch; their tips are rounded, and even a little enlarged, and frequently torn off, as if they had been attached to or buried in the flesh of the shark, in which the whole peduncle is imbedded. These filaments are formed of, and are continuous with the external transparent membrane of the peduncle, and they contain, up to the tips of every sub-branch, a hollow thread of corium, prolonged from the layer internally coating the whole peduncle. In all other Lepadidae, the peduncle increases in length, chiefly at the summit where joined to the capitulum, and in diameter, throughout nearly its whole length, except close to the base; but, owing to the constant disintegration of the outer surface, the old outside coat does not split in defined lines, like the membrane of the capitulum. In Anelasma, however, owing to the imbedded position of the peduncle, the old outer coats are preserved, the lines in which they have / split during continued growth being thus exhibited: those in the uppermost part almost symmetrically surround the peduncle, showing that here,

---

[44] Dr C. Schmidt in his 'Contribution to the comparative Anatomy of the Invertebrate animals', etc. (translated in Taylor's *Scientific Memoirs*, vol. v, p. 1), says that in young Crustacea, 'we find plain primitive fibres, which subsequently acquire the transversely striated aspect'.

as in other Lepadidae, has been one regular line of growth; but in the lower part the lines are extremely irregular; and what is almost unique, it appears that the blunt basal end is constantly increasing in length and breadth, and, apparently, at a greater rate than any other part. I judge of this latter fact, from the whole bottom of the peduncle being covered with numerous curved, or nearly circular, lines of natural splitting, the nature of which can be best understood by examining the much enlarged drawing (Pl. IV, fig. 3) of a small portion (taken by chance) of the membrane of the base, seen from the outside, and bearing some of the simplest branched filaments: other branches, as may be seen, have been cut off. This manner of growth explains the broad, blunt basal termination of the peduncle, so unlike that in other Lepadidae. New membrane is formed, not continuously as in other cases, under the whole surface of the old membrane, but in irregular patches; thus the portion marked a runs under b, but not under the little circles c, c, for these are the last formed portions and underlie the membrane a and b. I do not understand how the splitting of the old membrane is effected; but no doubt it is by the same process by which the membrane of the capitulum in other genera, as in Scalpellum, splits symmetrically between the several valves. In the branched filaments it is particularly difficult to understand their growth, for it is not possible, after examining them, to doubt that they continue to increase, and send off sub-branches, which it would appear probable, penetrate the shark's flesh like roots. I may remark that one, or more commonly two or three branched filaments stand nearly in the centre of each circular line of exuviation or splitting. The branched filaments first commence as mere little pustules, and these appear to be most numerous at the bottom of the peduncle. /

The final cause of the downward growth of the bottom of the peduncle, is obviously to allow of the animal burying itself in the shark's body, in the same way as Coronula and Tubicinella become imbedded by the downward growth of their parietes in the skin of Cetacea. The only other genus of Lepadidae, in which the growth of the peduncle is at all analogous, is Lithotrya, in this genus, however, the animal burrows mechanically into soft rock or shells.

I looked in vain for cement, or for the cement glands (but the specimen was in an extremely unfavourable state for finding the latter) or for the prehensile antennae of the larva. No doubt this cirripede at first becomes attached in the same way as others, but after early life, I suspect it is retained in its place, by being so deeply imbedded in the shark's body, and perhaps by the root-like branched filaments. The irregular growth and splitting of the membrane at the base of the peduncle, where the prehensile antennae of the larva must originally have been situated, would account for not finding them.

The inside of the peduncle (fig. 2g) was gorged, in the specimen examined by me, with immature ova. The innermost muscular layer consists of longitudinal bundles of unusual size, but placed rather far apart from each other; these do not extend to the very base of the peduncle, and at the upper end they curve inwards, almost to the middle of the underside of the diaphragm, separating the peduncle and capitulum. Outside these longitudinal muscles, there are delicate transverse ones, but apparently there are no oblique muscles in the upper part of the peduncle, as in other Lepadidae; near the bottom, the transverse muscles form a thicker layer with many of the bundles running in oblique lines.

Mouth. Lovén has not described this part quite accurately, owing to his not having used high enough magnifying powers. He states that the trophi are soft and functionless, which is far from the case. The whole mouth (fig. 2d), is unusually small; it is, to a certain / extent, probosciformed, and being curved a little downwards, projects slightly over the adductor muscle, to which it is closely placed. The labrum does not project more beyond the general surface of the body, than in many other cirripedes, but the probosciformed structure is caused by the elongation of the surface fronting the thorax. The summit of the mouth stands above the level of the top of the pedicels of the first pair of cirri. The labrum is slightly hollowed out in the middle of its upper margin; it can scarcely be called bullate, in which it differs from all other Lepadidae; on the other hand, the outer and inner folds of the labrum are not so close together as in Balanus. On each upper corner, there is, as usual, a small rounded prominence, close to which there is a second slight, rounded, spineless swelling; these latter represent the quite rudimentary *Palpi*.

The *mandibles* (figs 4, 5) are more highly developed than the other trophi; they are, however, very minute, the toothed edge being only about $16/1000$th of an inch in length, measured in its longest direction; the edge is unusually thick, with the teeth placed rather on one side; this organ, when viewed on the labrum side (fig. 5), shows two large teeth placed low down, with the inferior angle pectinated and broadly truncated; but when viewed on the other or maxillae side (fig. 4), several large and small teeth, placed alternately and irregularly in pairs, are seen extending along the whole edge. The mandibles are furnished, as usual, with three principal sets of muscles attached to the basal fold of the mouth.

The *maxillae* (fig. 7) are still smaller than the mandibles; the spinose edge being only $\frac{1}{100}$th of an inch in length; the edge, instead of being square, and furnished with a double row of long spines, as in all other cirripedes, is rounded, thick, club-shaped, and with the side facing the mandibles, thinly and irregularly strewed with short, thick, very minute spines; there is a large broad apodeme a, in the usual place, but it is much more transparent and flexible than common: there are also the usual muscles. / In other cirripedes, the mandibles alone seem to force the prey down the oesophagus; but here, the mandibles and maxillae equally stand over the orifice, and their adjoining spinose faces and edges, seem excellently adapted to force, by their united action, any minute living creature down the passage.

The *outer maxillae* are almost in as rudimentary a condition as the palpi; they are quite spineless; viewed externally, they appear like two smooth, blunt, very minute projecting points; but viewed internally, the membrane forming the supra-oesophageal hollow seems to be united actually to their tips, so that they do not project at all. I was surprised to find that the longitudinal muscles going to these organs were developed, in proportion to the other muscles, quite as fully as in ordinary cirripedes: hence, these two little outer maxillae, no doubt, serve as an underlip, and possess the usual backward and forward movement.

The surface of the prosciformed mouth facing the first pair of cirri, has a deep central longitudinal fold, and rather more than halfway down, a transverse fold; just above this latter fold, and therefore quite below the outer maxillae themselves, the two olfactory orifices are seated; these are unusually large, and the sack into which they lead, is most unusually large and deep. In this cirripede, I was first enabled to observe that the membrane lining the sack is tubular, and open at the bottom.

Cirri. There are, as usual, six pair, and not of very small size; they have a shapeless and rudimentary appearance; they are coloured, like the rest of the body, blackish purple: they are quite spineless, and not articulated, but their anterior faces are either obscurely or very plainly lobed, so that in some (for instance in the third pair, Pl. IV, fig. 6), nine or ten prominent steps could be counted, manifestly representing so many segments. The rami are equal in length in the first pair, and slightly unequal in the second and third pair; these two latter are longer than either the first or three posterior pair. / There is a small interspace as usual between the first and second pair of cirri. Internally, the cirri are occupied, even up to their tips, by delicate striae-less muscles. The external membrane of the thorax and limbs, when examined under a very high power, is seen to be covered with minute toothed scales, as in most cirripedes.

The thorax is articulated as usual: the posterior part, however, is smaller, and tapers more suddenly than in other species, and this corresponds with the smaller size and more rudimentary condition, of the three posterior pair of cirri, compared with the anterior pair. The prosoma is hardly at all developed. The orifice (Pl. IV, fig. 2e) of the acoustic [?] sack, beneath the first cirrus, is unusually large.

There are no filamentary appendages.

Alimentary canal. The membrane lining the oesophagus is unusually thin: it is furnished with the ordinary constrictor muscles, and others radiating from them like spokes of a wheel. The stomach is lined by unusually prominent biliary folds, which in the duodenum are transverse, sending forth, however, short folds at right angles; and these latter, in the proper stomach, become so much developed that the folds appear longitudinal. The rectum extends inwards, about as far as the base of the fourth pair of cirri, but is very short, owing to the little development of the three posterior segments of the thorax. The anus is seated in its usual place, at the dorsal basis of the penis, and is hidden by loose folds of skin; but there are no distinct caudal appendages. The stomach, in the specimen examined, was quite empty.

Reproductive organs. The penis (fig. 2c) is thick, short (about twice as long as the sixth cirrus), constricted at the base, ringed, spineless, with the terminal aperture large; internally it is well furnished with muscles. The two vesiculae seminales, appeared to be unusually small; and one was much smaller than the other; they do not (I believe) become united into a common tube, till near the apex of the penis. They were empty; and, I presume, / from the state of the ova, that their contents had lately been discharged. The whole thorax was filled with a white, fibrous and cellular mass, consisting perhaps of the testes in their undeveloped state. The individual dissected by me appeared to have been defective in its last act of reproduction, for there were only two or three ova attached to the fraenum on one side, and not very many on the other. The ova are much less elongated than is usual; they are of a remarkable size, namely $22/1000$ths of an inch in their longer diameter; the membrane by which they are united into a pair of lamellae is remarkably strong; the fraenum (Pl. IV, fig. 2*f*) on each side is large, strong, with rounded edges, pale coloured and hence conspicuous; on the side nearest the body, the whole surface is covered with club-shaped glands, having very short foot-stalks, and being in total length $5/6000$ths of an inch; these glands secrete a reticulated layer of gut-formed fibres, attached to the ovigerous lamellae. In the specimen described by Lovén, the lamellae (fig. 1, and fig.2*b*, *b*) appear to have been very large: and in that examined by myself, the peduncle was gorged with immature ova, showing that the female reproductive powers were ample, though at the foregoing period, only a few eggs had been formed.

Habits. According to Lovén, this species lives imbedded in the skin of *Squalus maximus* and *spinax*, in the North Sea: I suspect that it is not closely compressed in its cavity, otherwise, I do not see the use of the two layers of muscles round the whole peduncle; it probably adheres to the sides of the cavity by the tips of the branched, root-like filaments; owing to the flexible nature of the capitulum, this cirripede can offer little resistance to the water, and, therefore, is little likely to be torn out of its cavity. I have no doubt that it can fold the membrane of the capitulum, like a cloak, round its thorax and cirri; but it certainly can offer far less resistance, than other cirripedes, to any enemy. This creature must obtain its food, and considering its productiveness much food must be required, in a manner quite different / from nearly every other member of its order. As the whole of the peduncle is imbedded, and as the mouth is probosciformed, with the labrum a little curled over the adductor muscle, I conclude that this cirripede can reach minute animals crawling by on the surface of the shark's body.

It must be borne in mind that the mouth, as in all cirripedes, has the power of independent movement, and that the mandibles and maxillae are here beautifully adapted to catch and force down any small living creature into the muscular oesophagus; the rudimentary outer maxillae, moreover, no doubt have the power of scraping, like a lip, anything towards these prehensile organs. It will hereafter be seen, that the male of *Ibla Cumingii*, in which the cirri are quite rudimentary, obtains its food in a somewhat analogous manner, though in this case the whole peduncle moves, and not merely a prosbosciformed mouth: it deserves attention, that in the male Ibla and in Anelasma, in neither of which the cirri are prehensile, the palpi are rudimentary and useless. I am tempted to believe, that the largely developed olfactory sacks, and perhaps, likewise, acoustic [?] sacks, in Anelasma, replace, by giving notice of the proximity of prey, the loss of tactile cirri. It should be remembered that all cirripedes subsist on animals which happen to swim or float within reach of the cirri; but here it is only those which happen to crawl within reach of the prosbosciformed mouth. It would, however, be rash to assert that the cirri in Anelasma, considering their muscular though feeble structure, may not be of some slight use, when thrown over the prey, in preventing its escape.

Professor Steenstrup informs me that, from late observations, it appears that this animal always adheres to the shark's body in pairs. I regret extremely that I have not been able to examine a pair: that the individual examined by me was bisexual, I can hardly doubt, though the male organs certainly were feebly developed; it appears probable, that the individual described by Lovén was likewise / bisexual: but after the facts presently to be revealed regarding the sexes in Ibla and Scalpellum, it is quite possible that the male and female organs may be developed in inverse degrees in different and adjoining individuals.

The genus Anelasma is, I think, properly placed between Alepas and Ibla. In several of its characters, such as the absence of calcareous valves, the broad blunt end of the peduncle, the spineless cirri, the small size of the trophi, and more especially the absence of transverse striae in those muscles, which in mature cirripedes are thus furnished, we see that this genus is in some degree in an embryonic condition.

GENUS: IBLA

PLATES IV, V

IBLA. Leach, *Zoolog. Journal*, vol. ii, July, 1825
ANATIFA. Cuvier, *Mem. pour servir, ... Mollusques, Art. Anatifa*, 1817
TETRALASMIS. Cuvier, *Regne Animal*, 1830

(Foem. et Herm.) Valvae 4, corneae: pedunculus spinis corneis, persistentibus vestitus.
(Fem. and Herm.) Valves four, horny: peduncle clothed with persistent, horny spines.

Body partly lodged within the peduncle; mandibles with three teeth; maxillae with two obscure notches; outer maxillae pointed; olfactory orifices prominent; caudal appendages multiarticulate.

Male and complemental male, parasitic within the sack of the female or hermaphrodite; mouth and thorax seated on a long tapering peduncle, but not enclosed within a capitulum; mouth with normal trophi, but palpi small and almost rudimental; cirri rudimental, reduced to two pairs; penis reduced to a pore; caudal appendages rudimentary.

Attached to fixed littoral objects: Eastern Hemisphere.

General remarks. As there are only two species as / yet known, and as these resemble each other in every respect most closely, a generic description would be a useless repetition of the full details given under *Ibla Cumingii.* I have taken this latter species as the type, from having, owing to the kindness of Mr Cuming, better and more numerous specimens. Ibla and Lithotrya are the only two recent genera in which the body of the animal is lodged within the peduncle; but there is no distinction of any importance, though useful for classification, between the capitulum and peduncle; and these two parts, as we have seen, tend to blend together in some species of Conchoderma and Alepas. The entire absence of calcareous matter in the valves and spines of the peduncle, at first appears very remarkable; but we have seen a similar fact in Alepas, and there is an approach to it in some varieties of *Conchoderma aurita* and *C. virgata.* In all four valves of Ibla, the umbones, or centres of growth, are at their upper points. The horny spines on the peduncle, are the analogues of the calcareous scales in Scalpellum and Pollicipes; and in this latter genus, two of the species have their scales, almost cylindrical, placed irregularly, with new ones forming over all parts of the surface, and not exclusively at the summit – in which several respects there is an agreement with Ibla. The shape of the body (i.e. thorax and prosoma, Pl. IV, fig. 8a') is peculiar; but it is only a slight exaggeration of what we have seen in several genera, and shall meet again in some species of Scalpellum. The presence of hairs on the outer membrane of the prosoma is a peculiarity confined to this genus among the Lepadidae, though observed in the sessile genus,

Chthamalus. The caudal appendages in the *I. quadrivalvis* attain a greater length than in any other species of the family, being four times the length of the pedicels of the sixth cirrus. A far more important peculiarity is the fact of the oesophagus, in both species, running over or exteriorly to the adductor scutorum muscle, instead of, as in every other species, close under this muscle. I took great pains in ascertaining the truth of this singular / anomaly: the course of the oesophagus is approximately represented (in Pl. IV, fig. 8*a'*) by faint dotted lines. The stomach has no caeca; the biliary folds are longitudinal; there is a marked constriction at the line corresponding with the junction of the thorax and prosoma. There are no filamentary appendages.

The generative system gives the chief interest to this genus. We here first meet with males and females distinct; and, within the limits of this same restricted genus, the far more wonderful fact of hermaphrodites, whose masculine efficiency is aided by one or two complemental males. The complemental and simple males closely resemble each other, as do the female and hermaphrodite forms; but under the two following species I enter into such full and minute details on these remarkable facts, that I will not here dilate on them. I may add that, at the end of the genus Scalpellum, I give a summary of the facts, and discuss the whole question. The penis (Pl. IV, fig. 9*a*) in the hermaphrodite, *I. quadrivalvis*, is singular, from the length of its unarticulated support, and from the distinctness of the segments in the articulated portion.

As ovigerous fraena occur in the usual place in *I. quadrivalvis*, though much smaller than in any other species, I have no doubt that they occur in *I. Cumingii*, although I failed in observing them. The glands on the margin, in *I. quadrivalvis*, are singular, from not being borne on a long, hair-like footstalk.

Affinities. Ibla, though externally very different in appearance from Scalpellum, is more nearly related to that genus than to any other; in both genera some species have the sexes separate, the imperfect males being parasitic on the female, and other species are bisexual or hermaphrodite, but aided by parasitic complemental males. In Scalpellum, again, the oesophagus pursues a sinuous course, resembling that in Ibla, though it does not pass exteriorly to the adductor scutorum muscle. The disc of the prehensile antennae of the larva, in / both genera, has an unusual oblong form, like a mule's hoof; there is also an affinity between the two genera in the size and form of the ova, in the prominent orifices of the olfactory cavities, and in the peduncle not being naked; though, in these two latter respects, in the structure of the cirri, and in the multiarticulate caudal appendages, there is an equal affinity to Pollicipes and Lithotrya. I have already shown that Alepas is likewise related to Ibla.

1. IBLA CUMINGII

PLATE IV. FIG. *8*

I. (foem.) valvarum marginibus lateralibus, et superficie interiore, caeruleis: pedunculi spinis plerumque annulis caeruleo-fuscis.
Fem. – Valves coloured, along the lateral margins and on the upper interior surface, blue: spines on the peduncle, generally ringed with blueish-brown.

Caudal appendages barely exceeding in length the pedicels of the sixth cirrus: rami of the first cirrus unequal in length by about two segments.

Male – with scarcely a vestige of a capitulum: maxillae with fewer spines than in the female.

Habitat. Philippine Archipelago, Island of Guimavas; invariably attached to the peduncle of *Pollicipes mitella*, in groups of two or three together; Mus. Cuming. Tavoy, British Burmah Empire; Mus. A. Gould of Boston.

FEMALE

The capitulum is formed of four valves, but is hardly distinct from the peduncle. The latter includes, in its wide upper part, the animal's body. The valves, namely, a pair of scuta and terga, are composed of an extremely hard, horny substance, or properly chitine, and do not contain any calcareous matter; they are extremely flat or thin, and both pairs project freely, like curved horns, to a considerable height above the sack enclosing / the body: the terga project about twice as much as the scuta, and their flat apices generally diverge a little. The tips of the valves are frequently broken off; their surfaces are plainly marked or ribbed by the layers of growth, which are wide apart. The bases of the valves externally are hidden by the long spines of the peduncle.

Scuta. These are shorter and broader than the terga; their internal (Pl. IV, fig. 8*b'*) growing or corium covered surfaces are slightly concave, triangular, with the basal margin longer than the other margins and slightly excised in the middle: there is no depression for the strong adductor muscle: the internal surface of the free horn-like portion, has a small central fold (formed by an oblique crest) running from the summit of the triangular growing surface to the tip of the valve: in perfect specimens, the growing and the free horn-like portions (the latter represented much too long in fig. 8*a'* and *b'*) are about equal in length: the basal portion of one side of the scutum overlaps the tergum.

Terga. The internal growing surface (fig. 8*b'*) is almost diamond-shaped, and less in area than the scuta: external surface rounded; internal surface of the free horn-like portion, slightly concave.

Colour and structure of valves. The external surfaces of the scuta and terga are yellow along the middle, plainly marked by zones of growth, and finely ribbed longitudinally: the internal surfaces and sides of the horns of the two valves, are coloured fine blue or purple; in the terga, however, the internal surface is mottled with yellow. In some specimens, especially in one from Tavoy, each zone of growth was only very narrowly edged with blue. When a thin layer is removed from one of the valves, the dark blue or rather purple appears by transmitted light a beautiful pale blue; and it is a very singular fact, that this blue portion is permanently turned by very gentle pressure into a fiery red; the same singular effect is produced by muriatic and acetic acids. This blue part is much harder than the yellow; the latter exhibits, under / a high power, a folded structure, and is penetrated by a few tubuli, whereas the harder blue portion has a cellular or scaled appearance. The spines of the peduncle exhibit, in a smaller degree, similar phenomena.

Peduncle. This, as already remarked, cannot be distinctly separated from the capitulum; it is much compressed; it is composed of unusually thin and delicate membrane, transversely wrinkled and thickly clothed with long cylindrical horns or spines of chitine. These horns (fig. 8*c'*) are not the analogues of the spines which are articulated on the external membranes of many pedunculated and sessile cirripedes, but of the calcified scales on the peduncle of Scalpellum and Pollicipes; for they pass through the membrane (the underlying corium being marked by their bases) and are persistent, being added to, like the valves, during each successive period of growth. Their bases are concave, so that a section of the layers of growth exhibits a series of pointed cones, one within another. Each spine is nearly cylindrical, irregularly curled, and nodose or slightly enlarged at intervals: the apex smooth and pointed; the exterior surface longitudinally and finely ribbed, like the valves. The spines increase irregularly in size from the bottom to the top of the peduncle, those at the carinal and rostral ends being generally the longest; they point upwards and hide the bases of the valves. They are not arranged symmetrically, and new ones are formed over all parts of the peduncle. They are formed of the same substance as the valves, and do not contain any calcareous matter. These horns are yellowish, generally ringed with pale and dark blueish-brown, which on pressure becomes slightly opalescent with pale blue and fiery red: sometimes only the upper horns are thus ringed, and in rare instances all are simply yellowish. The muscles of the peduncle run up to the bases of the four valves.

Surface of attachment. The cement appears to proceed from only two points. In some specimens, a considerable length of one side of the peduncle was fastened / to the surface of attachment, the horns or spines being enveloped in the cement. The prehensile antennae of the larva will presently be described under the male.

The *length* of an average specimen, including the peduncle and valves is about half an inch, and the width across the widest part one-fifth of an inch. Mr Cuming has one specimen an inch in length, but this is owing to the peduncle being unusually tapering. In a specimen kept some years in spirits, the cirri, trophi, caudal appendages, and corium under the membrane between the scuta, were all dark purple; the sack and corium of peduncle clouded with purple, and the prosoma pale coloured.

The *body* (Pl. IV, fig. 8a') is small compared with the capitulum and peduncle; it is much flattened; the prosoma is of a very peculiar shape, being square, the sides of equal length, and, in an average-sized specimen, $75/1000$th of an inch long. The peculiar shape arises from the great distance between the first and second cirrus – from the mouth being far removed from the adductor scutorum muscle – and lastly, from the lower part of the prosoma being not at all protuberant. The thorax which supports the cirri is also unusually small, plainly articulated, and separated from the prosoma by a deep fold. The thin membrane of the prosoma is studded with some fine, pointed hairs, about ¾₀₀ths in length, and articulated on little circular discs.

Mouth, placed at a considerable distance from the adductor, and directed in an unusual manner towards the ventral surface of the thorax: the trophi are arranged, in a curved line, facing the thorax (see Pl. V, fig. 2, for this part in the male), and therefore less laterally than is usual.

Labrum (Pl. IV, fig. 8a' opposite c) highly bullate; the upper part produced into a blunt point: on its crest there are no teeth.

Palpi (fig. 8a' opposite d) small, blunt and rounded at their ends; inner margins slightly concave. /

Mandibles (Pl. X, fig. 4), with three teeth, of which the first is much larger than the second and third, and distant from them: inferior angle produced and pectinated; upper edges of the second and third teeth finely pectinated.

Maxillae (Pl. X, fig. 11) small, slightly but distinctly indented by two notches, supporting, besides the three upper great spines, three pairs of moderately long spines and some finer ones: apodeme short, thick.

Outer maxillae, unusually pointed, with the inner bristles not very numerous, continuously arranged; externally, the bristles are longer. Olfactory orifices, tubular, projecting, flattened, square on the summit smooth: they point upwards and obliquely towards each other: they arise more laterally than in the other genera, namely outside the bases of the outer maxillae, and between them and the inner maxillae.

Between the bases of the first pair of cirri, there is a conical prominence, clothed with bristles and coloured purple: it projects nearly as high as the top of the lower segment of the pedicel of the first cirrus: it lies over the infra-oesophageal ganglion, and serves, I suspect, to fill up a little interval between the outer maxillae.

Cirri long, little curved: the first pair (Pl. IV, fig. 8a') is situated at an extraordinary distance from the second; hence its basal articulation is on a level with the upper articulation of the pedicel of the second cirrus. In the three posterior cirri, the segments are laterally very flat, with their anterior surfaces not protuberant; each supports three pairs of thin, non-serrated bristles, of which the second pair is much shorter than the upper, and the lowest pair minute; between each pair there is a minute, rectangulary projecting bristle; dorsal tufts consist of two or three spines, of which one is longer than the others. The two bristles forming each pair are not of equal length; for in the rami of each cirrus, the inner row of bristles is much shorter than the outer; and this seems to be connected with the flatness of the whole animal, and the consequent little power of divergence in the rami / of the cirri. The first cirrus is rather short, with the rami unequal in length by about two segments: the anterior ramus is shorter and thicker than the other: segments numerous, each clothed with several rows of bristles. The second cirrus has the anterior ramus thicker and more thickly clothed with spines than the posterior ramus; this latter is rather more thickly clothed with spines than are the three posterior cirri: the third cirrus is in all these respects characterized like the second cirrus, but in a lesser degree. The pedicels of the second and third cirri are thickly and irregularly clothed with spines; in the three posterior pairs, the spines are placed in two regular rows, with some minute intermediate spines.

Caudal appendages (Pl. IV, fig. 8a', f), multiarticulate, thin, tapering, in one specimen equalling, in another just exceeding, in length the pedicels of the sixth cirrus. In the latter specimen there were thirteen segments, of which the basal segments were broader and shorter than the upper; these latter are slightly constricted round the middle, so that they resemble, in a small degree, an hour-glass. Their upper margins are surrounded by rings of bristles; the terminal segment being surmounted by one or two very fine bristles much longer than the others. The two appendages are closely approximate; each arises from a narrow elongated slip, attached to the side of the pedicel of the sixth cirrus.

Nervous system. I examined the upper part of the nervous chord, in order to ascertain whether the infra-oesophageal ganglion, which is of a globulo-oblong shape, was far separated from the second ganglion; and this I found to be the case, in accordance with the distance of the first cirrus from the second. I may here remark, that in *S. quadrivalvis* I discovered the eye, which, though in all probability really double, appeared to be single; it was situated near to the supra-oesophageal ganglion; and this ganglion was situated near to the adductor scutorum muscle, and at a considerable distance from the labrum. The aperture leading into the acoustic [?] sack, is situated / much lower down than is usual (Pl. IV, fig. 8a'), namely, at the length of the pedicel of the first cirrus beneath its basal articulation.

Generative system. The specimens here described, of which I examined six, are exclusively female; they have no trace of the external, probosciformed penis, or of the two great vesiculae seminales, or of the testes: on the other hand, the ovarian tubes within the peduncle are developed in the usual manner, and owing to the large size of the ova, are of large diameter, and hence very distinct: I detected, also, the true ovaria at the upper edge of the stomach.

MALE

PLATE V. FIGS 1–8

Of the above described *Ibla Cumingii* I dissected six specimens, four from the Philippine Archipelago,[45] and two from the Burmah Empire, and none of them, as we have just seen, possessed the probosciformed penis, the vesiculae seminales, or the testes, so conspicuous in other cirripedes; on the other hand, all were furnished with the usual branching ovarian tubes and sometimes with ova, and consequently were unquestionably of the female sex. Within each of these specimens there was attached within the sack, in a nearly central line, at the rostral end (Pl. IV, fig. 8a', *h*, magnified five times), a flattened, purplish, worm-like little body, projecting about ½₀th of an inch: in one of the six individuals, there was a second similar little creature attached at the carinal end of the sack. Before giving the reasons which I think conclusively prove that these little animals are the males of the ordinary form of the *Ibla Cumingii*, it will be convenient to describe their structure in detail.

The whole consists of a long, much flattened peduncle, separated from the mouth and thorax by an oblique fold (Pl. V, fig. 1*h*, *b*), which is conspicuous on the dorsal / margin under the cirri, and can be traced with difficulty to the ventral margin. The thorax, itself rudimentary, and supporting rudimentary cirri, is in some individuals, as in the one represented (fig. 1, *magnified 32 times*), covered by, or received in the oblique fold *h*, just mentioned: in other individuals the thorax is drawn out, and then the fold shows merely as a notch on the dorsal margin,

---

[45] I am deeply indebted to the liberality and kindness of Mr Cuming, in allowing me to cut up four specimens of this new species; and to Dr Gould, of Boston, U.S., for the examination of the Burmese specimens.

and the basal articulations of the cirri stand some little way above it. The basal edge of the large, well-developed mouth can be traced all round, and on the ventral margin b, is generally marked by a slight notch. The dimensions and proportions vary much: the longest specimen, including the imbedded portion, was $8/100$th, and the shortest barely $5/100$ths of an inch in length; the width of the widest portion varied from 1 to $2/100$ths of an inch: the specimen figured (Pl. IV, fig. $8a'$, and Pl. V, fig. 1), is a broad, short individual. Generally, the middle of the peduncle is rather wider than the upper part.

Peduncle. The main part of the animal, as may be seen in the drawing, consists of the peduncle, of which the imbedded portion tapers more or less suddenly in a very variable manner, and is of variable length – in one specimen being one-fourth of the entire length, and in another consisting of a mere minute blunt point. The free upper part of the animal is bent in various directions, in relation to the imbedded portion. The latter passes obliquely through the chitine membrane and corium, lining the sack of the female, and running along amidst the underlying muscles and inosculating fibrous tissue, is attached to them by cement at the extremity. The peduncle is often, but not in the individual represented, much constricted at the point where it passes through the skin of the female, and generally at several other points, especially towards the extremity (see fig. 1); the stages of its deeper and deeper imbedment being thus marked. The constrictions are, I believe, simply due to the continued growth of the male, whilst the hole through the membrane of the female does not yield. The imbedment, / which is considerable only when the lower part of the peduncle is almost parallel to the coats of the sack, seems caused by the growth and repeated exuviations of the female; I believe that the larva attaches itself to the chitine tunic of the sack, and that the cement, by some unknown means, affects the underlying corium, so that this particular portion of the tunic is not moulted with the adjoining integuments, and that the growth of the surrounding parts subsequently causes this portion to be buried deeper and deeper: it is, I believe, in the same way as the end of the peduncle in *Conchoderma aurita*, sometimes becomes imbedded in the skin of the whale to which it is attached.

The outer tunic of the peduncle is thin and structureless: in the fold (fig. $1h$) under the cirri, there is a central triangular gusset of still thinner membrane, corresponding in position to the membrane connecting the two terga in the female, and there subjected to much movement. I may here remark, that this fold, in its office of slightly protecting the thorax and in its position, evidently represents the capitulum with its valves, enclosing the whole body of the female. The outer tunic is lined by corium, mottled with purple, and within this there are two layers of striae-less muscles, transverse and longitudinal, as in all pedunculated cirripedes. The corium extends some way into the imbedded portion of the peduncle, and consequently, the outer tunic there continues to be added to layer under layer, and as it cannot be periodically moulted, it becomes much thicker than in the upper free part of the animal: the corium, however, does not extend to the extreme point, so that in it growth of all kind ceases.

Antennae. The peduncle terminates (Pl. V, fig. 1e) in the two usual, larval, prehensile antennae, which it is very difficult to see distinctly; they are tolerably well represented in fig. 5, greatly magnified. Their extreme length, measured from the basal articulation to the tip of the hoof-like disc, is $22/6000$ths of an inch, the disc itself being $7/6000$ths of an inch. The disc is slightly narrower / than the long basal segment, from which it is divided by a broad conspicuous articulation; its lower surface is flat and its upper convex, altogether resembling in shape a mule's hoof; its apex is fuzzy with the finest down; it bears a narrow ultimate segment, thrown, as usual, on one side; this segment supports on its rounded irregular summit, at least five, I believe, judging from the structure of the same part in the male larva of *Ibla quadrivalvis*, six or seven spines, longer than the segment itself: one long spine arises from the underside of the disc, near the base of the ultimate segment, and points backward: there is also a single curved spine on the outside, near the distal end of the basal segment. These organs were imbedded in a heart-shaped ball or cylinder of brown, transparent, finely laminated cement, and thus attached to the fibrous tissue of the female. The two cement ducts (fig. 1f) were very plain, each about $1/6000$th of an inch in diameter, containing the usual inner chord of opaque cellular matter. I traced them at the one end into the prehensile antennae as far as the disc; and at the other, up the peduncle for about one-fourth of its length, where I lost them, and could not discover with certainty any cement glands. I may, however, here mention, that I found in the lower half of the peduncle, numerous, yellowish, transparent, excessively minute, pyramidal bodies, with step-formed sides; of these two or three often cohered by their bases like crystals; I have never seen anything like these in other cirripedes, but it has occurred to me that they may possibly be connected with the formation of the cement: for in the last larval condition of Lepas, the cement ducts run up to the gut-formed ovaria, filled at this period with yellowish, grapelike, cellular masses, without the intervention of cement glands, and I can imagine that similar masses, not being developed into functional ovaria, might give rise to the yellow pyramidal bodies.

Mouth. The mouth is well developed; it is represented as seen vertically from above (in Pl. V, fig. 2, magnified / about 60 times); the positions of the cirri and the outline of the thorax are accurately shown by dotted lines; a lateral view is given in fig. 1. In the specimen figured, the longitudinal diameter of the mouth, including the labrum, was $5/400$th of an inch. The muscles of the several trophi have transverse striae, and are the strongest and most conspicuous of any in the body. The labrum is largely bullate, with its summit slightly concave; the trophi are arranged in a remarkable manner, in a semicircular line, so as to be opposed to the labrum rather than to each other: there are no teeth or spines on the crest of the labrum, which overhangs the oesophageal cavity.

The *palpi* (fig. 2b and fig. 3) are very small, dark purple, bluntly pointed, with a few small bristles at the point; they do not extend beyond the knob at each corner of the labrum, which is here present, as in all other Lepadidae; they are much smaller than in the female, though of a similar shape, and consequently, their points are much further apart: within their bases, the lateral muscles of the mandibles are, as usual, attached; they are represented (in fig. 3), as seen from the inside, with the eye on a level with the concave summit of the labrum. The rudimentary condition of the palpi is connected, as remarked under the *Anelasma squalicola*, with the absence of efficient cirri.

The *mandibles* (fig. 7) are well developed; they so closely resemble those of the female that it is superfluous to describe them: they are, however, smoother, without any trace of the teeth being pectinated, and with the inferior point smaller: measured in their longer direction, they are $7/2000$th of an inch in length, and, therefore, a little less than one-third of the size of those of the female. These organs have the usual muscles well developed, and the usual articulations.

The *maxillae* (fig. 8) have a rather rudimentary appearance; yet they have the same size relatively to the mandibles, as in the female, the spinose edge being ³⁄₂₀₀₀ths of an inch in length. These organs resemble to a certain / extent, those of the female, differing from them in being less prominent – in the outline being more rounded, with the notches even less distinct – and in the spines being fewer. The apodeme is short and broad.

The *outer maxillae* (fig. 6) are pointed, with a small tuft of bristles at the apex; they are much less hairy than in the female, but have nearly the same unusual shape. Outside their bases, and between them and the inner maxillae, the two well-developed, tubular, flattened, square-topped, olfactory orifices, project in exactly the same remarkable position as in the female; these are not represented (in fig. 2), though sometimes they can be very distinctly seen, when the mouth is viewed from vertically above.

Thorax and cirri. The thorax is in a rudimentary condition: I did not observe the usual articulations. The whole, as seen from vertically above, is of small size, compared with the mouth; the outline is accurately shown by dotted lines (in Pl. 5, fig. 2), together with the positions of the two pair of cirri, the caudal appendages, and anus. The posterior end of the thorax does not rise to the level of the summit of the mouth; and the thorax seems of no service, excepting perhaps as a sort of outer lip to protect the mouth. The cirri are in an extreme state of abortion, and evidently functionless; they are lined with purplish corium, without the vestige of a muscle; they are usually distorted and bent in different directions; they vary in size, and even those on opposite sides of the same individual, sometimes do not correspond, and do not arise from exactly corresponding points of the thorax. There are always two pair of cirri, which, as I conclude from the position of the excretory orifices, answer to the fifth and sixth pair in other cirripedes. Each cirrus (fig. 4) usually carries only one ramus, placed on a large basal segment, evidently corresponding to the pedicel of a normal cirrus. The posterior are larger than the anterior cirri, which latter spring from points a little lower down on the thorax. In the posterior cirrus / figured, the great basal articulation or pedicel, almost equals in length, and much exceeds in thickness, the four segments of the ramus; these segments are furnished on their upper dorsal edges with little brushes of spines, but have not even a trace of the normally larger and far more important anterior spines. In one specimen, the anterior cirrus had a large pedicel, carrying three segments, like those of the posterior pair; but in another specimen, one of the three segments showed traces of being divided into two, thus making four imperfect segments; whilst on the corresponding side of this same individual there were only two ill-formed segments, with their few spines differently arranged. Again, in a third specimen, the great basal segment of the anterior cirrus on one side, bore, exteriorly to the usual ramus, a single segment furnished with bristles, and evidently representing a second ramus; thus showing that the great basal segment certainly answers to a pedicel. I may here add, that on the integuments of these cirri, I observed with a high power, the serrated scale-like appearance common in other cirripedes. Directly between the bases of the sixth cirrus, there is a very minute papillus, which, under the highest power, can be seen to consist of two closely approximate, flattened points; these, I have no doubt, are the caudal appendages in an extremely rudimentary condition, for I traced the vesiculae seminales to this exact spot: close outside these rudimentary points, on a slight swelling, is the anus. It will presently be seen that in the male of the closely allied *Ibla quadrivalvis*, the nature of these caudal appendages admits of no doubt, for in this species they consist of more than one segment, are spinose, and close under them towards the mouth, there is a perfectly distinct papillus, representing the usual probosciformed penis.

Alimentary canal. The oesophagus is very narrow, and of remarkable length; from the orifice under the mandibles, it first runs back (in this respect not well represented in Pl. V, fig. 1), under the bullate labrum, and then straight down the peduncle, where it terminates / in the usual bell-shaped expansion, entering one side of the small globular stomach; the latter, at its lower end, is slightly constricted, and then is rather abruptly upturned. The rectum is of unparalleled length, and extremely narrow; it can be best detected after the dissolution by caustic potash of the softer parts, when its inner coat of chitine can be seen to be continuous, in the ordinary manner, with the outer integuments of the thorax. The anus, as already stated, is seated on a slight swelling, and consists of a small longitudinal slit (*f*, fig. 2), placed close outside the two very minute caudal appendages.

Organ of sight. In all the specimens, a little below the fold separating the mouth from the peduncle, and near the abdominal (or rostral) edge, a black ball (*c*, fig. 1), about $\frac{1}{1000}$th of an inch in diameter, is conspicuous. When dissected out, it is somewhat conical in form, and appears to consist of an outer coat, with a layer of pigment-cells of a dark purple colour, surrounding a transparent, rather hard lens, apparently leaving a circular orifice at the summit, and forming a short tube at the base, surrounding what I believe to be a nerve. I was not able to perceive that this eye consisted of two eyes united, which the analogy of other cirripedes makes me suppose probable, although in the ordinary and hermaphrodite *Ibla quadrivalvis*, the eye also appeared single. It is seated under the two transparent muscular layers, close upon the upper end of the stomach, and this is the exact position, as stated in the introductory discussion (p. 49), in which the eyes of pedunculated cirripedes are commonly situated.

Generative system. Within the muscular layer all round the upper part of the peduncle, and surrounding the stomach, there are numerous, little, rather irregular globular balls, with brown granular centres, so closely resembling the testes in other cirripedes, though of smaller size, that I cannot doubt that this is their nature: they were much plainer, larger, and more numerous in some specimens than in others. The vesiculae seminales / can seldom be made distinctly out; but having cut one specimen transversely across the thorax, they were as plain as could be desired, lying parallel and close to each other above the rectum (the animal being in the position as drawn), and therefore in their normal situation. Each had a diameter four times as great as that of the rectum. In this individual the contents seemed (whether from decomposition or state of development, or from my not having used high enough power, I know not), merely pulpy; but I have since found, in another specimen, masses of the most distinct spermatozoa, with the usual little knots on them, associated with numerous cells, about as large as and resembling those which I have examined in living cirripedes, and from which I have every reason to believe the spermatozoa are developed. The vesiculae seminales unite and terminate under the two extremely minute caudal appendages, and here I think I saw an orifice; but there is certainly no projecting, prosbosciformed penis.

Having dissected the six specimens with the utmost care, and having scrupulously examined the ovaria in other cirripedes during their early stages of development, even before the exuviation of the larval locomotive organs, and in specimens of smaller size than the male Ibla, I am prepared to assert that there are no ovaria, and that these little creatures are exclusively males. It should be borne in mind, that in some of the specimens there were perfect spermatozoa in the vesiculae seminales (as likewise in some of the males of *I. quadrivalvis*), and, therefore, if these individuals had been hermaphrodites, their ova would have been, at this period, well developed, and ready for impregnation: in this state it is almost impossible that they could have been overlooked. Moreover, it is probable that such ova would not have been very small, for the larvae whence the parasitic males are derived, attain (as might have been inferred from the known dimensions of their prehensile antennae, and as we shall show actually is the case in *I. quadrivalvis*), the size common among ordinary Cirripedia. /

Concluding remarks. That these animals are true cirripedes, though having so different an external appearance from others of the class, admits of not the least doubt. The prehensile antennae, enveloped in cement and including the two cement ducts, would have been amply sufficient, without other parts – for instance, the mouth, by itself perfectly characteristic with each organ, together with the whole alimentary canal, constructed on the normal plan – to have proved that they were Cirripedia. Under the head of the closely allied *Ibla quadrivalvis*, we shall, moreover, see that the males are developed from larvae, having every point of structure – the peculiar quasi-bivalve shell, the two compound eyes, the six natatory legs, etc. – characteristic of the order. But in some respects, the males are in an embryonic condition, though unquestionably mature, as shown by the spermatozoa; thus, in the thorax and mouth opening throughout their whole width into the cavity of the peduncle, that is, homologically into the anterior part of the head, and in the viscera being there lodged instead of in the thorax and prosoma, there is a manifest resemblance to the larva in its last stage of development: the absence of a prosbosciformed penis, the spineless peduncle, the food being obtained without the aid of cirri, and the length of the rectum, are likewise embryonic characters. Not only are these males, as just remarked, Cirripedia; but they manifestly belong to the pedunculated family. If a specimen had been brought to me to class, without relation to its sexual characters, I should have placed it, without any hesitation, next to the genus Ibla; if the mouth alone had been brought, I should assuredly have placed it actually in the genus Ibla: for let it be observed how nearly all the parts resemble those of *Ibla Cumingii*, excepting only in size and in being less hairy. The trophi are arranged in the same peculiar position as in the female; the labrum is largely bullate, without teeth on the crest; the palpi, though relatively smaller, are of the same shape; so are the mandibles; the maxillae are more rounded and less prominent, but have the same / exact size relatively to the mandibles; the outer maxillae have the same, quite peculiar pointed outline, and the olfactory orifices are tubular, and hold the same unusual position. It is most rare to find so close a resemblance in the parts of the mouth, except in very closely allied genera, and often species of the same natural genus differ more. Again, in the long oesophagus and constricted stomach there is a resemblance to Ibla. In the male of *Ibla quadrivalvis*, the caual appendages are multiarticulate; now, this is a character confined to four genera, namely, Ibla, Alepas, Pollicipes, and Lithotrya. I may add, that large tubular olfactory orifices are confined to the same genera, together with Scalpellum. Lastly, it particularly deserves notice, that the prehensile antennae, in having a hoof-like and pointed disc, with a single spine on the heel, much more closely resemble these organs in Scalpellum, certainly the nearest ally of Ibla, than in any other genus; they differ from the antennae in Scalpellum, only in the ultimate segment not having a notch on one side. These organs, unfortunately for the sake of comparison, were not found in the female and ordinary form of Ibla. The full importance of the above generic resemblance in the antennae, will hereafter be more clearly seen, when their classificatory value is shown in the final discussion on the sexual relations of Ibla and Scalpellum.

Here, then, we have a pedunculated cirripede *very much* nearer in all its essential characters to Ibla than to any other genus, and exclusively of the male sex; and this cirripede in six specimens, from two distant localities, adhered to an Ibla exclusively of the female sex. May we not, then, safely conclude that these parasites are the males of the *Ibla Cumingii*? Considering that, in the same class with the Cirripedia, there is a whole family of crustaceans, the Lerneidae, in which the males, compared with the females to which they cling, differ as much in appearance as in Ibla, and are even relatively smaller, I should not have added another remark, had there not been under the head of the following species, and of the next genus / Scalpellum, a class of allied facts to be advanced, which in some respects support the view here taken, but in others are so remarkable and so hard to be believed, that I will call attention to the alternative, if the above view be rejected. The ordinary *Ibla Cumingii* must have a male, for that it is not a hermaphrodite can hardly be questioned, seeing how easy it always is to detect the male organs of generation; and we must consequently believe in the visits of a locomotive male, though the existence of a locomotive cirripede is improbable in the highest degree. Again, as the little animal, considered by me to be the male of *I. Cumingii*, is exclusively a male (for there were no traces of ova or ovaria, though the spermatozoa were perfect), we must believe in a locomotive cirripede of the opposite sex, though the existence in any class of a female visiting a fixed male is unknown:[46] in short, we should have hypothetically to make two locomotive cirripedes, which, in all probability, would differ as much from their fixed opposite sexes, as does the cirripede, considered by me to be the male of *I. Cumingii*, from the ordinary form. This being the case, I conclude that the evidence is amply sufficient to prove that the little parasitic cirripede here described, is the male of *Ibla Cumingii*.

If we look for analogies to the facts here given, we shall find them in the Lerneidae already alluded to, but in these the males are not permanently attached to the females, only cling, I believe, to them voluntarily. The extraordinary case of the Hectocotyle, originally described as a worm parasitic on certain Cephalopoda, but now shown by Kölliker to be the male of the species to which it is attached, is perhaps more strictly parallel. So again in the entozoic worm, the *Heteroura androphora* the / sexes cohere, but are essentially distinct: 'this singular species, however', according to Professor Owen,[47] 'offers the transitional grade to that still more extraordinary Entozoon, the *Syngams trachealis*, in which the male is organically blended by its caudal extremity with the female, immediately anterior to the slit-shaped aperture of the vulva. By this union a kind of hermaphroditism is produced; but the male apparatus is furnished with its own peculiar nutrient system; and an individual animal is constituted distinct in every respect, save in its terminal confluence with the body of the female. This condition of animal life, which was conceived by Hunter as within the circle of physiological possibilities, has hitherto been exemplified only in the single species of Entozoon, the discovery of the true nature of which, is due to the sagacity and patient research of Dr

---

[46] It deserves notice, that in the class Crustacea, both in the Lerneidae and in the Cirripedia, the males more closely resemble the larvae, than do the females; whereas among insects, as in the case of the glow-worm in Coleoptera, and of certain nocturnal Lepidoptera, it is the female which retains an embryonic character, being worm-like or caterpillar-like, without wings. But in all these cases, the male is more locomotive than the female.

[47] *Cyclopaedia of Anatomy and Physiology*, p. 142.

C. Th. Von Siebold.' In Ibla, the males and females are not organically united, but only permanently and immovably attached to each other. We have in this genus the additional singularity of occasionally two males parasitic on one female.

I have used the term parasitic, which perhaps ought strictly to be confined to cases where one creature derives its nutriment from another, inasmuch as the male is invariably and permanently attached to and imbedded in the female – from its being protected by her capitulum, so that its own capitulum is not developed – and from its feeding on minute animals infesting her sack. The male Ibla must seize its prey, guided probably by its well-developed olfactory organs, through the movement of its long, flexible body, furnished with muscles, and with the mouth seated on the summit. We have already seen one instance of a cirripede, the Anelasma, obtaining its food without the aid of cirri, by means of its prosciformed, flexible mouth. The eye can serve only to announce to the male when the female opens her valves, allowing occasionally some minute prey to enter. In / ordinary cirripedes the penis is long, articulated, and capable of varied movements, I presume for the purpose of impregnating each separate ovum: the male Ibla has no such organ; and no doubt the whole body, furnished like the penis with longitudinal and transverse muscles, serves the same purpose! I may remark, that it seems surprising that so small a male should secrete sufficient semen to impregnate the ova of the female, but the ova are not nearly so numerous in Ibla as in most genera of cirripedes; and the smallness of the males in some parasitic Crustacea has already been alluded to. The male must always be younger than the female, for the latter must first grow large enough for the larva of the male to crawl into her sack. Whether the male lives as long as the female I know not, but he certainly lives for a considerable period and increases in size, as shown by the depth to which the end of the peduncle is imbedded. Moreover we shall see, under the next species, that the male is metamorphosed from a larva, not one-sixth of its own size.

In the male Ibla, abortion has been carried to an extraordinary and, I should think, almost unparalleled extent. Of the twenty-one segments believed to be normally present in every Crustacean, or of the seventeen known to be present in cirripedes, the three anterior segments are here well developed, forming the peduncle: the mouth consists as usual of three small segments: the succeeding eight segments are represented by the rudimentary and functionless thorax, supporting only two pair of distorted, rudimentary and functionless cirri: the seven segments of the abdomen have disappeared, with the exception of the excessively minute caudal appendages; so that, of the twenty-one normal segments, fifteen are more or less aborted. The state of the cirri is curious, and may be compared to that of the anthers in a semi-double flower; for they are not simply rudimentary in size and function, but are monstrous, and generally do not even correspond on opposite sides of the / same individual. As males in other classes of the animal kingdom often retain some female characters, so here (though the case is not strictly analogous)[48] the male possesses the cementing apparatus, which homologically is part of an ovarian tube modified.

The individuals in every other genus (with the exception of Scalpellum), in the several families, in the three orders of Cirripedia, are hermaphrodite or bisexual. Why, then, is Ibla unisexual; yet, becoming, in the most paradoxical manner, from its earliest youth, essentially bisexual? Would food have been deficient, and was the seizure of infusoria by another and differently constructed individual, necessary for the support of the male and female organs? The orifice of the sack of the female is unusually narrow; would the presence of testes and vesiculae seminales have rendered her thorax and prosoma inconveniently thick? Seeing the analogous facts in the six, differently constructed species of the allied genus Scalpellum, I infer there must be some profounder and more mysterious final cause.

---

[48] Certain plants offer a closer, though not perfect, analogy. Thus, in the florets of some compositous flowers, the pistil, besides it proper female functional end, serves to brush the pollen off the anthers; while, in the florets of some other compositae (see the account of Silphium in *Ch. K. Sprengel Das entdeckte Geheimniss der Natur*), the pistil is functionless for its proper end, the flower being exclusively male, but its style is developed, and still serves as a brush. So in the male Ibla, part of the ovaria, in a modified condition, is still present, and serves as a cementing apparatus.

2. IBLA QUADRIVALVIS

PLATE IV. FIG. 9

ANATIFA QUADRIVALVIS. Cuvier, *Mém. pour servir . . . Mollusq.*, Art. Anatifa, Plate, figs 15, 16, 1817
IBLA CUVIERIANA. J. E. Gray, *Annals of Philosophy*, vol. x, New Series, August, 1825
IBLA CUVIERIANA. J. E. Gray, *Spicilegia. Zoolog.*, Pl. iii, fig. 10
TETRALASMIS HIRSUTUS. Cuvier, *Regne Animal*, vol. iii, 1830
ANATIFA HIRSUTA. Quoy et Gaimard, *Voyage de l'Astrolabe*, Pl. xciii, figae. 7–10, 1834 /

I. (Herm.), valvis et pedunculi spinis sub-flavis: basali tergorum angulo, introrsùm spectanti, hebete, quia margo carinalis inferior longiùs quam margo scutalis prominet.

Hermaph. Valves and spines on the peduncle yellowish: basal angle of the terga, viewed internally, blunt, owing to the lower carinal margin being more protuberant than the scutal margin.

Caudal appendages four times as long as the pedicels of the sixth cirrus: rami of the first cirrus unequal in length by about six segments.

Complemental male, with a notched crest on the dorsal surface, forming a rudiment of a capitulum: maxillae well furnished with spines.

Kangaroo Island, South Australia (Mus. Brit., given by Cuvier to Leach); Adelaide, South Australia (Mus. Stutchbury); King George's Sound, Voyage of Astrolabe; New South Wales, attached to a mass of the Galeolaria decumbens (Mus. Hancock).

HERMAPHRODITE

All the external parts so closely resemble those of *I. Cumingii*, that it would be superfluous to describe more than the few points of difference. The horny substance of both scuta and terga is uniformly yellow; though in dried specimens, from the underlying corium being seen through the valves, these generally have a tinge of blue.

The *scuta*, viewed internally, are less elongated transversely; they have their basal margins slightly more hollowed out, and the fold on the upper free and horn-like portion rather deeper.

The *terga*, viewed internally, have the apex of the growing or corium-covered surface higher relatively to the scuta than in *I.*

Cumingii; and the basal angle is much broader, owing to the lower carinal margin being much more protuberant than the scutal margin. The spines on the peduncle are all yellowish-brown, and are rather longer than in *I. Cumingii*. I observed in three or four specimens, that the lowest part of the peduncle / had become *internally* filled up with the usual, brown, transparent, laminated cement, cone within cone, so that this lower part was rendered rigid and stick-like; this, latter effect, I apprehend, is the object gained by the formation of cement within the peduncle, of which I have not observed any other instance. The entire length of the largest specimen was one inch; some other specimens were only half this size.

The thorax and prosoma are of the same shape as in *I. Cumingii*, and in the largest specimen, about one-tenth of an inch square; the prosoma, as in that species, is hairy. In the *mouth*, all the parts are closely similar to those of *I. Cumingii*, but one-third larger; the crest of the labrum is a little roughened with minute points: the palpi are squarer and blunter at their extremities: the mandibles have their second and third teeth nearly equal in size to the first, and they do not appear pectinated: the maxillae have their spinose edge very nearly straight: the outer maxillae are pointed. The olfactory orifices are similarly situated, and of similar shape; they are dark coloured.

Cirri. These also are similar to those of *I. Cumingii*; the segments, however, of the three posterior cirri have each four pair of spines, placed very close together in a transverse direction. First cirrus has its two rami unequal in length by about six segments. The anterior rami of the second and third cirri are thicker, and more thickly clothed with spines, than the posterior rami, to perhaps a greater degree than in *I. Cumingii*. In the posterior cirri, the upper segments of the pedicels are nearly as long as the lower segments.

Caudal appendages, four times as long as the pedicel of the sixth cirrus, and three-fourths of the length of the rami of this same cirrus: segments thirty-two in number, and therefore as many as those forming the sixth cirrus: the upper segments are much thinner and longer than the basal segments; each furnished with a circle of short bristles; whole appendage excessively thin and tapering: the two closely approximate. /

Colour. From some well-preserved dried specimens in Mr Stutchbury's possession, it appears that the sack, cirri and trophi, were dark blue, as in *I. Cumingii*; after being long kept in spirits, these parts become brown.

Generative system. The penis (Pl. IV, fig. 9a) is very singular in structure; it is of the ordinary length, but of small diameter; it tapers but little; it consists of a movable articulated, and a fixed unarticulated portion; this latter is smooth, much flattened, not divided into segments, and projects straight out under the caudal appendages; it is about one-third of the length of the entire penis; it corresponds with a part present in all cirripedes, but here surprisingly elongated. The articulated portion consists of separate segments, twenty in number, quite as distinct as those of the cirri; each one is oblong, being longer by about a third part than broad; each has a few short bristles round its upper margin; the terminal segment has a circular brush of bristles. The vesiculae seminales are easily seen, though they are narrow; they are slightly tortuous; they enter the prosoma, and lie on each side of the stomach; their outer case has a ringed structure, but is not fibrous; the contents in the best specimen consisted of a mass of spermatozoa, which I saw with perfect distinctness. The testes are unusually large and egg-shaped.

Ova, spherical, $5/400$ths of an inch in diameter, united as usual into two ovigerous lamellae. The ovigerous fraena are extraordinarily small, and might be very easily overlooked; their length, in a full-sized specimen, was only $7/400$ths of an inch, and they projected only $2/400$ths from the inner surface of the sack. The glands on their margin, to which the lamellae adhere, are pointed oval, with an extremely short footstalk, and that rather thick; the entire length of gland and footstalk, being only $2/3000$ths of an inch. The larvae, in their first stage of development, offer the usual characters, and closely resemble those of Scalpellum; the prosciformed mouth, however, is remarkably prominent, and the limbs unusually thick. /

Affinities. This species most closely resembles *I. Cumingii*, and cannot be distinguished externally, except by the absence of the blue colour on the marginal and interior portions of the valves; and this can hardly be ascertained without separating and cleaning them, owing to the blueness of the underlying corium. Internally some slight differences may be perceived in the form of the valves. Considering these so slight differences, it is highly remarkable that this species should be hermaphrodite, whilst *I. Cumingii* is unisexual. There is a greater, though still slight, difference in the included animal's body; the palpi in *I. quadrivalvis* are blunter, the mandibles smoother, the olfactory orifices darker coloured; the rami of the first cirrus more unequal, the spines more numerous on the segments of the posterior cirri, and lastly and most conspicuously, the caudal appendages are very much longer relatively to the length of the sixth cirrus, than in *Ibla Cumingii*.

COMPLEMENTAL MALE

I have examined one specimen of the hermaphrodite *I. quadrivalvis*, preserved in spirits from Kangaroo Island, and one dry from Adelaide, both places in South Australia, and four from an unknown locality, purchased from Mr Sowerby; and within five out of these six specimens, males were attached. In one of them, two males of different ages were included, one adhering to the peduncle of the other: in *I. Cumingii*, also, it may be remembered, there was a case of two males parasitic on one female. I may add that I opened another quite young specimen, from Adelaide, not counted with the above, and it was without a male. The males in the five specimens were attached low down, at the rostral end, almost in a horizontal position, stretching across the bottom of the sack; one of them, however, was placed considerably on one side. One individual which I measured, was $16/100$ths of an inch in length, and $5/100$ths in width in the widest part, / namely, about half down the peduncle. I may state, for the sake of comparison, that the hermaphrodite to which this individual was attached, was, including the peduncle and capitulum, one inch in length, that is, six times as long as the male, and one-fifth of an inch in width, that is, four times as wide. The above measurements show that the male of this species is rather more than twice as large as that of *I. Cumingii*. In consequence of this greater size, I dissected, with the utmost care, the one specimen which was excellently preserved in spirits, and found every part, with a few exceptions, so exactly the same as in the male of *I. Cumingii*, only larger and more conspicuous, that it will be sufficient to indicate the few points of difference.

The most conspicuous difference is, that the oblique fold separating the thorax and peduncle is more plainly developed, projecting at the point (corresponding to *h* in fig. 1, Pl. V), $5/1000$ths of an inch; in the middle the fold is notched; it can be traced more easily than in *I. Cumingii*, running beneath and parallel to the basal edge of the mouth, to the ventral margin of the body. In the mouth there is hardly any difference; the maxillae, however, have two notches even plainer than in the hermaphrodite *I. quadrivalvis*, or than in the male *I. Cumingii*, but the depth of such notches is always a variable character; there are also more spines on the edge in the male of the present species, than in *I. Cumingii*. Both mandibles and maxillae in the male *I. quadrivalvis*, are larger than in the male *I. Cumingii*, to a greater degree than the larger proportional size of the body in the former will account for; and this, likewise, is the case with these same organs in the hermaphrodite *I. quadrivalvis* compared with the female *I. Cumingii*. The tubular olfactory orifices are situated in the same peculiar position as in the hermaphrodite, and as in both sexes of *I. Cumingii*: they are $1/500$th of an inch in diameter, and about as thick as one of the lower segments in the rami of the sixth cirrus.

The thorax, as in the male of *I. Cumingii*, is quite / rudimentary, and serves as a mere flap to protect the mouth. In the three specimens carefully examined, the posterior cirri had each only one ramus, whilst the anterior cirri generally had two: in one specimen, one of the rami in the anterior cirrus was formed of five segments, and the other ramus of three segments, both rami being supported on a uni-articulated pedicel; but on the opposite side of the same individual, the anterior cirrus was represented by a mere knob. The longer ramus of the anterior cirrus, in the best developed individual, barely exceeded in length the mandibles measured along the line of the teeth! In one specimen between the bases of the posterior cirri, there were two perfectly distinct caudal appendages; these, like the cirri, are in a quite rudimentary condition; one was $5/1000$ths of an inch in length, and consisted of three segments, the upper edges of which had short spines; the other was shorter, uni-articulated, but spinose. In a second specimen, these appendages were quite aborted. Close under them, on the inside or towards the mouth (that is, in the normal position), there was a rudimentary but quite distinct penis, with the apex projecting freely, and with the sides distinguishable from the ventral surface of the thorax, for the length of $1/1000$th of an inch: the corium lining this little penis made the terminal orifice plainly visible. The vesiculae seminales lie in the usual position, and are conspicuous; they are slightly tortuous, with their ends blunt: in the specimen so well preserved in spirits, they were filled with a mass of spermatozoa, perfectly distinct; and the whole cavity of the body was lined with globular and pear-shaped testes. Assuredly there was no vestige of ovarian tubes. From the greater size and excellent preservation of this specimen, which rendered the examination of the generative system so easy, I was able to examine the contents of the stomach, in which I found the delicate epithelial coat, separated as usual, and containing cellular matter, on which the animal had preyed, but the nature of which I was unable to / make out. The anus was much plainer than in the male of *I. Cumingii*. I saw the eye distinctly. I could not distinguish the orifices of the acoustic [?] sacks; and I think I should have seen them, if they had existed.

Prehensile antennae. I examined these in the larvae presently to be mentioned, and therefore they were in better condition than in the mature animal when cemented. Their total length, measured along the outside, from the basal articulation to the end of the disc, is $32/6000$th or $33/6000$ths of an inch – that is, one-third longer than in *I. Cumingii*; whilst the hoof-like disc itself is $8/6000$ths, or only $1/6000$th of an inch longer than this same part in *I. Cumingii*: the apex of the disc is downy, or bears some excessively minute spines. The ultimate segment has its end irregularly rounded, with the spines obscurely divided into two groups, the outer group consisting of two or three longer and thinner spines, and the inner group of, as I believe, five rather shorter spines: the longer spines equal in length the whole ultimate segment. I could not perceive that they were plumose, as in many other genera. A single, rather thicker and long spine, pointing backwards, is attached to the underside of the disc, nearly opposite to the point where the ultimate segment is articulated on the upper convex surface. Another single, curved spine is attached on the outer side of the basal segment, near its distal end.

Development of the male. In the specimen before alluded to, which included two males, one of these was only the $30/1000$ths of an inch in length, and therefore between one-fifth and one-sixth of the size of the mature male. It had, probably, undergone only one exuviation since its metamorphosis, for the larva is nearly as long, namely, $25/1000$ths of an inch. In this young male, the mouth formed one-third of the entire length: it was attached, not as in every other case to the sack of the hermaphrodite, but low down to the peduncle of the other male.

In the sack with these two males, there were certainly / four, I believe five, larvae, which in every main point of structure resembled the larvae of other pedunculated cirripedes. From the peculiar form of their prehensile antennae, differing in no respect, except in the proportional lengths of the segments, from the same organ in the male

I. Cumingii, I can feel no doubt that these were the larvae of the male *I. quadrivalvis*; for a moment's reflection will show how excessively improbable it is, that several larvae of some other cirripede, and that a cirripede intimately allied to the parasitic male Ibla, should have forced themselves, without any apparent object, into the sack of the hermaphrodite Ibla. The larvae, though not yet attached, were on the point of attachment, so that the single eye of the mature animal could be distinctly seen, lying near to the two great compound eyes of the larva. We have also just seen, that one male quite recently here had undergone its metamorphosis. The larvae are $25/1000$ths of an inch in length, and rather more than $19/1000$ths in width in the widest part: they are boat-shaped, the dorsal edge forming the keel of the boat; the anterior end is only a little blunter than the posterior end; the quasi-bivalve carapace is smooth. All the essential points of structure in the larvae of other cirripedes at this stage, could be distinctly here seen – such as the two compound eyes, with the apodemes to which they are attached, and the two oblong sternal plates whence the apodemes spring – the adductor muscle – the six natatory legs, with long plumose spines – the abdomen, with its three small segments and the caudal appendages – the prehensile antennae already described – and, lastly, the two little auditory [?] sacks at the antero-sternal edges of the carapace, but not so near the anterior extremity as in Lepas. The four or five larvae, after having undergone in the open sea the several preparatory metamorphoses common to the class, must have voluntarily entered the sack of the hermaphrodite: ultimately would they, on finding two males already attached there, have retired, and sought another individual less well provided; or / would they all have remained, and so formed a polyandrous establishment, such as we shall presently see occurs sometimes in Scalpellum? This must remain quite uncertain.

In this same hermaphrodite specimen of *I. quadrivalvis*, the two ovigerous lamellae contained some hundreds of larvae in the first stage of development, which were liberated from their enveloping membranes by a touch of a needle: they were about $16/1000$ths of an inch in length, and presented all the usual characters of larvae at this period. What a truly wonderful assemblage of beings of the same species, but how marvellously unlike in appearance, did this individual hermaphrodite present! We have the numerous, almost globular larvae, with lateral horns to their carapaces, with their three pair of legs, single eye, probosciformed mouth and long tail: we have the somewhat larger larvae in the last stage of development, much compressed, boat-formed, with their two great compound eyes, curious prehensile antennae, closed rudimentary mouth and six natatory legs so different from those in the first stage: we have the two attached males, with their bodies reduced almost to a mouth placed on the summit of a peduncle, with a minute, apparently single eye shining through the integuments, without any carapace or capitulum, and with the thorax as well as the legs or cirri rudimentary and functionless: lastly, we have the hermaphrodite, with all its complicated organization, its thorax supporting six pairs of multi-articulated two-armed cirri, and its well-developed capitulum furnished with horny valves, surrounding this wonderful assemblage of beings. Unquestionably, without a rigid examination, these four forms would have been ranked in different families, if not orders, of the articulated kingdom.

Concluding remarks. If the creature which I have considered as the male of *Ibla Cumingii* be really so, and the evidence formerly given seems to me amply conclusive, then the animal just described, from its close affinity in every point of structure with the former, assuredly is the / male of *Ibla quadrivalvis*. But feeling strongly how improbable it is, that an additional or complemental male should be associated with an hermaphrodite, I will make a few remarks on the only possible hypothesis, if my view be rejected – namely, that the two parasites considered by me to be exclusively males, are not so, but are independent hermaphrodite cirripedes, the female organs and ova (which, if present, would have been nearly mature, judging from the presence of spermatozoa in both species) having been overlooked by me in every specimen: and again, that in the animal described as the female *I. Cumingii*, I have, though minutely dissecting several specimens, and finding far smaller parts, such as the organs of sense and nervous system, entirely overlooked all the conspicuous male organs, though when I came to *I. quadrivalvis*, and naturally expected to find it likewise exclusively female, a single glance showed me the great prosciformed penis, and by the simplest dissection the vesiculae seminales and testes were exhibited. Such an oversight is scarcely credible; but even if assumed, we have to believe in the extraordinary circumstance of the two parasites being species of an independent genus, not only the very next in alliance to the animals to which they are attached, but in certain most important points, namely, the organs of the mouth, actually deserving a place in the very same genus. Moreover, the two parasites differ from each other, not only in about the same slight degree, but in a corresponding manner, as do the two Iblas to which they are attached; thus the mouths of *Ibla quadrivalvis* and *I. Cumingii* are closely similar (the difference being barely of specific value), so are the mouths of the two parasites; but the parts are larger in the hermaphrodite *I. quadrivalvis*, than in *I. Cumingii*, so are they in the parasites. Again, the most conspicuous character in *I. quadrivalvis*, is the number of segments in the caudal appendages, far exceeding those in the other species of Ibla, as well as of every other pedunculated cirripede, and the parasite of this species has articulated spinose / appendages, far larger than the barely visible, non-articulated pair in *I. Cumingii*.

Considering the whole case, there seems no room to doubt the justness of the conclusion arrived at, under the former as well as under the present species, namely, that these little parasites are the males of the two species of Ibla to which they are attached; wonderful though the fact be, that in one case, the male should pair with a hermaphrodite already provided with efficient male organs. It is to bring this fact prominently forward, that I have called such males, complemental males; as they seem to form the complement to the male organs in the hermaphrodite. We look in vain for any, as yet known, analogous facts in the animal kingdom. In the genus Scalpellum, however, next in alliance to Ibla, in which, consequently, if anywhere, we might expect to find such facts, they occur; and until these are fully considered, I hope the conclusions here arrived at, will not be summarily rejected. Although the existence of hermaphrodites and males within the limits of the same species, is a new fact among animals, it is far from rare in the vegetable kingdom: the male flowers, moreover, are sometimes in a rudimentary condition compared to the hermaphrodite flowers, exactly in the same manner as are the male Iblas. If the final cause of the existence of these complemental males be asked, no certain answer can be given; the vesiculae seminales in the hermaphrodite of *Ibla quadrivalvis*, appeared to be of small diameter; but on the other hand, the ova to be impregnated are fewer than in most cirripedes. No explanation, as we have seen, can be given of the much simpler case of the mere separation of the sexes in *Ibla Cumingii*; nor can any explanation, I believe, be given of the much more varied arrangement of the parts of fructification in plants of the Linnean class, Polygamia. /

GENUS: SCALPELLUM

PLATES V, VI

SCALPELLUM. Leach, *Journ. de Physique*, t. lxxxv, July, 1817
LEPAS. Linn, *Systema Naturae*, 1767
POLLICIPES. Lamarck, *Animaux sans Vertebres*, 1818
POLYLEPAS. De Blainville, *Dict. des Sc. Nat.*, 1824
SMILIUM (pars generis). Leach, *Zoolog. Journal*, vol. 2, July, 1825
CALANTICA (pars generis). J. E. Gray, *Annals of Philosophy*, vol. x (new series), August, 1825.
THALIELLA (pars generis). J. E. Gray, *Proc. Zoolog. Soc.*, 1848
ANATIFA. Quoy et Gaimard, *Voyage de l'Astrolabe*, 1826–34
XIPHIDIUM (pars generis). Dixon, *Geology of Suffolk*, 1850

(Herm. et Foem.) *Valvis 12 ad 15: lateribus verticilli inferioris quatuor vel sex, lineis incrementi plerumque convergentibus: sub-rostrum rarissime adest: pedunculo squamifero, rarissime nudo.*

(Herm. and Fem.) Valves 12 to 15 in number: latera of the lower whorl, four or six, with their lines of growth generally directed towards each other: sub-rostrum very rarely present: peduncle squamiferous, most rarely naked.

Filamentary appendages, none: labrum, with the upper part highly bullate: trophi, various: olfactory orifices, more or less prominent: caudal appendages, uniarticulate and spinose, or none.

Males, parasitic at or near the orifice of the sack of the female or of the hermaphrodite: thorax enclosed within a capitulum, furnished with three or four rudimentary valves, or with six perfect valves: peduncle either short and distinct, or confounded with the capitulum: sometimes mouth and stomach absent, and cirri non-prehensile; sometimes mouth and cirri normal.

Generally attached to horny corallines, in the warmer temperate seas over the whole world.

I have felt much doubt in limiting this genus: the six recent species which it contains, differ more from each other than do the species in the previous genera. Mr / Gray has proposed or adopted generic names for four of the species, and a fifth certainly has equal claims to this same rank. These genera have been founded almost exclusively on the number of the valves; and oddly enough, the numbers have generally been given wrongly, namely, in Scalpellum, Calantica,

Thallella, and Xiphidium. Scalpellum blends through *S. villosum* into Pollicipes; and this latter genus has an equal right with Scalpellum, to be divided into subgenera, three in number. Hence, no less than eight genera might be made out of the twelve recent species of Scalpellum and Pollicipes, and their formation, in some degree, be justified; but, in my opinion, this inordinate multiplication of genera destroys the main advantages of classification. At one time, I even thought that it would be best to follow Lamarck, and keep the twelve recent species in one genus; but considering the number of fossil species, I believe the more prudent course has been followed, in retaining the two genera Scalpellum and Pollicipes; more especially as I can hardly doubt, that several other species will be hereafter discovered.

Having so lately described in the *Memoirs of the Palaeontographical Society*, the fossil species, I will not here further allude to them, than to state, that out of the fifteen species therein described, *S. magnum* comes very close to the recent *S. vulgare*, and that several Eocene and Cretaceous species, such as *S. quadratum, S. fossula*, and *S. maximum*, are allied to *S. rutilum* and *S. ornatum*. *Scalpellum villosum*, a recent species, has stronger claims than any other species to be generically separated; and its habits, in not being attached to horny corallines, are also different, but the identity of its complemental male with that of *S. Peronii*, and its numerous points of resemblance in structure with the other species, have determined me not to separate it. *Scalpellum Peronii, villosum*, and *rostratum*, in having a subcarina – in the rostrum being pretty well developed – and in the complemental male being pedunculated, and furnished with / a functional mouth and prehensile cirri, may be separated from *S. vulgare, ornatum* and *rutilum*; but even between these two little groups, *S. rostratum* is in some respects intermediate, namely, in having three pairs of latera, and more especially in the rudimentary condition of the valves of its complemental male, and in the position in which the male is attached to the hermaphrodite. The three species in the second little group, namely, *S. vulgare, S. ornatum*, and *S. rutilum*, are more nearly allied to each other in all their characters, especially in the characters drawn from their males, than are the other three species. *S. ornatum* and *S. rutilum* are considerably nearer to each other than any other two of the species. Upon the whole I conclude that the six species must be thrown either into five or into four genera (the first three species making one genus), or all into one genus, and this latter has appeared to me the preferable course. The separation even of Scalpellum and Pollicipes, as already stated, is hardly natural. The fact of these genera having existed from a remote epoch, and having given rise during successive periods to many species now extinct, is probably the cause that the few remaining species are so much more distinct from each other, than is common in the other genera of Lepadidae. Whenever the structure of the whole capitulum in the fossil species is well known, and as soon as more species, recent and fossil, shall have been discovered, then probably the genus Scalpellum will have to be divided into several smaller genera.

Description. The *capitulum* is much compressed, and generally produced upwards; it is formed of from twelve to fifteen valves, which are rather thin, and with the exception of S. *ornatum*, almost entirely covered by membrane, bearing spines: the valves are seldom locked very closely together. A sub-rostrum exists only in S. *villosum*, which species leads on to Pollicipes: in S. *vulgare* the rostrum is rudimentary and hidden. The scuta, terga and carina, are much larger than the other valves: these five valves seem to differ essentially from / the others in being at first developed under the form of the so-called primordial valves: the other valves commence by a small indistinct brown spot, very different from the hexagonal tissue of the primordial valves: I saw this very clearly in young specimens of S. *vulgare*. At first, the scuta, terga and carina, grow exclusively downwards (and permanently so in most fossil species), and therefore the growth of the scuta and carina is in an absolutely opposite direction to what it is in Lepas, Paecilasma, and Dichelaspis. After a short period the scuta are added to at their upper ends; the portion thus added, stands at a rather lower level, and projects in a rather different direction from the first-formed part of the valve, giving to it, in some respects, the appearance of having been broken and mended. This structure is common to S. *vulgare*, S. *rostratum* and S. *Peronii*. The upper Latera (except in S. *villosum*) grow in the same manner, namely, at first exclusively downwards, and then both upwards and downwards. The rostral and carinal latera (with the same exception of S. *villosum*) have their umbones seated laterally, at opposite ends of the capitulum – the umbones of the rostral latera being close to the rostrum, and those of the carinal pair close to the carina, and consequently their chief growth is directed towards each other. The carina in all the species, except S. *villosum*, is either bowed or angularly bent; in the latter case the lower half is parallel to the peduncle, and the upper half, extending far up between the terga, is parallel to their longer axes. In some of the species the carina is added to almost equally at both ends; in *S. ornatum* it grows but little at the upper end, and to a varying degree in different individuals according to their age; in *S. rutilum* the umbo is at the apex, and there is consequently no upward growth; lastly, in *S. villosum* the carina widening much from the apex to the basal margin, grows exclusively downwards, and a portion of the apex projects freely – characters all common to the carina in the genus Pollicipes. The upper latera occur / in all the species; in the lower whorl there are either two or three pair of latera, in the former case the infra-median pair being absent. The latera differ considerably in shape in the different species.

The *peduncle* is generally rather short, and, with the exception of *S. Peronii*, is covered with calcified scales. These scales are generally small, and placed symmetrically in close whorls, in an imbricated order, with each scale corresponding to the interspace between two scales in the whorls above and below. In *S. ornatum*, the scales are so wide, transversely, that there are only four in each whorl. In *S. villosum*, the scales are spindle shaped and arranged somewhat irregularly in transverse rows, not very near to each other. New calcareous scales originate only round the top of the peduncle, and they continue to grow only in the few upper whorls; and as the peduncle itself continues to increase in diameter by the formation of new inner membranous layers and the disintegration of the old outer layers, the calcareous scales come in the lower part of the peduncle to stand further and further apart. In the earliest stage of growth there are no calcareous scales on the peduncle in *S. vulgare*; they first appear under the carina. Spines are articulated in great numbers on the surface of the peduncle in *S. vulgare*, *S. Peronii*, and *S. villosum*, and very short ones on that of *S. rostratum*.

Attachment. All the species, except *S. villosum*, are attached to horny corallines: the singular means of attachment in *S. vulgare* will be described under that species, and is probably common to several of the other species. The larva in most, or in all cases, when it proceeds to attach itself, clings head downwards to the branch, and hence the capitulum comes to be placed upwards, with its orifice fronting the branch and the carina outwards. The sucking disc of the prehensile antennae of the larva, in the five species examined, was a little pointed, and in shape resembled the hinder hoof of a mule: this may perhaps be accounted for by the / narrowness of the branches of the corallines, to which it has to adhere: a large circular disc, as in Lepas, would have been worse than useless: the ultimate segment in most or all the species, has on its inner side (the segment being supposed to be extended straight forward) a notch or step, bearing, I believe, two spines.

Size and colour. Some of the species attain a medium size, others are small. The valves are generally clouded red or pink, but sometimes white.

Mouth. The various parts vary far more than in any genus hitherto described. The labrum is highly bullate, with the upper part forming a rounded overhanging projection, and with the lower part much produced, so that the mouth is placed far from the adductor scutorum muscle, and consequently the orifice is directed more towards the ventral surface of the thorax than in most other cirripedes: on the crest of the labrum there are some very small teeth in several of the species, but not in all. The mandibles have either three or four main teeth, generally with either one or two small teeth intermediate between the first and second large teeth, and in the case of *S. Peronii*, with small teeth between all the larger ones. The maxillae have their edges furnished with many spines, and are either straight or have the inferior part prominent and step-formed. The outer maxillae have the spines on their inner edges either continuous or divided into two groups, of which latter structure we have not hitherto had any very well characterized example. The olfactory orifices are either highly or moderately protuberant.

In most of the species the prosoma is little developed, and the first cirrus is placed far from the second. The *cirri* are generally but little curled, and have elongated segments, with long, generally serrated spines: the first cirrus varies in proportional length; the second and third cirri have both their rami more thickly clothed with spines than are the three posterior cirri, the spines being generally arranged in three or four longitudinal rows: / the cirri, however, of *S. villosum* in all respects resemble closely the cirri of *Pollicipes sertus* and *P. spinosus*.

The *caudal appendages* are uniarticulate, small, and clothed with spines: in *S. villosum*, however, differently from in all other allied forms, there are no appendages.

The *stomach*, in those species which I opened, is destitute of caeca. There are no filamentary appendages.

Generative system. The ova are nearly spherical, and remarkably large, as was stated to be the case in the introductory discussion, in which the larva of *S. vulgare*, in the first stage of development, was described: the ovigerous fraena are small. The testes are large, but the vesiculae seminales in some of the species extraordinarily small. *Scalpellum ornatum*, and perhaps *S. rutilum*, are unisexual; the other species are hermaphrodite, but most or at least some of the individuals, are furnished with complemental males. These latter are fully described under each species, so I will here only remark, that *S. ornatum*, which alone (excepting perhaps *S. rutilum*) is unisexual, has less claim than the other species to be generically separated: we have seen also, in Ibla, that similar sexual differences occur in two most closely allied species. It is very singular how much more some of the males and complemental males in Scalpellum differ from each other, than do the female and hermaphrodite forms; this seems due to the different stages of embryonic development at which the males have been arrested. In the males, however, of *S. rostratum*, *S. Peronii*, and *S. villosum*, compared one with another, but not with the males of the other species, the parts of the mouth and apparently the cirri, resemble each other more closely, than do the same organs in the hermaphrodites. At the end of this genus I shall give a summary on the highly remarkable sexual relations both in Scalpellum and Ibla.

Distribution. The species seem distributed over the whole world, but as far as we can trust our present scanty materials, are most common in the warmer temperate regions. The *S. vulgare* ranged from the Norwegian seas to Naples. Most of the species are inhabitants of deep water. /

Affinities. In the preliminary remarks, we have seen how this genus blends into Pollicipes; and under the head of Oxynaspis, I have shown its close affinity to that genus. If, indeed, we take *Pollicipes spinosus*, and destroy all but six of the already minute and almost rudimentary latera, we shall, as far as the capitulum is concerned, convert it into a Scalpellum, closely similar to *S. villosum*. If we take any species of Scalpellum (excepting *S. villosum* and *S. rutilum*), and destroy all the valves, but the scuta, terga and carina, we shall convert it into an Oxynaspis. Lastly, I have shown under Ibla, that in several most remarkable peculiarities of structure, there is a manifest affinity between Scalpellum and that genus.

Geological history. Full details on this subject have been given in the *Memoirs of the Palaeontographical Society*. I will here only state, that the oldest known form of Scalpellum occurs in the Lower Green Sand.

[†SUB-CARINÂ NULLÂ]

1. SCALPELLUM VULGARE

PLATE V. FIG. 15

SCALPELLUM VULGARE. Leach, *Encyclop. Brit. Suppl.*, vol. iii, 1824
LEPAS SCALPELLUM. Linn, *Systema Naturae*, 1767
LEPAS SCALPELLUM. Poli, *Test. utriusque Siciliae*, Pl. vi, fig. 16, 1795
POLLICIPES SCALPELLUM. Lamarck, *An. sans Vertebres*, 1818
POLYLEPAS VULGARE. De Blainville, *Dict. Sc. Nat.*, Plate, fig. 4, 1824
SCALPELLUM LAEVE, var. Leach, *Zoolog. Journal*, vol. ii, p. 215, 1825
SCALPELLUM SICILIAE, var. Chenu, *Illust. Conch.*, Pl. iv, fig. 9
SCALPELLUM VULGARE (et var.). Brown, *Illust. of Conch.*, Pl. li, figs 7 to 20, 1844

S. (Herm.) valvis 14, si rostrum paene rudimentale includatur: lateribus superioribus inaequaliter ovatis. /
(Herm.) Capitulum with 14 valves, including the rudimentary rostrum: upper latera irregularly oval.

Mandibles, with four or five teeth: maxillae, with the edge straight, bearing numerous spines.

Complemental male flask-formed, with four rudimentary valves; no mouth; cirri not prehensile; attached to the occludent margin of the scutum, near the umbo.

Great Britain, Ireland, France, Norway, Naples. Attached to horny corallines, at from twenty to thirty, sometimes even to fifty fathoms in depth, according to Forbes and MacAndrew.

HERMAPHRODITE

Description. Capitulum much flattened with the apex produced, of a pale brown colour, sometimes faintly tinted purple, composed of fourteen valves, of which the rostrum is rudimentary and barely visible externally; valves thin, white, translucent, smooth, slightly marked by the lines of growth, separated from each other by rather wide interspaces of colourless membrane, which is thickly clothed by small, articulated spines of unequal length. The valves, excepting sometimes their umbones, are also covered with membrane, bearing spines, placed in rows parallel to the lines of growth; the spines are particularly numerous round the orifice of the sack.

Scuta slightly convex, thrice as long as broad; upper part much acuminated; occludent margin almost straight; basal margin nearly at right angles to the occludent margin; the tergal margin is separated from the lateral margin by an angle more or less prominent; a slight curved ridge runs from the umbo to this angle, and this deserves especial notice, inasmuch as it indicates the outline which the valve assumed in its earliest growth, and which is permanently retained in most of the older fossil species. Along the occludent margin, there is a trace of a ledge, developed in a variable degree, and which is noticed only on account of the plainly visible ledge along this same margin, in the allied genus Oxynaspis. The / umbo, or centre of calcification, is seated close to the occludent margin, and at about one-fourth of the length of the valve from the apex. Internally (fig. 15a', Pl. V), the part above the umbo is flat; and beneath this upper part, there is a large rounded hollow (d) for the adductor muscle: a fold or indentation (a) running downwards from the umbo, extends in a very oblique line across the occludent margin. this fold is of high interest as giving lodgment to the complemental males, and will hereafter often be referred to.

Terga, triangular, flat; occludent margin, very slightly arched.

Carina, much bent, with the umbo placed at barely one-third of the entire length of the valve from the apex. Two very slight ridges can be perceived, one on each side, running from the umbo to the basal margin, and separating the roof from the parietes of the valve; these ridges are of great use in distinguishing the fossil carinae of Scalpellum, from the carinae of Pollicipes. The part above the umbo is formed by the upward production of a marginal slip along each side of the valve, which slips in the fossil species (C in the woodcut, fig. 1, given in the Introduction), I have designated as the intra-parietes. The lower part of the valve gradually widens from the umbo downwards; internally, the whole is deeply concave, and continuously curved. The angle varies at which the upper and lower portions externally meet each other; but is never less than 135°. The upper part of the carina runs up between the terga for three-quarters of their length; the basal margin does not extend down low enough to pass between the carinal latera.

Rostrum (fig. 15b', seen externally, and highly magnified), minute, almost hidden by the enveloping membrane and by the small prominent umbones of the rostral latera; in area equalling about one-fourth of the rostral latera; externally pyramidal, with the upper side rather longer than the lower; internally slightly concave, square, with the upper margin and sometimes with the lower / margin, slightly hollowed out. Umbo of growth nearly central.

Upper latera, flat, irregularly oval, with an almost rectangular shoulder under the basal angle of the terga; in area, about one-third larger than the largest valve of the lower whorl; the exact degree of elongation of the oval figure varies a little. Umbo seated a little above the central point.

Lower whorl – rostral latera, nearly twice as long as broad, lying under the basal margins of the scuta: umbo seated over the rostrum; opposite end, towards which the valve widens either sensibly or but little, is either square or rounded; in area, less than any of the other valves, excepting the rostrum; in breadth, equalling either half or one-third of the height of the infra-median latera; growth, directed chiefly towards the infra-median latera. The freely projecting umbo is about one-sixth part of the entire length of the valve.

Infra-median latera, rather larger than the carinal latera; their shape varies from elongated pentagonal with the angles rounded, to oval, with the longer axis directed upwards. The umbo is seated a little above the middle of the basal margin, so that there is some little growth downwards, but the main growth is upwards. The upper point generally stands a little above that of the carinal latera.

Carinal latera, flat, less in area than the infra-median latera; basal margin nearly straight; carinal margin slightly hollowed out, terminal margin arched and protuberant. The umbones of the two valves almost touch each other under the middle of the carina; main growth towards the infra-median latera and upwards; umbones projecting not above one-fifth of the entire length of the valve.

Peduncle, much flattened, rarely as long as the capitulum, with the upper end nearly as wide as it; the lower end is either blunt, or tapers to a very fine point. The calcareous scales are transversely elongated, and are about / four times as wide as high; their internal surfaces are slightly concave, and their external, convex; the two ends are pointed. Viewed internally, the scales approach in shape to rhomboids. There are, in a medium-sized specimen, about twenty scales in each whorl, their tips overlapping each other: the whorls are placed not very near each other and at rather unequal distances, except round the uppermost part, where, being in process of formation, they are packed closely together. The membrane uniting the scales, supports numerous transverse rows of articulated spines, varying from 1/100th to 1/500th of an inch in length, and each furnished with a long sinuous tubulus, 1/10000th of an inch in diameter, running through the membrane to the underlying corium.

Attachment. Specimens are attached to various horny corallines, and occasionally to the peduncles of each other.[49] In both cases, supposing the coralline to be erect, the capitulum is placed upwards, with its orifice towards the branch to which it is attached, and consequently with its carina outwards. Where several are crowded in a group, their peduncles often become twisted and their positions irregular, with their orifices facing in any direction. This uniform position is simply the consequence of the larva attaching itself head downwards, and from the position of the prehensile antennae, necesarily with its sternal surface parallel and close to the branch of the coralline; hence the dorsal surface, which afterwards is converted into the carina, faces outwards. The peduncle, as already stated, often tapers, at its basal extremity, to a sharp point. In very young specimens, for instance in one with a capitulum only 1/20th of an inch in length, the method of attachment is the same as in Lepas and many other genera, namely, by cement proceeding exclusively from the antennae of the larva; but in older and full-grown specimens, instead of the whole / bottom of the peduncle becoming flattened and broadly attached, which would be here impossible, the cement is poured out through a straight row of orifices along the rostral edge, thus causing, by an excellent adaptation, a narrow margin to adhere firmly to the thin and cylindrical branches of the coralline. These orifices are represented, magnified seven times (in Pl. IX, fig. 7), in which the lower attached portion of the
peduncle is split open and exhibited; they are circular, and stand at regular intervals, in a straight line; the higher orifices are larger, but further apart from each other than the lower ones; in one full-grown specimen, I counted ten of these orifices in a length of exactly a quarter of an inch. At each period of growth, the corium recedes a little from the attached portion of the peduncle; of which portion, the

[49] Mr Peach (*Transact. Brit. Assoc.*, 1845, p. 65), states that this is sometimes the case in Cornwall; and I have seen a similar instance in a fine group from Naples.

greater part is thus left empty and as incapable of further growth, as are the larval antennae at the extreme point: in the specimen figured, the corium extended a little below the upper orifice. The prehensile antennae, however, I must remark, do not strictly rise from the extreme point of the peduncle, but at a little distance from it, on the rostral surface; this simply ensues from the antennae in the larva, being situated on the sternal surface, close to, but not actually on the front of the head. The two cement glands are seated high up on the sides of the peduncle, and remote from each other; they are small, unusually globular and transparent. The two cement ducts (fig. 7a a) proceeding from them, are 3/2000ths of an inch in diameter, and run in a zigzag line; at the point where they pass through the corium to enter the lower attached portion of the peduncle, they become closely approximated, and partially imbedded in the membrane of the peduncle. Together they run along the rostral edge, giving out through each orifice a little disc of brownish cement, and finally they enter the larval antennae. The peduncle, just above the attached portion, where still lined by corium, no doubt increases in diameter at each period of growth, and must, I presume, / become pressed against the almost parallel branch of the coralline. The corium, at this same period, shrinks, or is absorbed, and the two cement ducts come in contact with, and adhere to, the inner surface of the outer membrane of the peduncle; and then, by a process which I do not understand in this or any other cirripede, apertures are formed both in the ducts and through the membrane, so that the cement passes through, firmly fastening the outer surface of the peduncle with its calcareous scales and spines, to the coralline.

The structure of the larval prehensile antennae will be most conveniently described when we come to the complemental male; and figures (10–12, Pl. V) will be given.

Size and colours. Montagu states (*Test. Brit.*, pl. 18) that British specimens rarely have a capitulum 0·62 of an inch in length; I have, however, seen an Irish specimen, 0·7 long; and several specimens, from the Bay of Naples, 0·8 long, and including the peduncle, 1·3 in length. The valves in all the specimens are white, and the membrane connecting them either nearly white, or dirty pale yellowish, or purplish-brown. Within the sack the corium under the valves is tinted pale purple, and two very faint bands of the same colour can generally be distinguished running down the two sides of the peduncle. Body, coloured yellowish-white, with the upper segments of the pedicels of the cirri, tinted in front with purple.

Body, much flattened, the prosoma is very little developed; the mouth placed far from the adductor muscle, and is directed in a remarkable manner towards the ventral surface of the thorax: the first pair of cirri stands far separated from the second pair.

Mouth. Labrum with the upper part highly bullate, forming an overhanging projection equalling the longitudinal axis of the mouth; basal margin much produced; crest with a row of bead-like teeth.

Palpi rather small, with their external margin straight, and internal margin oblique: the bristles on the two palpi just meet each other. /

Mandibles, with five or six teeth, with the second (or second and third, when there are six teeth), smaller than the others; in two specimens, there were five teeth on one side and six on the other; inferior angle rather broad and strongly pectinated.

Maxillae with the edge nearly straight, without any notch, but with the inferior portion very slightly projecting; there are twelve or thirteen pairs of unequal spines, of which some of the middle ones are rather longer than the others, and almost as long as the two upper great spines.

Outer maxillae. On the inner margin the bristles are divided into two separate tufts; exteriorly, near the base, there is a distinct rounded swelling with bristles. The olfactory orifices are highly protuberant, approximate, flattened, scarcely tapering towards their upper ends.

Cirri. The five posterior pair are elongated, very little curled, with short pedicels; their segments are long, not at all protuberant in front, bearing five or six pairs of long, slightly serrated spines, with a very minute tuft of bristles between each pair, and with some short lateral spines on the inner side of each segment; on the fourth pair of cirri, these lateral spines are considerably developed; dorsal tufts consist of fine spines, with one much longer than the others. *First pair* short, separated by a wide interval from the second; rami unequal in length, by between two and four segments; longer ramus having nine segments, scarcely half as long as the rami of the second cirrus; shorter ramus with seven segments; in the same individual there were twenty segments in the sixth cirrus. The segments in the shorter ramus of the first cirrus are oblong in a transverse direction, and may be compared to a set of shields placed transversely and strung together; in the longer ramus the segments are longitudinally oblong; in both they are thickly covered with spines. *Second cirrus*; the anterior ramus is a little broader than the posterior ramus, with the segments bearing about five rows of bristles; / fifteen segments in the shorter ramus. *Third pair*, with the two rami equal in thickness, and with the segments differing very little from those of the posterior cirri, excepting that the serrated spines in the external lateral rows are rather larger. The fourth pair is remarkable by having, on the inner side of the upper edge of each segment, a little tuft of minute smooth spines, flattened, and a little enlarged near their ends, so as to be spear-shaped; I could not see these singular spines on the other cirri. The lower segments of the pedicels of all the cirri, excepting the sixth pair, are remarkable from having their inner edges, in the middle, produced into a considerable, abrupt, rounded projection, irregularly covered with spines.

Caudal appendages (Pl. X, fig. 21), very small, flattened, of nearly the same width throughout; in a medium-sized specimen, only $1/100$th of an inch in length; each bears from ten to twenty small bristles placed distantly from each other, of which those on the rounded apex are the longest.

Generative system. The penis is remarkably acuminated; the vesiculae seminales are unusually small, and enter only for a short distance into the prosoma; the testes are large. The ovarian tubes are of large diameter; the ova are nearly spherical and large, namely, $9/400$ths of an inch in diameter; they are not numerous, and lie in single layers in the two lamellae. The ovigerous fraena are well developed, and lie under the scuta; one I measured was $5/100$ths of an inch in length and $2/100$ths in width; the margin is obliquely truncated and slightly sinuous. This species breeds late in the autumn, and even in mid-winter; I have examined a specimen from Cornwall with ova containing larvae; taken on the 26th of October; again, in another specimen from Belfast, sent to me by Mr Thompson, taken in January, there were ova in the lamellae, and therefore no doubt impregnated; and on February the 12th I received from Mr Peach from Cornwall, specimens so very young that they must have become attached during the first days of the month. /

Varieties. The specimens from near Naples (which I owe to the kindness of the Rev. F. W. Hope), are somewhat larger, and differ slightly from those of Britain: they form, I imagine, the *S. Siciliae* of Chenu. After carefully examining them internally and externally, I think it is quite impossible to consider them specifically distinct, for although in several specimens, the valves were placed a little further apart from each other – the upper latera a little more elongated – the carinal latera rather narrower in their upper half – the infra-median latera rather more rounded – and, lastly, in the scuta, the tergal margin extended almost in the same line with the lateral margin; nevertheless in other specimens, I could perceive no difference whatever. It is, however, remarkable that in several full-grown Neapolitan specimens there were no complemental males, whereas I have never seen a single full-grown British specimen without such being present. In some specimens in the British Museum, without any given locality, I have observed considerable variation in the breadth of the carinal and rostral latera.

COMPLEMENTAL MALE
PLATE V. *FIGS 9–14*

When first dissecting *Scalpellum vulgare*, I was surprised at the almost constant presence of one or more very minute parasites, on the margins of both scuta, close to the umbones: these are represented, but rendered darker and therefore more conspicuous than in nature, in the drawing (Pl. V, fig. 15), which is three times the natural size. I carelessly dissected one or two specimens, and concluded that they belonged to some new class or order among the Aticulata; but did not at that time even conjecture, that they were cirripedes. Many months afterwards, when I had seen in Ibla, that a hermaphrodite could have a complemental male, I remembered that I had been surprised at the small size of the vesiculae seminales in the hermaphrodite *S. vulgare*, so that I / resolved to look with care at these parasites; on doing so, I soon discovered that they were cirripedes, for I found that they adhered by cement, and were furnished with prehensile antennae, which latter, I observed with astonishment, agreed in every minute character, and in size, with those of *S. vulgare*: the importance of this agreement will not at present be fully appreciated. I also found, that these parasites were destitute of a mouth and stomach; that consequently they were short-lived, but that they reached maturity; and that all were males. Subsequently the five other species of the genus Scalpellum were found to be present more or less closely analogous phenomena. These facts, together with those given under Ibla (and had it not been for this latter genus, I never probably should have even struck on the right track in my investigation), appear sufficient to justify me, in provisionally considering the truly wonderful parasites of the several species of Scalpellum, as Males and Complemental Males. When these parasites are fully described, will be the proper time to discuss and weigh the evidence on their sexual relations and nature. I will now describe the parasite of S. *vulgare*.

General appearance. Shape, flask-like, compressed (Pl. V, fig. 9, magnified 36 times), with a short neck: the outline is usually symmetrical, but sometimes is a little distorted on the underside. The creature is imbedded more than half its length or depth in the transparent, spine-bearing chitine border of the scutum of the hermaphrodite. Its length, or longer axis, varies from $^{10\text{ to }11}/_{400}$ths; its breadth, or transverse axis, is $^{6\text{ to }7}/_{400}$ths; and its thickness, for it is much flattened, is only $^{4}/_{400}$ths of an inch. On the summit, there is a fimbriated orifice *a* the size of which can rarely be made out quite distinctly, owing to the extreme thinness of the membranous edges. A little way beneath the orifice, there are four little blunt, bristly points *b*, generally rather more than the $^{1}/_{1000}$th of an inch in length; they are rather variable in size, and seem to be of no functional importance; directly beneath them, *l* there are four little calcareous beads (as may be known by their dissolving with effervescence in any acid, and breaking easily under the needle); these are $^{3}/_{2000}$ths of an inch in their larger external diameter; they are rather deeply imbedded in the outer integument, and taper a little downwards ending in a concave terminal point, into which a minute tubulus enters, like those passing into and through the valves of ordinary Cirripedia: along the axis of imbedment, there are often $^{4}/_{2000}$ths of an inch in length. These calcareous beads of rudimentary valves are seated in pairs, at the two ends of the flattened animal, so that when the animal is laid on one side, the upper bead in each pair exactly covers and hides the lower one. The outer integument is composed of chitine, as may be inferred from boiling caustic potash having no effect on it; the upper part is thicker than the imbedded portion and is wrinkled transversely; it is covered with minute spines $^{4}/_{10,000}$ths of an inch in length, either single or in groups of two and three (Pl. V, fig. 14). This outer tunic is lined by corium, sometimes slightly mottled with dull purple; and this by delicate longitudinal, striaeless muscles, running from the base up to the under edge of the orifice; these longitudinal muscles are crossed, at least, in the upper part, by still finer transverse muscles.

Thorax and abdomen. When the external integument is cut open, the thorax (Pl. V, fig. 13) is found lodged within an inner sack or rather tube, extending from near the bottom of the animal, up to the external orifice. The whole thorax is sometimes forced through the orifice, owing perhaps to the action of the spirits of wine and consequent endosmose, and is thus well displayed without dissection. The thorax tapers a little, is much flattened and straight; its length, together with the terminal abdominal lobe, is about 9/400ths of an inch; it is formed of very thin, most finely hirsute membrane, transversely wrinkled and so extensible, that when everted by the internal muscles being seized, it stretches to twice its former length; in this condition, five transverse articulations are / displayed. The abdominal lobe is smooth, and cannot be stretched, or turned inside out by pulling the above muscles. On the thorax, corresponding with the interspaces between the five transverse articulations, there are four pair of short limbs, but their bases, I believe, are prolonged across the inner or ventral surface of the thorax, so as almost to touch each other. These limbs, I believe, have no articulations, except, perhaps, where united to the thorax. The anterior or lowest limb, on each side, supports two or sometimes only a single spine; this pair is rather smaller than the second, and is placed a little more distant from it, than are the upper pairs from each other. The second pair differs from the upper two, only in having its three spines a very little shorter. The two upper or posterior pair exactly resemble each other; each has two spines on the summit, and a third seated lower down, on a little notch on the outer side, but with its point on a level with the others. The points of the spines of the two upper limbs, stand on a level with the external spines at the end of the abdomen. All the spines are of excessive tenuity and sharpness; they are straight, long, and not plumose.

The abdominal lobe is square, and from not being wrinkled, has a different appearance from the thorax: on each of the posterior angles, there are three moderately long, very sharp spines, with the tips of the outer pair bent a little inwards; in the middle between them, there are two little spines, and a little below and outside these latter, on the ventral surface, there are two other longer spines with their tips bent inwards; and again, lower down, two other pair, one beneath the other, of short spines. Perhaps, the three pair of spines on the ventral surface, mark the three segments, which are distinct on the abdomen of the larva in the last stage of its development, in Lepas and other genera. In the same way, it is probable that the lateral spine on the notch in each limb, marks the point where, in the larva, there is an articulation. Altogether, there are seven / pairs of spines on the abdomen, and eleven pairs on the thoracic limbs.

A little way beneath the lower or anterior pair of limbs, the thorax is abruptly bent, and becomes confluent with the lower internal parts of the whole animal. Here, the very delicate membrane of chitine which lines the sack or tube, extending from the external orifice, can be seen to be continuous, as in all cirripedes, with the outer tunic of the thorax. Within the thorax, there are some longitudinal muscles, without transverse striae, which, I believe, enter the short limbs, but not the abdomen, as I infer from the latter not being everted when they are pulled. At their lower ends these muscles terminate abruptly, and from being contracted are often a little enlarged. They extend a short way beneath the lower pair of limbs, and are, I suspect, attached to the outer integument of the animal, near the base.

After the most careful dissection of very many specimens, and their examination in many different methods (as by caustic potash, etc.), I can venture positively to assert that there is no vestige of a mouth, or masticatory organs, or stomach: I did not see any anus, but I will not affirm that such does not exist.

In the upper part of the animal, lying under the superficial muscles, and close beneath the upper line of their attachment, I found in all the specimens, an eye, of a pointed oval form, rather less than the $1 1/12,000$ths of an inch in diameter, formed of an outer capsule, lined with purple pigment cells, and surrounding, as it appeared, a lens. The eye is not introduced (in fig. 9), for I could not see it, except by dissection, and therefore do not know its exact relative position.

Generative system. The contents of the animal, between the sack containing the thorax and the outer integuments, and directly under the thorax, varied much in condition: in young and lately attached specimens the whole consisted of a pulpy mass with numerous oil globules; in other specimens, apparently more mature, / there were vast numbers of cells, sometimes cohering in sheets, about $3/10,000$ths of an inch in diameter, and having darkish granular centres; these I believe to be the testes, for in a specimen presently to be mentioned, in which the vesicula seminalis was gorged with spermatozoa, I found adhering to its outside, a mass of cells of exactly the same diameter, but now empty and transparent instead of having brownish centres. Lastly, in several other specimens, at the very bottom of the sack-formed animal, there was a brownish, pear-shaped bag, of different sizes in different individuals, and occasionally broader even than the thorax. This bag contained either pulpy matter, or a great mass of spermatozoa. Before being disturbed, these spermatozoa lay parallel to each other in flocks, and they yielded to the needle in a peculiar manner, so that I found (having had experience with these bodies in living Cirripedia) I could almost tell before examination under the compound microscope, whether or not I should see spermatozoa. Many had distinct heads,[50] which were two or three times as broad as the filamentary bodies; the latter when placed between glass were the 1/20,000th of an inch in diameter. I compared these spermatozoa with others taken out of the vesiculae seminales of the individual hermaphrodite *S. vulgare*, to which the parasite was attached, and could not perceive the slightest difference in them. The brownish pear-shaped bag, or vesicula seminalis, the coat of which seems fibrous, could sometimes be distinctly traced, sending a chord or prolongation far up the thorax: at the end of the abdominal lobe, no doubt there is an orifice; and this, I believe, I once distinguished. Owing to this chord, the bag often / adheres to the thorax, when the latter is dissected out of the general integuments; in this condition, I twice clearly made out that it was single: in one other specimen, however, there appeared to be two small vesiculae seminales. By using a condenser and very brilliant light, the outline of the vesicula seminalis could sometimes be distinguished before dissection, at the bottom of the sack-formed animal; and such was the case in the specimen drawn (in fig. 9).

Although I have dissected, at least, thirty specimens, taken at different times of the year, and from different localities, and when many of the specimens were mature and ready for the impregnation of ova, as clearly shown by the presence of innumerable spermatozoa, I have never seen even a trace of an ovum or ovaria.

[50] I do not understand the development of the spermatozoa in Cirripedia: in a recent Chthamalus and Balanus, I found the greater number had a little filament in front of the head or nodular enlargement, which latter varied in size and in shape from globular to that of a spindle. The filament before the head, also, varied in proportional length; it did not project in exactly the same straight line with the hinder part, and some of the spermatozoa were entirely without this filament in front; such is the case with the spermatozoa here described.

Antennae and attachment. The prehensile antennae (Pl. V, fig. 10), are seated a little above the very base of the sack-like animal; and this might have been expected from the antennae in the larva, being seated on the ventral surface, not at the very extremity of the head. By a very strong light, they can sometimes just be seen whilst the parasite is attached to the hermaphrodite (the scutum of the latter having been cleaned on the underside), and are thus represented (in fig. 9). They are formed of thicker membrane than the general integument of the body: the second segment, or disc, is pointed and hoof-like; when seen in profile (fig. 11), the upper convex surface has a uniform slope with the upper surface of the basal segment; it is furnished with a single backward pointing spine, attached, I believe, on the underside, nearly opposite the articulation of the ultimate segment: at the apex, there are some excessively minute hairs or down. The ultimate segment projects rectangularly outwards as usual, and has on its inner side, rather beneath the middle, a conspicuous notch (fig. 12), which bears two or three long, non-plumose spines; on the summit there are three or four rather shorter spines. On the outside of the great basal segment there is a single spine curving backwards. / The importance of the following measurements (in fractions of an inch) will hereafter be seen.

Length of whole organ, from end of disc to the further margin of the oblique basal articulation	$38\text{--}39/6000$
Length of whole organ, to the inner margin of the oblique basal articulation	$1/6000$
Breadth of basal segment, measured halfway between the basal and second articulations – the limb being viewed from vertically above	$8/6000$
Length of hoof-like disc, measured from the apex to the middle of the articulation with the basal segment	$9\text{--}10/6000$
Breadth of hoof-like disc, measured from the apex to the middle of the articulation with the basal segment	$5/6000$
Length of ultimate segment	$6/6000$
Breadth of ultimate segment beneath the notch	$7/20{,}000$
Breadth of ultimate segment above the notch	$5/20{,}000$

I did not see the cement ducts, which, perhaps, was owing to the corium extending from the inside of the whole animal some way into the antennae, thus rendering them rather less transparent than in common cirripedes. That the ducts and cement glands exist, is certain, for the antennae in every case were enveloped in a little irregular mass or capsule of the usual, brown, transparent, laminated cement. When several of these parasites were attached close together, the cement ran up between them.

I may here state, that I found on one Scalpellum, three males very lately attached, and not as yet imbedded in the chitine border; they were white, opaque, pulpy, and full of oily globules; the lower part was considerably more pointed, and extended further beyond the prehensile antennae, than in the older and imbedded specimens. There were distinct remnants of two great reddish-brown eyes, showing that in this respect the larvae of the male in their last stage of development, are characterized like the larvae of other Lepadidae. The male larva would, probably, be a little larger than the male itself; but yet compared with the larva in the earliest stage, there can have been unusually little increase of size during the several intermediate metamorphoses; I judge of this from the dimensions of the larva of the hermaphrodite in the first stage, namely, $9/400$ths of an inch, exactly the size of some of the smaller males. In the allied genus Ibla, / the increase is also less than is usual, namely, from $15/1000$ths of an inch, the diameter of the ovum, to only $25/1000$ths of an inch, the length of the boat-shaped larva, just before its final metamorphosis.

Habits and concluding remarks. The males are imbedded in the spinose chitine border of the occludent margin of the scuta, exactly over an oblique fold or notch (fig. 15á *a*), close by the umbo. This fold has no direct relation to the males, but being present is taken advantage of by them; for it occurs in the young hermaphrodite, before the attachment of the males, and in species of the genus in which the males are attached to other parts. It occurs, also, in fossil species of Pollicipes, and in these it seems caused by the upper inner part of the valve being rendered more and more prominent during growth: in the present species, I suspect, its origin is connected with the formation of a ridge bounding the outer side of the pit for the adductor scutorum muscle: we shall see in the next species, that this fold is of the highest importance in relation to the position of the males. The transparent chitine border of the scuta is broad, and fills up the fold in the shell, so that the outline of the occludent margin is not affected by it: in the drawing (fig. 9) some of the inner layers of chitine *e e*, which dipped into and filled up the fold, have been removed, that the lower part of the animal might be more plainly exhibited. The chitine bears numerous spines of various lengths, which must afford some protection to the males, rudely arranged in lines, parallel to the edge of the valve, indicating the successively formed layers of chitine; each spine has a fine, tortuous tubulus connecting its base with the underlying corium. The extreme outer edge of the border is thin, forming a kind of lip, close beneath which the delicate tunic lining the sack is attached. During continued growth, the valve is added to in thickness, and so is the chitine border, and likewise in breadth. It appears that the larva of the male must attach itself on the underside / of this border, on the edge of the tunic of the sack, and that by the action of the cement, the corium beneath is killed (as I believe always is the case with other parasitic Cirripedia), whereas on both sides, the chitine continues to be added to, so that the male, excepting the upper and always projecting portion, becomes imbedded at first laterally, and ultimately all round: I have seen specimens in several different stages of imbedment. Hence, in old specimens, with a thick and broad chitine border, it might and does come to pass that one male is imbedded (the valve being laid flat) directly beneath another.

I have examined a great number of specimens from various localities, taken at different times of the year – some dozen specimens from Cornwall,[51] and several from unknown localities in various collectons; some from Ireland, from the Shetland Islands, from Norway, and from near Naples. Every one of these specimens, with the exception of some of the Neapolitan ones, had parasitic males attached to them: I must also except very young specimens, on which they never occur. On a Cornish specimen, with a capitulum a little more than one-fifth of an inch in length, it may be mentioned as unusual that there were three males. In young specimens there is generally one male on each scutum, but sometimes there are two, and sometimes none on one side. In large old Cornish specimens I have counted on the two sides together, six, seven, and eight males, and in one Irish specimen no less than ten, seven all close together on one valve and three on the other, but I do not suppose that all these were alive at the same time. In the Neapolitan specimens, however, which are the largest that I have seen, there was in no case more / than two; and out of seven or eight specimens, four had not any male; so that it would appear there is something in this locality hostile to the development of the parasitic males. I have noticed only one instance (that given in fig. 9) in which the males were imbedded a little way apart; generally they touch each

[51] I am greatly indebted to Mr Peach for his unwearied kindness in procuring me fresh specimens. Mr W. Thompson allowed me to dissect one, possessing particular interest, out of his three Irish specimens. Professor Forbes procured me a specimen from the Shetland Islands, and Professor Steenstrup was so kind to take pains to send me some Scandinavian specimens.

other, and are cemented together: where there are several males, they occur at different levels, as measured from the under or upper surface of the chitine border: in one instance of four males adhering to one valve, I distinctly perceived that the lowest one was white, pulpy, and recently attached; the two above, which were placed close together and between the same laminae of chitine, were mature; and the third still higher up, was dead, empty, transparent, and half decayed: in some other instances, I have found the uppermost parasites dead, and, together with the surrounding chitine, partially worn away.

The larva of the male must have a different instinct from the larva of the hermaphrodite; for the latter attaches itself head downwards to a coralline, whilst the male larva crawling on the scuta of the hermaphrodite, discovers, I presume by eyesight, the fold in the shell beneath the translucent border of chitine, and there invariably attaches itself. Its object in choosing this particular spot, I believe, simply is that the depth or thickness of the chitine is there greater, and sufficient for its imbedment, which would hardly be the case elsewhere. This parasite has, as we have seen, no mouth or stomach, and indeed, considering its fixed position and the non-prehensile condition of its limbs or cirri, a mouth would have been of no service to it, without it had been extraordinarily elongated. The male must live on the nourishment acquired during its locomotive larval condition; and its life no doubt is short, but yet not very short, as I infer from the depth to which mature specimens are buried in the chitine border. The full development of the spermatozoa consumes, I suppose, some considerable lapse of time. The thorax and limbs, though / furnished with muscles, are obviously, as already remarked, of no use for prehension; these parts serve, probably, to defend the little creature, when its eye announces the passing shadow of some enemy, and for this purpose they are well adapted from the extreme sharpness of the spines. The thorax, into which I traced the vesicula seminalis, no doubt also serves for the emission and first direction of the spermatozoa; and hence, perhaps, its singularly extensible structure. I have already remarked, that in specimens preserved in spirits, the thorax is often largely protruded, and bent down at right angles to the orifice. I presume this is caused by endosmose; nevertheless it deserves notice, that it was in these protruded specimens that the vesicula seminalis was most conspicuously gorged with spermatozoa. I suspect the longitudinal and transverse muscles lining the upper part of the outer integuments of the whole animal, can be of little use to the creature, without it be to aid in the protrusion of the thorax, and perhaps in the violent expulsion of the spermatozoa, thus causing them to reach the ovigerous lamellae within the sack of the hermaphrodite. It is also probable, that the action of the cirri of the hermaphrodite, would tend to draw inwards the spermatozoa in the right direction. In one specimen, the spermatozoa in the hermaphrodite and in the male were mature at the same time; in another this was not the case; and as the males, apparently, become attached at all periods of the year, this want of coincidence in maturity must often occur. Can the males retain their spermatozoa, till told by some instinct, that the ova in the sack of the often fecundated hermaphrodite are ready for impregnation; or are the spermatozoa sometimes wasted, as must annually happen with such incalculable quantities of the pollen of many dioecious plants?

This little cirripede is, in many respects, in a partially embryonic condition. There is no separation between the capitulum and peduncle; there is no mouth; and the thorax, throughout its whole width, opens into the anterior / part of the animal: the limbs differ greatly from those both of the mature cirripede and of the larva, but come closest to the latter: the preservation of the abdomen is a well-marked embryonic character. On the other hand, the four rudimentary calcareous valves, the narrow orifice, the hirsute outer integument, the two muscular layers, the single eye, and male internal organs, are all characteristic of the fully developed condition. The four little valves, as I believe, represent the scuta and terga, though they are placed considerably below the orifice: the little bristly points have no homological signification, and are absent in the male of the following closely allied species. The four pairs of limbs answer to the four posterior cirri, as may be inferred from their proximity to the abdominal lobe, and from the three posterior pairs closely resembling each other, and differing a little from the first pair; this latter pair corresponds with the third pair in the hermaphrodite form of Scalpellum. If I am right in believing that only a single vesicula seminalis is ordinarily developed in the male, this is a special and singular character.

As stated in the beginning of this description, from the one great fact of the absolute correspondence of the prehensile antennae of the parasite, with those of the hermaphrodite *Scalpellum vulgare*, together with its fixed condition, its short existence, and exclusively male sex, I have thought myself justified in provisionally considering it as the complemental male of the cirripede to which it is attached; but I hope final judgement will not be passed on this view, until the whole case is summed up at the end of the genus.[52] /

2. SCALPELLUM ORNATUM

PLATE VI. FIG. 1

THALIELLA ORNATA. J. E. Gray, *Proc. Zoolog. Soc.*, p. 44, Annulosa, Plate, 1848.

S. (*Foem.*) *valvis 14, sub-rufis: lateribus superioribus quadranti-formibus, arcu crenâ profundâ notato.*

(Fem.) Capitulum with 14 reddish valves: upper latera quadrant-shaped with the arched side deeply notched.

Mandibles with three teeth; maxillae narrow, bearing only four or five pair of spines.

Males, two, lodged in cavities on the undersides of the scuta; pouch-formed, with four unequal, rudimentary valves: no mouth: cirri not prehensile.

Algoa Bay, South Africa. Attached to Sertularia and Plumularia. British Museum.[53]

FEMALE

Capitulum oblong, with the upper portion much produced; valves, 14, thick, naked, closely locked together, irregularly clouded with pale crimson; the membrane connecting the valves is not furnished with spines. On most of the valves there are furrows and ridges diverging from the umbones, and the lines of growth are plainly marked: in the valves of the lower whorl, the umbones are slightly protuberant.

Scuta, convex, unusually thick, oblong, quadrilateral, with the

---

[52] I trust, before long, that some naturalist, with more skill than I possess, will examine these parasites on *Scalpellum vulgare*, which unfortunately is the only species of the genus that can be easily obtained. Fresh specimens, or those preserved in spirits of wine, are necessary. The action of boiling caustic potash is very useful in cleaning the prehensile antennae. If these latter organs are sought in the hermaphrodite for the sake of comparison, young specimens, adhering to clean branches of a coralline, should be procured, and caustic potash used.

[53] I am greatly indebted to Mr Bowerbank for specimens of this extremely interesting species; also to Mr Morris, to whom Mr Bowerbank had given some of the original specimens.

occludent margin the longest; lateral margin slightly hollowed out. The umbo (and primordial valve) is situated at the uppermost point of the valve, and consequently the growth is exclusively downwards. On the under side (Pl. VI, figs 1b' and 1c'), in about the middle of the valve, there is a pit a for the adductor scutorum muscle, the depth and distinctness of which varies a little; / above the pit, and between it and the apex, there is a transverse, oblong, deeper depression b, within which, the male is lodged. A small portion of the apex of the valve projects over the terga.

Terga, large, nearly equalling the scuta in area, flat and subtriangular; the scutal margin is not quite straight. The apex of the valve is thick and solid, and must have projected freely for a length equalling one-third of the occludent margin.

Carina, laterally broad, angularly bent; slightly widening from the apex to the base; internally, deeply concave. The position of the umbo varies, in young specimens it is seated at the uppermost point, and consequently in such there is no upward growth; in older specimens, from the junction and upward production of that part on each side of the valve, which I have called in fossil specimens the intra-parietes, the valve is added to above the umbo, but to a lesser degree than in *S. vulgare*. Slight ridges separate the roof from the parietes, and the parietes from the intra-parietes.

Rostrum, minute, narrow, widening a little from the apex downwards, inserted like a wedge between the umbones of the rostral latera, and hardly projecting above their upper margins, so as to be easily overlooked: internally concave.

Upper latera (fig. 1a'), a quadrant-shaped, with a deep square notch cut out of the arched margin, which notch receives the upper point of the carinal latera; the surface of the valve between the notch and the umbo is depressed.[54]

Rostral latera, small, gradually widening from the umbo to the opposite end, which is obliquely rounded.

[54] The only valve which I have seen at all like this, is a fossil specimen from the Upper Chalk of Scania; this is described in my memoir on the Fossil Lepadidae (Palaeontographical Society), under the name of *Scalpellum solidulum* (Pl. 1, fig. 8, e, f) and is perhaps erroneously there considered as a carinal latus.

Infra-median latera, approaching to diamond-shaped, placed obliquely to the longer axis of the capitulum; or the upper part may be described as spear-shaped. /

Carinal latera: these appear as if formed of two valves united together; the upper portion, widening as it ascends in a curved line, terminates in a rounded margin, which enters the deep notch in the upper latera; the other and lower portion is shorter, and terminates in a square margin abutting against the infra-median latera; the umbones of the carinal latera project beyond the line of the carina.

Direction of the lines of growth in the valves. This should always be carefully observed, on account of the great diversity there is in this respect between the different species, especially when the recent are compared with the older fossil species; moreover, one of the chief characters between the genus Scalpellum and Pollicipes, depends on the direction of the lines of growth. In the scuta, terga, rostrum, and upper latera of the present species, the chief growth is downwards; in the carina, in mature specimens, it is both upwards and downwards; in the carinal latera, both upwards and towards the infra-median latera; in the infra-median latera chiefly upwards; and, lastly, in the rostral latera, towards the infra-median latera.

Peduncle, short, not half as long as the capitulum; calcareous scales imbricated as usual, tinged red, almost crescent-shaped, acuminated at both ends, of remarkable length, so that in each whorl there are only four scales: a full-sized scale equals in length one of the rostral latera. The tips of two scales, in one whorl, lie under the middle points of the carina and rostrum; and in the whorl, both above and below, a single much curved scale occupies this same medial position. The peduncle does not seem to have been attached in any definite position to the horny coralline, as is the case with *S. vulgare*.

Length of capitulum in the largest specimen 0·2 of an inch.

The *Mouth* is directed towards the ventral surface of the thorax. The *labrum* is far removed from the adductor muscle, with the upper part forming an overhanging projection; I believe there are some very minute bead-like / teeth on the crest. *Palpi*, small, narrow, thinly clothed with bristles.

Mandibles, with three teeth, of which the first is distant from the second; inferior angle not much acuminated, pectinated on both edges.

Maxillae, small, narrow, produced, without any notch, with two large upper spines, of which one is much thicker than the other; on the convex upper margin there are some minute tufts of very small hairs.

Outer maxillae, with few bristles, arranged in a continuous line on the anterior surface; on the external surface there is a tuft of long bristles. Olfactory orifices situated laterally, forming two flattened, tubular projections.

Cirri. First pair placed not far from the second; the three posterior pair not very long, with their segments elongated, not protuberant, bearing four pair of non-serrated spines, with a single short bristle between each pair; dorsal tufts small, with one spine longer than the others. First cirrus rather short, segments not very broad; second cirrus with the rami nearly equal in length, anterior ramus rather thicker than the posterior ramus, with three longitudinal rows of spines.

Caudal appendages. These are minute, rather broad, not half as long as the lower segments of the pedicels of the sixth cirrus, with four very long spines at the tip.

Penis. There is no trace of a prosciformed penis in the four specimens examined; and as this organ is present in every ordinary cirripede, with the exception of *Ibla Cumingii* which we know to be exclusively female, so we may infer with some confidence that the form here described is female, although it is impossible in specimens once dried to demonstrate the absence of the vesiculae seminales and testes.

Affinities. This is a very distinct species; it is, however, much more nearly related to *S. rutilum*, than to any other species; and next to this, to *S. vulgare*; from this latter species it chiefly differs in the large scales of the / peduncle, in the scuta not being added to at their upper ends, and in the membrane covering and connecting the valves being spineless; but there is a greater difference in the trophi and in the cirri. The peduncle of *S. ornatum* presents some resemblance to that of the singular cretaceous genus, *Loricula*.

MALE

All the specimens, as already stated, were dry, but in an excellent state of preservation, so that after having been soaked in spirits, they could be minutely examined. In the four which I opened, I found, in a transverse pouch on the underside of each scutum, a male lodged; in a fifth dead and bleached specimen, the cavities in the shell for the reception of the males, were present; and in a sixth young specimen, also dead, cavities were in process of formation. As compared with plants, the relation of the sexes in this species may be briefly given, by saying that it belongs to the class *Diandria monogynia*. I will first describe the males themselves, and then the cavities in the shell of the female. The males differ in every point of detail, from the complemental males of *S. vulgare*, but yet present so close a general resemblance, that a comparative description will be most convenient.

The general shape of the whole animal is rather more elongated, and I suspect flatter, but this latter point could not be positively ascertained in dry specimens. The entire length is greater, being in the largest specimen $13/400$ (instead of at most $11/400$), and the width, $7/400$ of an inch. The orifice is not fimbriated; the four bristly points over the calcareous beads are absent. The whole outer integument is much thinner, owing evidently to its protected position, and is not covered by little bristles, but with an extremely high power, minute points arranged in transverse lines can be distinguished. The calcareous beads, or rudimentary valves, are thin and regularly oval. It is remarkable that in all the specimens, two on one / side were smaller than the two on the other side – the smaller beads being $16/6000$, and the larger, $22/6000$ of an inch in diameter; therefore more than twice the size of one of the beads in *S. vulgare*, which are only $9/6000$ externally in diameter. From the position of the eye, close to one margin, near the upper end of the flattened animal, and from the manner in which the little limbs and spines lay between two of the beads at the opposite end, it was manifest that these latter, one large and one small, corresponded with the terga of the other cirripedes, and that the other two, near the eye, answered to the scuta. The valves being of unequal sizes on the right- and left-hand sides of the animal, is probably connected with one side being pressed against the hard, shelly valve of the female; in the same way as the valves in certain Paecilasmas, are smaller and flatter on the side nearest to the crustacean to which they are attached. The eye, in being slightly notched on the upper and lower edge, shows signs of really consisting of two eyes, which I believe is always normally the case; it is rather larger, in the proportion of 13 to 11, being $13/12,000$ of an inch in diameter, than in *S. vulgare*; and from the almost perfect transparency of the integuments, is far more conspicuous than in that species. Hence when the valves of the female are opened, the black little eye is the first part of the male which catches the attention. No vestige of a mouth could be discovered.

Thorax and abdomen. The thorax, as in *S. vulgare*, is highly extensible, and when stretched exhibits the same five transverse folds or articulations; when contracted, it is broader, so that even the truncated end of the abdomen is wider than the lower (properly anterior) end of the thorax in *S. vulgare*. Its thin outer integument is studded with excessively minute points in transverse rows. The four pair of limbs are longer than in *S. vulgare*, but the spines on them much shorter and thicker; each limb (including the first) supports three spines, of which one is seated on a notch low down on the outside, and / is longer than the other two; of these two, the one on the same side with the notch, is a little longer than the other. The spines on the first and second pair of limbs are considerably shorter than those on the third pair, and these latter, are a little shorter than those on the fourth or posterior pair. Hence, the spines on the thoracic limbs, compared with those of *S. vulgare*, present considerable differences, both in their relative and absolute dimensions. The abdominal lobe is in proportion rather shorter; its end is less abruptly truncated, and supports a row of, I believe, six moderately long, and basally thick spines; these spines are not so long as those surmounting the fourth pair of limbs. On both lateral margins of the abdomen, rather on the ventral face, there is a row of, I believe, seven long spines, but it is very difficult to count the spines in specimens which have been once dried. I was able to distinguish that the two lower pair of spines on the ventral surface, are seated a little way one below and within the other, as in *S. vulgare*. The abdominal spines altogether form quite a brush, and there are certainly several more than in *S. vulgare*, and those on the two sides are much longer.

Antennae. The disc is hoof-like, with the upper surface forming a straight line with the upper edge of the basal segment; the apex is pointed and clothed with some fine down; there is a single spine pointing backwards, which rises from the lower flat surface. The ultimate segment was hidden in laminae of cement; and I was not able to make out its structure. There is a single spine on the outer edge of the basal segment, in the usual position. The entire length of the limb, measured from the end of the disc to the further margin of the basal articulation, is $36/6000$ths of an inch; measured to the inner margin, it is $21/6000$ths of an inch; the disc itself is $12/6000$ths of an inch long; these measurements differ a little both absolutely and proportionally, compared with those of the antennae of S. *vulgare*.

Cavities in the scuta of the female for the reception / of the males. These extend nearly parallel to the tergal margin, transversely across the valves, for three-fourths of their width; they are seated above the depression for the adductor muscle, and are more conspicuous than it; they are deep and well defined, and each exactly contains one male. The males are placed with their orifices in a little notch in the occludent margin, and their prehensile antennae at the further end. The distance to which the cavities extend across the valve, and their distance from the upper or tergal margin, varies a little, but chiefly in accordance with the age of the specimens; for the valve continues to increase in width, whilst the size of the cavity remains the same. The occludent margin of the scutum in the largest female, was 0·1 of an inch in length; of another, in which there was a fully developed cavity, 0·084; of a third, in which there was no cavity, only a slight concavity, with a preparatory impression, the length of the occludent margin was 0·062. The larger and smaller of these three valves, are drawn of their proper proportional sizes (in Pl. VI, figs 1*b'*, 1*c'*). The preparatory impression (fig. 1*c'*, *b*), consists of a narrow, not quite straight, extremely slight furrow, of slightly irregular width, bordered on each side by a very minute ridge, which is distinctly continuous with the inner edge of the occludent margin, both above and below the cavity. The furrow appears to have been formed by calcareous matter not having been deposited along this line, during the thickening or growth of the internal surface of the valve: I suspect, that it originates at a single period of growth, for I could see no signs of successively formed transverse lines. I believe that it is strictly homologous with the fold, over which the complemental male is attached in *S. vulgare*, but carried, for a special purpose, much further across the valve and rectangularly inwards, for in structure and position both are identical. In comparing the internal views of the scuta in *S. vulgare* and *S. ornatum* (Pl. V, fig. 15*a'*, and Pl. VI, fig. 1*c'*), it must be borne in mind, that the latter should be compared, as clearly shown by / the lines of growth, with that portion alone of the scutum in *S. vulgare*, which lies under the curved ridge connecting the umbo and tergo-lateral angle. The deep cavity in which the male is lodged, is formed subsequently to the preparatory furrow, simply by the gradual thickening of the surrounding surface of the valve, more especially of a ridge just above the pit for the adductor muscle, and of another broad ridge just beneath the tergal margin. The deepest part of the cavity lies parallel to the tergal margin along the upper side, and here, in the older valves, the preparatory furrow can by care be distinctly traced. In conformity with the shape of the cavity, the orifice or notch in the occludent margin of the scutum, is situated at the point where the preparatory furrow sweeps round and enters. I believe that the cavity is lined by membrane, and that between the cavity and the body of the female, there is a complex membranous layer – a pouch or bag being thus formed. An imaginary section of this pouch (with the thickness of all the parts extremely exaggerated and in a reversed position) is given (in Pl. VI, fig. 1d'): a is the shell; x the cavity, converted, as I believe, into a pouch by, firstly, the delicate tunic c lining the sack of the female; secondly, a double layer d of corium; and, thirdly, by a special, rather thick membranous layer b, which thinning out round the cavity coats only part of the under surface of the scutum. This latter membrane I have not seen in any other cirripede, and I believe it is nothing but the tissue, here not calcified, which, in a calcified condition, ordinarily forms the valves. On this view, the males may be said to be lodged in pouches, formed in the thickness of the valves.

Concluding remarks. The males from the absence of a mouth (and no doubt of a stomach), must necessarily be short-lived, and, I suppose, are periodically replaced by fresh males.[55] In one instance, the remnants of the two / great compound eyes of the larva, could be seen at the end of the pouch, opposite the orifice. The larvae, I conclude, crawl in at the orifice, one side of which is formed, as we have seen, of yielding membrane, and scratch out the dead exuviae of the former occupant: certainly, the males are less firmly attached to their pouches, though some small quantity of cement is excreted, than are other cirripedes to the objects to which they are attached. The small size of the female, and her valves not being thickly edged with chitine, accounts for the males having pouches specially formed for them, instead of being, as in *S. vulgare*, laterally imbedded in the chitine border of the scuta. In hereafter weighting the evidence on the nature of the parasites in Ibla and in Scalpellum, the fact of the valves of the supposed female being here modified for the special purpose of

---

[55] It is possible, though opposed to all analogy, that the females may be short-lived, and breed only once, in which case the males would not have to be periodically replaced.

lodging the males, will be seen to be important. If we imagine the male parasites to be extraneous animals, and that by adhering to the sack of the Scalpellum, they injure the corium and thus prevent the growth of the shell over an area exactly corresponding to their own size, and so form for themselves cavities; yet what can be said regarding the preparatory furrows? surely these narrow lines cannot have been produced by the pressure of the much broader parasites. Must we not see in the furrows, the first marking out, if such an expression may be used, of the habitation for the male, which has to be specially formed by the independent laws of growth of the female?

3. SCALPELLUM RUTILUM

PLATE VI. FIG. 2

S. (Foem. an Herm.) valvis 14 subrufis: carinae tecto plano, utrinque cristâ rotundatâ instructo; margine basali truncato: lateribus superioribus latitudine duplo longioribus.

(Fem. or Herm.) Capitulum with 14 reddish valves: carina with the roof flat, bordered on each side by a rounded ridge; basal margin truncated: upper latera twice as long as broad. /

Mandibles with three teeth: maxillae narrow, bearing only four or five pair of spines: segments of the second and third pair of cirri with one side wholly covered with spines.

Males, two, lodged in hollows, on the undersides of the scuta; pouch-formed, with four [?] rudimentary valves; no mouth; cirri not prehensile.

Habitat unknown; associated with *Dichelaspis orthogonia*. British Museum.

FEMALE OR HERMAPHRODITE

There is only a single specimen in the British Museum, and this had nearly all its valves separated, and many of them in fragments: from its state of decay, I think the specimen must have been dead, when originally collected.

Description. The capitulum consists of fourteen valves, including from analogy a rostrum.[56] Valves, apparently covered with membrane, bearing some thin spines on the margins; clouded with a fine, though pale, orange tint; surfaces plainly marked with lines of growth.

Scuta, elongated, nearly three times as long as broad; apex, pointed; basal margin extremely oblique, forming an acute angle with the occludent margin; the lateral margin is slightly hollowed out, and is separated from the tergal margin by a large rectangular projection or shoulder. The occludent margin is nearly straight; externally, there is a slight ridge running down the middle of the valve, from the apex to the baso-lateral angle; and a second / ridge running from the apex to the tergo-lateral angle. The lines of growth do not end abruptly at the tergo-lateral angle, as is the case with *S. ornatum* and several fossil species, but run up a little way along the tergal margin. The umbo is seated at the uppermost point, and, therefore, the main growth is downwards. There is a large rounded depression for the adductor muscle (*a*, fig. 2*a*′), and higher up, opposite the tergo-lateral angle, there is another hollow *b*, for the lodgment of the males; this latter is of nearly the same shape as the hollow for the adductor muscle, but rather more conspicuous than it. From the appearance of the under surface of the scuta, it might readily have been thought, that there had been two adductor muscles.

Terga, of large size, longer than the scuta, flat, triangular, with the whole inferior part much produced and spear-like. A portion of the apex must have projected freely above the sack.

Carina (Pl. VI, fig. 2*b*′), simply bowed (i.e., not rectangularly bent), with the umbo (and primordial valve) seated at the upper point; rather massive, narrow, only slightly increasing in width from the upper to the lower end; the two sides are flat, and at right angles to the roof, which is bordered on each side by a rather broad, square-topped ridge (*see section* fig. 2*c*′); or the roof may be said to have a square-edged furrow running from the apex to the basal margin, and widening

---

[56] In my first, and as I thought, careful examination of the separated valves (my only materials) of this species, I mistook one of the triangular rostral latera for the rostrum, and hence was unfortunately led into an error in my *Monograph on the Fossil Lepadidae of Great Britain*, in which I state that the present species has only twelve valves in the capitulum; and I inferred from this, that *S. quadratum, S. fossula*, etc., had only twelve valves; I still believe this to be correct, but the existence of fourteen valves in *S. rutilum* and *S. ornatum*, the recent species to which the above fossils are most closely allied, no doubt is a strong argument in favour of this higher number.

downwards; these two ridges have their lines of growth oblique, and hence have a twisted appearance; the central depressed portion of the basal margin, which is square or truncated, descends lower down than the two ridges. The sides of the valve close to the apex are broad, and consist, as I believe, of intra-parietes, as well as of parietes, but these parts are not separated from each other by ridges, as is commonly the case, more especially with the fossil species. I have described the carina in some detail, on account of its resemblance to that of the cretaceous *S. fossula*, *S. trilineatum*, and *S. quadricarinatum*. /

Rostrum, unknown; but one probably existed.

Upper latera, of large size, elongated, quadrilateral, approaching to diamond-shaped, with the angles rounded, nearly twice as long as broad; almost flat; upper half acuminated, lying between the scuta and terga; the lower half broad, forming a rectangular projection lying between two latera of the lower whorl. The umbo is near the apex, the greater part of the growth being downwards, but the valve is added to a little, round the two sides of the apex; these additions do not take place in the early stages of growth (as explained under *S. vulgare*), and, therefore, they form a depressed rim.

Rostral latera, almost exactly triangular, curved; basal margin furnished with a just perceptible rim.

Infra-median latera, quadrilateral, sides unequal in length; the carino-basal margin being the longest; in area not quite twice double the rostral latera; directed obliquely upwards.

Carinal latera, sub-triangular, produced upwards, with the apex rounded, and the two lateral margins hollowed out; the basal margin exceeds a little in length the basal margin of the rostral latera. The umbones of these two latera are seated at their basal outer angles, so that the growth of the valves is towards each other and upwards. The umbo of the infra-median latus is seated at the baso-rostral angle, and hence the growth is obliquely upwards. The umbones of the rostral latera must have been close together, over the unknown rostrum.

Length of capitulum about 4/10th of an inch.

Peduncle, only small fragments are preserved; the calcified scales are small, closely imbricated, several of them together only equalling in length the basal margin of the rostral latera. Each scale is thin, transversely elongated; basal imbedded portion straight; upper margin rounded.

Mouth. Labrum with the upper part highly bullate, forming an overhanging projection; palpi apparently small and narrow. /

Mandibles, narrow, produced, with three teeth; inferior angle pectinated, as is sometimes the third tooth; the distance between the tips of the first and second teeth equals that between the second tooth and the inferior angle.

Maxillae, extremely narrow, produced, without any notch; spinose edge exactly one-third of the length of the mandibles: beneath the two upper great spines there are only three or four pair of spines; on the convex upper margin there are some minute tufts of the smallest hairs.

Outer maxillae, rounded with the inner margins very sparingly but continuously covered with bristles. I could not ascertain whether the olfactory orifices were tubular.

Cirri. These consisted, in the one specimen, of merely small fragments. The segments of the posterior cirri are elongated, not protuberant, and support, I believe, five pair of non-serrated spines, and an exterior row of very minute spines, dorsal spines fine and long. Either the second or third cirri, or probably both, are remarkable for having the whole of one side of each segment covered with irregular rows of long spines. Moreover, in the upper segments of these same cirri, between each separate dorsal tuft, there is placed one or two long bristles. The first cirrus appears to have had very broad segments, and these are singular from the spines in the dorsal rows, being extremely long. In some of the cirri, several of the basal segments are soldered together.

Caudal appendages, lost.

From the state of the specimen, it was quite impossible to ascertain whether the individual here described was a hermaphrodite or female; from the analogy of its nearest congener, *S. ornatum,* the latter is the most probable; but the genus Ibla shows how the sexes may differ in the most closely allied forms.

Affinities. From the hollows on the undersides of the scuta, for the lodgment of the males; from the umbones of the scuta and of the carina being situated on the apices / of these valves; and from all the characters of the mouth, S. *rutilum* is much more closely allied to S. *ornatum* than to any other species.

MALE, OR COMPLEMENTAL MALE

In the concavity or hollow above the depression for the adductor muscle (Pl. VI, fig. 2a'), I found males, but in so extremely decayed a condition, that they could hardly be examined. On one side, however, I distinctly saw the larval prehensile antennae, with pointed, hoof-like discs; and part of the thorax, with its small limbs and long spines, as in S. *vulgare* or S. *ornatum*. I also saw clearly the eye. The four calcified beads or rudimentary valves, I believe, were present; but in removing the specimen, the whole fell to pieces and was lost. The outer integument was covered with rather thick, very minute bristles, each about $2/10,000$th of an inch in length, and therefore only half the length of those on the complemental males of S. *vulgare*. The cavities for the males are not formed, as in S. *ornatum*, by the thickening of the internal surface of the valve round a defined space, but by the scutum being externally convex and internally concave down the middle, hollows being thus produced both for the lodgment of the males and for the attachment of the adductor muscle. These hollows are separated from each other by a slight transverse ridge. I do not know at which point of the margin of the valve the orifice of the male is situated, but I presume close under the apex. In this species, as in S. *ornatum*, there can be no question that the scuta of the female are specially modified by their own growth for the reception of the males. It must be added that, as it was not possible to ascertain whether the ordinary form of S. *rutilum* was hermaphrodite or female, so it must remain doubtful whether the parasites are males or complemental males; but the former, I think, is most probable. /

††*Sub-carinâ presente*

4. SCALPELLUM ROSTRATUM

PLATE VI. FIG. 7

S. (Herm.) valvis 15: rostro permagno: laterum paribus quatuor: pari superiore pentagono.

(Herm.) Capitulum with 15 valves: rostrum very large: four pair of latera; upper latera pentagonal.

Mandibles with four teeth; maxillae with the inferior angle prominent.

Complemental male, attached between the mouth and adductor scutorum muscle; pedunculated; capitulum bearing a pair of elongated scuta and a rudimentary carina; mouth and cirri prehensile.

Philippine Archipelago; Island of Bantayan. Attached to a horny coralline: twenty fathoms. Mus. Cuming.

HERMAPHRODITE

Capitulum, with the upper part narrow and produced.

Valves, fifteen in number, placed close together, clouded pale red, covered with membrane, which is thickly clothed with minute points.

Scuta rather small, oval, with the upper end pointed; rather convex; basal and lateral margins blending into each other; the upper produced portion above the umbo is small; there is a deep pit for the adductor muscle, and there is a fold on the occludent margin in the usual position; occludent margin not straight.

Terga large, one-third of their own length longer than the scuta; flat, subtriangular; the three margins are not quite straight; the carinal margin projects a little above the apex of the carina, and the scutal margin is excised to fit the upper part of the scuta.

Carina bowed, internally deeply concave; upper portion above the umbo, about one-fourth of the total length, extending between the terga for two-thirds of their length, / up to the slight prominences on their carinal margins: a ridge separates, on each side, the parietes from the tectum.

Rostrum (fig. 7a) unusually large, about two-thirds of the length of the scuta, and twice as long as the rostral pair of latera; internally concave, externally carinated; outline of the upper portion acutely triangular, of the lower portion rounded; umbo seated at the upper end.

Upper latera pentagonal, with the apex rounded.

Rostral latera flat, four-sided, with the basal margin the longest, and the baso-carinal angle produced.

Infra-median latera nearly equalling in area the upper latera; not descending so low down as the rostral and carinal latera; outline of lower half semi-oval, of upper half rectangular.

Carinal latera flat, four-sided, with the basal margin the longest, and slighly protuberant; baso-rostral angle produced; whole valve larger than the rostral latus, but closely resembling it in form.

Sub-carina minute, not above one-third of the size of the rostral latera, which are the smallest of the other valves; internally deeply concave; externally solid, pyramidal, standing out beyond the surface of the carina, with the umbo at the apex.

The umbones of the four pair of latera are seated a little above the centre in each valve, on the summit of a raised triangular portion; this arises from the valve at first growing only downwards, and when added to at the upper end, the new part forms a ledge at a lower level round the old part, which had already acquired some thickness.

Peduncle, short, about half the length of the capitulum; narrow; thickly clothed with minute, longitudinally elongated, spindle-shaped, calcareous scales or beads, which project but little.

Length of the capitulum, rather under 3/10ths of an inch.

In a *young specimen*, with its capitulum, together with the peduncle, only 1/10th of an inch long, the scuta, terga, and carina are very large in proportion to the valves of the / lower whorl. The latter project more, and are externally more pointed, as in the genus Pollicipes. The rostrum is well developed; the infra-median latera, in proportion, are the least of all the valves. The carina is straight and pointed, and not, relatively to the scuta, quite so long. The scuta are rather broader in proportion to their length, which would naturally follow from less having been added to their apices – these valves at first growing only downwards. The membrane covering and connecting the valves is furnished with long thin spines.

Mouth. Labrum placed far from the adductor scutorum muscle, with the upper part exceedingly prominent; apparently there are no teeth on the crest. Palpi blunt.

Mandibles, narrow, with four teeth, of which the second is not smaller than the others; inferior angle sharp and produced, barely pectinated.

Maxillae. Under the two or three great upper spines, there is a tuft of fine bristles; the inferior part of the edge is step-like, and much upraised.

Outer maxillae, with the inner edge deeply notched, and the bristles arranged in two quite distinct tufts; the bristles on the outer surface are long. Olfactory orifices, thin, tubular, and projecting.

Cirri. The first pair is placed far from the second; the three posterior pair are long and straight, with their segments much elongated, not protuberant, bearing four or five pair of long spines, with little intermediate tufts of minute spines, and with the minutest spines on the lateral upper edges. Dorsal tufts with one spine extremely long, equalling a segment and a half in length; the others very short. Spines all serrated. First cirrus not very short; rami nearly equal, with the four terminal segments of both tapering; all the basal segments much thicker, and thickly covered with bristles. Second cirrus (as well as the third in a less degree), with the anterior ramus thicker than the posterior ramus, and with all the lower segments in both rami thickly clothed with three or four longitudinal rows of spines. /

Caudal appendages, spinose, uniarticulate; but the specimen was injured, and I could not exactly make out their shape: I believe it was oval, and thickly fringed with fine spines.

Penis, very small, almost rudimentary, narrow, and hairy, scarcely exceeding in length the pedicel of the sixth cirrus.

COMPLEMENTAL MALE
PLATE VI. FIG. 5

Before describing the parasite of the present species, which departs entirely from the character of the males of the three preceding species, it is proper to state that I consider it to be a complemental male simply from analogy, as will hereafter be more fully shown at the end of the genus. Had a specimen of the parasite been brought to me without any information, I should have concluded that it was an immature individual of a new genus of pedunculated cirripedes, remarkable from the rudimentary condition of the valves, and exhibiting, in one important character, namely, in the form of the larval prehensile antennae, an alliance to Scalpellum. Had I been then told that three individuals in a group, had been found attached to *S. rostratum*, not outside the valves, but to the integument, in a central line, between the labrum and the adductor scutorum muscle, in such a position that when the Scalpellum closed its valves, these parasites were enclosed within the capitulum, my surprise would have been great; for it is very improbable that this singular and unparalleled position was accidental in this one group of specimens, inasmuch as there seems to be a relation between the naked condition of the capitulum of the parasite, and the protection afforded to it by the capitulum of the Scalpellum. It further becomes apparent on reflection, that these minute parasites, though having the appearance of immaturity, can not increase in size, or but little, for if they did grow, and acquired an ordinary size, they would / either be killed by the pressure of the scuta of the Scalpellum, or they would destroy the latter, and in doing so soon lose their own support, and thus necessarily perish!

The one full-grown specimen of *S. rostratum*, in Mr Cuming's collection, was in a good state of preservation, but dry. The three parasites were attached, as stated, close under the labrum, between it and the adductor muscle. They are constructed like ordinary Cirripedia, and have a mouth, thorax and cirri, enclosed in a capitulum, supported on a peduncle of moderate length and narrow. The entire length of the capitulum and peduncle, as far as could be ascertained in the shrivelled condition of the specimens, was $35/1000$ths, and the greatest width of the capitulum $11/1000$ths of an inch. Both capitulum and peduncle are hirsute with spines, nearly $1/1000$th of an inch in length, mingled with shorter hairs in little rows of three and four together. The figure 5 (in Pl. VI) is merely a restoration, as accurate as could be made from the much shrivelled specimens. There are only three valves – namely, an oval carina *a*, seated rather high up on the capitulum, in a rudimentary condition and only $1/1000$th of an inch in length, and a pair of scuta; these latter consist of a narrow, slightly curved plate, $8/1000$ths in length, broadest at the lower end, where the breadth is $2/1000$ths of an inch. The prehensile antennae, at the end of the peduncle, have pointed hoof-like discs: I was not able to make out the other parts. It deserves notice, that in the young specimen of the ordinary form of *S. rostratum*, $1/10$th of an inch in length, and therefore only thrice as long as the parasites, all the valves were perfect, and seemed to have followed the ordinary law of development.

Mouth. The largely bullate labrum is placed far from the adductor, in the same manner as in the hermaphrodite. The mandibles have three large sharp teeth, with the inferior point very sharp and small, so that there is one less tooth than in the hermaphrodite. The maxillae have two or three large upper spines, the others being very thin; I believe the lower part is upraised and step-like, / as in the hermaphrodite. The outer maxillae are bilobed in front, with a few short bristles on the outer side near the bottom. I was not able, from the dried state of the specimens, to discover whether the olfactory orifices were tubular. Altogether it was apparent, from this imperfect examination, that there was a close similarity between the mouth of the parasite and of the hermaphrodite.

The *thorax* is unusually elongated.

Cirri. The first pair is very short, and is distant from the second. All have the appearance of immaturity, with their pedicels very long in proportion to their rami; the latter are slightly unequal in length, even in the sixth pair. There appeared to be six segments in the rami of the sixth pair, each segment bearing two or three pair of long spines.

Caudal appendages, with two or three little spines on their summits.

Penis, short, blunt, thick at the apex, with one or two spines on it. I did not see any ovaria, but this could hardly have been expected in specimens in a dried condition, without they had happened to have been in a gorged condition. Certainly there were no ova.

In the general summary at the end of the genus, I shall give my reasons for believing this parasite to be the complemental male of the *Scalpellum rostratum.*

5. SCALPELLUM PERONII

PLATE VI. FIG. 6

SMILIUM PERONII. J. E. Gray, *Annals of Philosoph.*, new series, tom. x, 1825
SMILIUM PERONII. J. E. Gray, *Spicilegia Zoologica,* tab. iii, fig. 10, 1830
ANATIFA OBLIQUA. Quoy et Gaimard, *Voyage de l'Astrolabe,* Pl. xciii, fig. 16, 1823–34
POLLICIPES OBLIQUA. Lamarck, *An. sans Vertebres* (2nd edition)

S. (Herm.) valvis 13: laterum paribus tribus; pari superiore multùm elongato: pedunculi squamis calcareis nullis. /

(Herm.) Capitulum with 13 valves: three pair of latera; upper latera much elongated: peduncle without calcareous scales.

Mandibles with 10 or 11 unequal teeth: maxillae with the edge nearly straight, bearing numerous spines.

Complemental male, attached externally, between the scuta and below the adductor muscle; pedunculated; capitulum formed of six valves, with the carina descending far beneath the basal angle of the terga; mouth and cirri prehensile.

Swan River, Australia, attached to a coralline; Mus. Cuming. Port Western, Bass's Straits, as stated in the *Voyage of the Astrolabe*. Mus. Brit.

HERMAPHRODITE

Capitulum formed of thirteen valves; namely, two scuta, two terga, a carina and sub-carina, a rostrum, a pair of upper latera, and two pair of lower latera; these latter valves, with the sub-carina and the rostrum, make a whorl of six pieces. The upper part of the capitulum is, as usual, produced. The upper valves are separated (in specimens which have not been dried) by rather wide interspaces of membrane; they are covered (excepting, generally, their umbones), by membrane, which in the interspaces is clothed with fine spines. The spines, or the marks where they were once articulated, are visible over nearly the entire surface of the membrane covering the valves. The spines are particularly numerous round the orifice of the sack. The whole capitulum (in a dried condition), is coloured dull purplish-red, which is only in part due to the underlying corium, for the valves themselves are pale red. After having been long kept in spirits, the whole capitulum becomes colourless. The valves are smooth, faintly marked by lines of growth. The umbones of the lower valves project outwards, giving a denticulated appearance to the base of the capitulum.

Scuta, slightly convex, oblong, breadth about two-thirds of the length, almost quadrilateral, with the / upper portion produced into a flat projection; this projection is almost spear-shaped, being constricted a little on each side below the apex. There is a deep pit for the adductor muscle. The umbo is near the apex, the part above not being above one-fifth of the whole length of the valve. As in *S. vulgare*, the growth is at first downwards, and subsequently a little upwards and downwards, thus producing the upper, small, spear-like projection, which lies at a lower level than the umbo. There is a fold on the occludent margin.

Terga, large, flat, triangular; carinal margin slightly hollowed out; occludent margin slightly arched, with a small portion protuberant to a variable amount. The apex is slightly curved towards the carina.

Carina, long, internally deeply concave, angularly bent, the lower portion slightly longer and wider than the upper part; the two halves meet each other at about an angle of 135°; the upper half is parallel to the longer axis of the terga, between which it extends for three-fourths of their length. The external surface is rounded, except near the umbo, where the edge is carinated; growth almost equally upwards and downwards; the parietes and tectum are not separated by ridges.

The *sub-carina* lies close under the carina, and is placed almost transversely to the longer axis of the capitulum; external surface arched and smooth, the whole having the shape of half of a cone, with the apex a little curved outwards; seen internally, it may be said to be formed of two triangular wings placed at right angles to each other; basal margin straight; in size equalling the carinal latera.

Rostrum, lying almost transversely to the longer axis of the capitulum, under the basal margins of the scuta; in shape (fig. 6a) closely resembling the sub-carina, but about one-third larger than it; larger also than either the rostral or carinal latera; seen externally, appears like a half cone; seen internally, is formed of two triangular wings (with curved edges), placed at right-angles to each other. /

Upper latera, internally flat, oblong, twice as long as broad; upper end square, truncated; upper half rather wider than the lower half; fully twice as large as either of the lower latera. The basal points extend below the basal margins of the scuta. The umbo is placed a little above the centre.

Rostral latera, minute, scarcely exceeding one-third of the size of the carinal latera, and very much less than the rostrum; they are placed transversely under the basal point of the upper latus, or rather between it and the baso-lateral angle of the scutum; basal margin, as seen internally, straight; upper margin arched; rostral angle produced; internally flat; the whole valve is very thick and solid, so that the umbo which lies at the rostral end, projects rectangularly outwards.

Carinal latera, oblong, nearly quadrilateral, with the upper angle produced; placed obliquely, parallel to the lower half of the upper latera; umbo slightly prominent, seated near the apex, with three rounded ridges proceeding from it; internal surface very slightly concave.

Peduncle and attachment. The peduncle is short, not equalling the capitulum in length. The whole surface is most thickly clothed with minute spines, which are not visible when the specimen is dry; I think it probable that they may sometimes all drop off before a new period of exuviation. The peduncle does not (at least in the specimens which I have examined, which were grouped in a bunch) taper at the lower end to a point; and after careful examination, I feel sure that the cement does not debouch from several successively formed orifices, as in *S. vulgare* and as in some Pollicipes, but only from the two original orifices in the prehensile antennae of the larva. In these latter organs, the sucking disc is hoof-like and pointed, and is narrower than the basal segment. The ultimate segment has on its inner side (supposing this segment stretched straight forwards), a notch or step bearing at least three spines. The proportions of the different parts differ slightly from those / in *S. vulgare*; but, as I shall hereafter have to give all the measurements, I do not think them worth repeating here. In the one large group of specimens examined by me, in Mr Cuming's possession, all were attached symmetrically to the coralline, as in the case of *S. vulgare*, capitulum upwards, and their carinas outwards.

Length of capitulum about three-quarters of an inch; width about half an inch; entire length, with peduncle, a little more than one inch.

The *mouth* is placed far from the adductor muscle.

Labrum, with its basal margin much produced; upper part highly bullate, forming a rounded projection equalling the longitudinal axis of the rest of the mouth; crest without any teeth.

Palpi, triangular, with the two margins, thickly clothed with bristles; on each side of the mouth, near where the palpi are united to the mandibles, there is a slight, orbicular, shield-like swelling.

The *mandibles* (Pl. X, fig. 3) have nine or ten very unequal teeth, with the inferior angle rather broad and pectinated; of these, there are four main teeth, of which the second is always the smallest, and between the four, one or two small teeth are interpolated; so that the total number is either nine or ten, and often varies on the two sides of the same individual, as likewise does the shape of the inferior angle.

Maxillae, with the edge nearly half as long as that of the mandibles, supporting from seventeen to twenty pairs of spines; the upper pair is only slightly larger than the others; a part near the inferior angle projects slightly beyond the rest of the nearly straight edge. The apodeme, at its base or point of origin, is unusually broad and flat.

Outer maxillae, large and triangular. The inner margin is slightly concave, and continuously covered with short spines. The outer margin is bilobed, as in *S. vulgare*, with the basal part supporting a great tuft of long bristles, of which the greater number turn outwards, and almost / cover the olfactory orifices. The latter are slightly prominent, placed some way apart from each other, with the above-mentioned tufts of bristles between them. All the spines of the trophi are in some degree doubly serrated.

Cirri. The first pair is seated rather far from the second pair, and the prosoma being little developed, the shape of the body nearly resembles that of *S. vulgare*. The posterior cirri are elongated, very little curled, with the segments much flattened, not at all protuberant, bearing from five to seven pair of long serrated spines, with a few small spines in an exterior row; between each pair there is a very minute tuft of small bristles; the upper lateral rim of each segment is toothed with small spines; spines of the dorsal tufts, long, serrated. *First pair*, elongated, having numerous segments, namely, seventeen, whilst the sixth pair in the same individual had only twenty-one segments; rami nearly equal; segments short, nearly cylindrical, thickly clothed with long serrated spines. The *second* and *third* pair are nearly equal in length; they have their anterior rami slightly thicker than their posterior rami, both being much more thickly clothed with spines, than are the three posterior pair of cirri. Pedicels, rather short, with their inner edges not forming a projection, as in *S. vulgare*.

Caudal appendages (Pl. X, fig. 20), uniarticulate, flat, rounded at their ends and moderately long; clothed most thickly, like brushes, with very fine bristles, which latter are serrated, and are longer than the appendages themselves.

Penis, of small size, narrow, pointed, and thickly clothed with delicate hairs; in length equalling only one-fourth of the sixth cirrus.

Ovigerous fraena, small, semicircular; entire edge thickly covered with glands. Ovarian tubes, within the peduncle, fully developed as usual.

THE LEPADIDAE

Affinities. This species differs from all the others in the absence of calcareous scales on the peduncle; but it / has no other character which at all justifies its generic separation. In the shape of the scuta and carina it comes nearest to *S. vulgare*. Taking all the characters together, it is scarcely possible to say to which of the other species it is most closely allied, having close affinities with all. In the entire structure, however, of the complemental male, immediately to be described, this species certainly comes nearer to *S. villosum* than to any other species. I may add, that in *S. villosum* the latera are almost rudimentary, and therefore tend to disappear, whereas in *S. Peronii* it is the calcareous scales on the peduncle which have actually disappeared.

COMPLEMENTAL MALE
PLATE VI. FIG. 3

I examined, owing to the great kindness of Mr Cuming, six dry specimens of the hermaphrodite *S. Peronii*, from Swan River, and one in spirits from another locality, in the British Museum. Out of these seven specimens, only three appeared to have had parasites attached to them, and these I infer, from reasons to be more fully given at the end of the genus, are complemental males. One of the three specimens, however, had two males close together. These parasites were firmly cemented to the integument of the hermaphrodite, in a fold, in a central line between the scuta, a little below (the animal being in the position in which it is figured) the adductor scutorum muscle, and therefore some way below the umbones of these valves. When the scuta are closed, the parasites, from their small size, are enclosed and protected. In every detail of structure, they are obviously pedunculated Cirripedia.

The *capitulum* (Pl. VI, fig. 3) has six valves; namely, a pair of scuta and of terga, a carina, and a rostrum, all united by finely villose membrane, furnished near the orifice with some much longer and thicker spines. The capitulum is truncated in a remarkable manner, the orifice not being, as in the hermaphrodite, in the same line with the peduncle, but almost transverse to it, and therefore almost / parallel to the surface of attachment. The largest specimen measured transversely, through the scuta and terga, was $30/1000$ths of an inch in breadth; another was only $26/1000$ths to $27/1000$ths: this latter specimen, measured longitudinally, from the base of the carina to the tips of the terga, was $15/1000$ths of an inch. A scutum of the largest specimen was

17/1000ths in length. The scuta and terga are broadly oval, with the primordial valves very plain at their upper ends. I may here mention, that in a central line between the scuta, I observed the *apparently* single, minute, black eye, as in ordinary Cirripedia.

The *carina* is straight, triangular, and internally slightly concave; its basal margin descends far below the basal points of the terga.

The *rostrum* is shorter, and internally more concave than the carina: I believe it projects more abruptly outwards than is represented in the figure.

The *peduncle* commences some little way below the scuta: it is narrow and very short: it is finely villose: it is lined by delicate transverse striaeless muscles, within which there are the usual stronger, longitudinal muscles. The base is flat and truncated. I examined, and carefully compared, the prehensile antennae with those of the hermaphrodite, and found every part and every measurement the same. The full importance of this identity will hereafter be more fully insisted on. The antennae are represented of their proper proportional size in fig. 3.

Mouth. The labrum, as in the hermaphrodite, is highly bullate, and far removed from the adductor scutorum muscle. The *palpi* are small and triangular, with their blunt apices clothed with a very few scattered bristles.

Mandibles, with only three teeth, and the lower angle minute, slightly pectinated; the first tooth is distant from the second, and larger than it. Width of the whole organ, 0·0021 of an inch.

Maxillae, bearing only a few spines, furnished with a long apodeme; beneath the upper large pair there is a notch, under which there are two spines of considerable size and / a small tuft of fine bristles; width 0·001 of an inch, and therefore only 1/16th of the size of the same organ in the hermaphrodite: the relative sizes of the maxillae and mandibles are the same in the male and hermaphrodite.

Outer maxillae blunt, triangular, with a few thinly scattered bristles on the inner face; those on the outside being longer.

Cirri. The first pair is far removed from the second; the rami are very short, barely exceeding the pedicel in length; they are formed of only four segments, each bearing a pair of spines; but on the end of the terminal segment, there are three spines, of which the central one is very long. Second pair also short. In the sixth pair there are five or six elongated segments, each bearing three pair of long spines; dorsal tufts large. The cirri are furnished with transversely striated muscles.

The *caudal appendages* exist as two very minute plates, with a few bristles at their apices.

The *penis* is not acuminated, with four bristles at the end; it is short, equalling only the lower segment of the pedicel of the sixth cirrus. In the one specimen preserved in spirits, I unfortunately omitted to search for the vesiculae seminales; I cannot doubt that such existed, but it would have been important to have ascertained whether they contained spermatozoa. I made out, most distinctly, that there was no trace of ovarian tubes within the peduncle; and my assertion may be believed when I state, that I traced the two much finer and more transparent cement ducts, from the prehensile antennae up to the body of the animal: in Lepas I have *repeatedly* detected, with ease, the ovarian tubes within the peduncle, before the calcification of the valves had even commenced, and therefore at a much earlier period of growth than in these parasites. Consequently I am prepared to affirm, that these parasites are not females, but that, as far as can be judged from external organs, they are exclusively males.

Concluding remarks. In comparing the capitulum of the hermaphrodite with that of the complemental male / (Pl. VI, figs 6 and 3), we must be struck with the differences in their shape, in the number, relative sizes, and forms of the several valves. It should, however, be borne in mind, that the scuta and carina in the hermaphrodite at first grow exclusively downwards; so that if we remove the upper portions subsequently added, the difference in shape in these valves is not so great as it at first appears. The rostrum in the male is of much larger relative size; whilst of the upper latera there is not a trace, although in the hermaphrodite these valves are larger than the rostrum. The terga, compared with those of the hermaphrodite, differ more essentially than do the other valves; and the manner in which the primordial valves project, shows that from the first commencement of calcification, the lines of growth have followed an unusual course. The great breadth and shortness of the terga is evidently related to the shortening of the whole capitulum, and the transverse position of the orifice; and this shortening of the capitulum, no doubt, is rendered necessary for its reception and protection within the shallow furrow between the scuta of the hermaphrodite. Finally, if we compare the internal parts of the hermaphrodite and male, the differences are considerable, though partly to be accounted for by the youth of the latter: the form and position of the labrum, and the distance between the first and second pair of cirri, is the same in both; but the mandibles and maxillae differ considerably.

To put the case as I have before done, if a specimen of one of these parasites had been brought to me to class without any information of its habits – the downward direction of growth in all the valves, the presence of a rostrum, the villose outer integument, all the details of the prehensile antennae, the form of the animal's body, and the position of the labrum, would have convinced me that, though a quite new genus, it ought to have stood close to Scalpellum, and nearer to it than to Ibla. /

6. SCALPELLUM VILLOSUM

PLATE VI. FIG. *8*

POLLICIPES VILLOSUS on Plate (TOMENTOSUS in text). Leach, *Encyclop. Brit. Suppl.*, vol. iii, Pl. lvii, 1824

POLLICIPES VILLOSUS.[57] G. B. Sowerby, *Genera of Shells, Pollicipes*, fig. 3, 1826

CALANTICA HOMII. J. E. Gray, *Annals of Phil.*, vol. x, p. 100, 1825

S. *(Herm.) valvis 14: sub-rostro praesente: carinâ paene rectâ: laterum paribus tribus; pari superiore triangulo.*

(Herm.) Capitulum with 14 valves: sub-rostrum present: carina nearly straight: three pair of latera; upper latera triangular.

Mandibles with four teeth, of which the second is the smallest: maxillae with a projection near the inferior angle: no caudal appendage.

Complemental male, attached externally between the scuta, below

---

[57] As Mr Sowerby has adopted the name *villosus*, I have followed him; though as *tomentosus* is used through some mistake by Leach in the text, both names have equal claims as far as priority is concerned.

In Lamarck, *Animaux Sans. Vert.*, the *P. villosus* of Sowerby is made synonymous with *Anatifa villosa* of Brugière, which is certainly incorrect, although the *A. villosa* of this latter author is not positively known.

the adductor muscle; pedunculated; capitulum formed of six valves, with the carina not descending much below the basal angles of the terga: mouth and cirri prehensile.

Eastern Seas[58] [?] attached to shells and rocks. Mus. Brit.; College of Surgeons; Cuming.

HERMAPHRODITE

Capitulum with fourteen valves, consisting of a pair of scuta and of terga, a carina (which five valves are much / larger than the others), a rostrum, sub-rostrum, sub-carina, and three pair of small latera. All the valves are covered by membrane, as are the calcareous scales on the peduncle; and this membrane everywhere is densely clothed with spines. The upper valves are not very thick; they stand rather close together. The eight valves of the lower whorl are more solid, and are placed far apart; they are small, tending to become rudimentary. None of the valves are added to at their upper ends, in which respect this species differs remarkably from the others of the genus, and approaches in character to Pollicipes.

Scuta, with a deep hollow for the adductor muscle, triangular, with the basal margin elongated, and protuberant.

Terga, large, flat, triangular, basal point blunt, with the carinal margin slightly hollowed out, and the scutal margin protuberant. Apex solid.

Carina, rather longer than the terga, straight, gradually widening from the upper to the basal end, deeply concave. In young specimens the upper part is slightly bowed inwards. Apex solid.

Sub-carina, with the inner surface crescent-shaped; the umbo points transversely outwards; in width it exceeds the largest of the latera.

Rostrum, triangular, internally (fig. 8a) concave; basal margin slightly hollowed out, and deeply notched; rather less in width than the carina; short, with the umbo pointing upwards and outwards. In young specimens the apex curves a little inwards.

[58] No habitat is attached to any of these specimens; but Mr Sowerby informs me that he has seen specimens attached to the *Modiola albicostata* of Lamarck, which shell is said by the latter author to be found in the seas of India, Timor, and New Holland.

Sub-rostrum, with the inner surface transversely elongated (fig. 8*b*), slightly crescent-shaped, about two-thirds as wide as the rostrum. The apex points transversely outwards.

Latera, three pairs; the middle pair apparently corresponds with the upper latera of the other species of the genus. The two other pair of latera, together with the rostrum and sub-carina, form a whorl. The sub-rostrum / lies by itself, a little beneath this whorl. The latera are smaller than the rostrum or the sub-carina. They are placed far distant from each other; their inner surfaces are triangular; their umbones point upwards; the rostral pair is smaller than the other two pair, which are of equal size. The exact position of the rostral latus differed on the two sides of the specimen examined; apparently its normal position is at the baso-lateral angle of the scuta.

Peduncle, wide at the summit, longer than the capitulum; calcified scales small, not arranged very regularly; flattened, spindle-shaped, rather far separated from each other; imbedded in membrane, so that even their summits are rarely uncovered. The surface of the membrane is thickly clothed with spines, which are strong, thick, yellow, pointed, and furnished with large tubuli running to the underlying corium. These spines are arranged in groups of from three or four, to five or six. Besides these larger spines, the whole surface is villose with very minute colourless spines, not above $\frac{1}{20}$th of the length of the larger ones. The surface of attachment is broad. This species, not being symmetrically attached to a coralline, the peduncle does not curve, as in most of the other species, towards the rostrum.

The capitulum is above half an inch in length.

Mouth. The labrum is much produced downwards, but yet the mouth is not very far distant from the adductor muscle: the upper part is bullate, forming a small overhanging point, and in longitudinal diameter equals the rest of the mouth. *Palpi* blunt.

Mandibles with four teeth, strong, short, thick, the second tooth much smaller than the others; inferior angle broad, pectinated.

Maxillae with a long, rather sinuous edge, which, near the inferior angle, has a narrow projecting point, bearing rather finer spines; there is, also, apparently, a very minute tuft of small spines close under the two large upper spines: there are, altogether, about twenty pair of spines, without counting the smaller ones. /

Outer maxillae, with the inner edge slightly concave, continuously covered with bristles; exteriorly, with a prominence covered with longer bristles. Olfactory orifices prominent, protected by a slight punctured swelling between the bases of the first pair of cirri.

Cirri. Prosoma moderately developed; first pair of cirri rather far removed from the second pair. The segments of the three posterior pair are not elongated, short, slightly protuberant in front, bearing four or five pairs of strong spines; a little below each pair, there is an intermediate tuft of very fine straight bristles, of which the upper tuft is the largest; on the lateral upper rims there are some short, strong spines; dorsal tufts rather small and thick; spines all more or less serrated, especially on the broad basal segments of the three anterior cirri. Pedicels of the cirri not particularly protuberant in front. First cirrus with rami, slightly unequal in length; not short; basal segments much thicker and more protuberant than the upper segments. Second cirrus; anterior ramus with six or seven basal segments highly protuberant, and crowded with spines; posterior ramus with about six segments, similarly characterized. Third cirrus with the anterior ramus having six, and the posterior ramus five segments, also similarly characterized.

Caudal appendages absent, there being only a slight swelling on each side of the anus.

The *oesophagus* runs parallel to the labrum, and enters obliquely the summit of the stomach, which is destitute of caeca: the biliary envelope is longitudinally plicated.

There are no *filamentary appendages*.

Testes large, branched like a stag's horns, attached in a sheet to the ventral surface of the stomach: the vesiculae seminales enter the prosoma, and have their reflexed ends not very blunt. The *penis* is rather narrow, with the terminal half plainly ringed, and bearing tufts of fine bristles arranged in circles, one tuft below the other; on the basal half there are only a few scattered minute bristles. /

Affinities. In the downward growth of all the valves, in the presence of a sub-rostrum, in the shape of the scuta, carina, and more especially of the triangular latera, in the form of the peduncle, with its irregularly-scattered calcified scales, in the shape of the animal's body, in the structure both of the mandibles and maxillae, in the arrangement of the spines, both on the anterior and posterior cirri, *Scalpellum villosum* most closely resembles, or rather is identical with, Pollicipes. Had it not been for the fewness of the valves forming the capitulum, and from the presence of complemental males, I should have placed this species alongside of *Pollicipes spinosus* and *sertus*. In not having caudal appendages, *S. villosum* differs from all the species of Scalpellum and Pollicipes; but this organ is variable to an unusual degree in Pollicipes.

COMPLEMENTAL MALE

PLATE VI. FIG. 4

From the kindness of Professor Owen, Mr Gray, and Mr Cuming, I have been enabled to examine six specimens of this species; and on two of them I found complemental males. They were attached in the same position as in *S. Peronii*; namely, beneath the adductor muscle, in the fold between the scuta, so as to be protected by the latter when closed. This parasite is six-valved, and has a close general resemblance with that of *S. Peronii*, but differs in very many points of detail. It is represented of the natural size (at a' fig. 4). The capitulum is $43/1000$ths of an inch, measured across the scuta and terga; and the same measured from the base of the carina to the top of the capitulum; hence it is broader, by a quarter of the above measurement, and considerably higher than the male of *S. Peronii*. From the capitulum being higher, that is, not so much truncated, the orifice is placed more obliquely. The membrane connecting the valves is finely villose, and is besides furnished with spines, conspicuously thicker and longer than those on the male *S. Peronii*. The scuta and terga are much more elongated, / a scutum being here $85/1000$ths of an inch in length. The carina descends only just below the basal points of the terga, instead of far below them. The rostrum is a little broader and more arched than the carina; it is $2/1000$ths in length, and therefore more than two-thirds of the length of the carina, the latter being $28/1000$ths of an inch from the apex to the basal margin. The primordial valves, with the usual hexagonal tissue, are seated on the tips of the scuta, terga, and carina, but not on the rostrum; so that these valves follow the same law of development, as in the ordinary and hermaphrodite form of Scalpellum. The scuta (a, fig. 4, greatly enlarged), the terga b, and carina c of the male, resemble the same valves in the hermaphrodite, much more closely than do these valves in the male and hermaphrodite *S. Peronii*. The rostrum has not its basal margin hollowed out, and is very much larger relatively to the carina, than in the hermaphrodite. The large relative size of the rostrum in the complemental male both of this species and of S. *Peronii*, is a remarkable character, which I can in no way account for.

The peduncle is narrow and short, but in a different degree in the two specimens examined. It is naked. The prehensile antennae were not in a good state of preservation: the disc is narrower than the basal segment, and only slightly pointed, in which important respect it differs from the same part in the foregoing species; at its distal end, rather on the inner side, there are two or three spines, apparently in place of the excessively minute hairs, which are found at the same spot in some or in all the other species of Scalpellum, and in Ibla: similar strong spines occur in Pollicipes. Unfortunately, for the sake of comparison, I was not able to find the prehensile antennae in the hermaphrodite S. *villosum*.

Mouth. Labrum bullate, with teeth on the crest. *Palpi* blunt, spinose.

Mandibles, with three teeth; inferior point rather strongly pectinated.

Maxillae, with a considerable notch under the upper pair of large spines; inferior part of the edge not prominent. /

Outer maxillae, with the spines on the inner edge arranged into two groups. Olfactory orifices tubular and prominent, with some long bristles near their bases. In the mandibles having only three teeth, in the maxillae being notched and in the lower part not being prominent, and, lastly, in the bristles on the inner face of the outer maxillae being arranged in two groups, these several organs differ from those in the hermaphrodite.

Cirri. First pair short, with only three or four segments in each ramus: second cirrus, with the basal segments not very thickly clothed with spines: sixth cirrus with seven segments, not protuberant in front, each bearing four pairs of spines, without intermediate tufts.

Caudal appendages, none. This is an interesting fact, considering that these organs are likewise absent in the hermaphrodite S. *villosum* – an absence highly remarkable, and confined to the genus Conchoderma and the one species of Anelasma.

Penis thick, not tapering, rather exceeding in length the pedicel of the sixth cirrus, square at the end, and furnished with some spines. In one specimen, I believe I distinguished the vesiculae seminales: if so, they contained only pulpy matter, and not spermatozoa. There were no ovarian tubes within the peduncle, which was lined by the usual muscles; I traced the two delicate cement ducts, running from within the antennae close up to the animal's body. Hence in this case, as in that of *S. Peronii*, I dare positively affirm that ovarian tubes do not occur; for it is out of the question that I could have traced the cement ducts, and, at the same time, overlooked the far larger and more conspicuous ovarian tubes, into which, moreover, the ducts, had they existed, would have run. Consequently, these parasites are not females; but judging from the probosciformed penis, and from the presence, as I believe, of vesiculae seminales, they are males.

The complemental males of the present species, and of *S. Peronii*, so closely resemble each other, that what I / have stated regarding the affinities of the latter, are here quite applicable. It is singular how much more alike the parts of the mouth and the cirri of these two complemental males are, than the corresponding parts in the two hermaphrodites: this no doubt is due to the two males having been arrested in their development, at a corresponding early period of growth. Several of the characters, by which the hermaphrodite *S. villosum* so closely approaches, and almost blends into the genus Pollicipes – such as the thicker cirri, with the intermediate tufts of bristles, the small second tooth of the mandibles, and the little brush-like prominence on the maxillae – are not in the least apparent in the complemental male.

SUMMARY ON THE NATURE AND RELATIONS OF THE MALES AND COMPLEMENTAL MALES, IN IBLA AND SCALPELLUM

Had the question been, whether the parasites which I have now described, were simply the males of the cirripedes to which they are attached, the present summary and discussion would perhaps have been superfluous; but it is so novel a fact, that there should exist in the animal kingdom hermaphrodites, aided in their sexual functions by independent and, as I have called them, complemental males, that a brief consideration of the evidence already advanced, and of some fresh points, will not be useless. These parasites are confined to the allied genera Ibla and Scalpellum; but they do not occur in Pollicipes – a genus still more closely allied to Scalpellum; and it deserves notice, that their presence is only occasional in those species of Scalpellum which come nearest to Pollicipes. In the genera Ibla and Scalpellum, the facts present a singular parallelism; in both we have the simpler case of a female, with one or more males of an abnormal structure attached to her; and in both the far more extraordinary case of an hermaphrodite, with similarly attached complemental males. In the two species of Ibla, the complemental and ordinary males resemble each other, as closely as do the corresponding hermaphrodite and female forms; so it is / with two sets of the species of Scalpellum. But the males of Ibla and the males of Scalpellum certainly present no special relations to each other, as might have been expected, had they been distinct parasites independent of the animals to which they are attached, and considering that they are all cirripedes having the same most unusual habits. On the contrary, it is certain that the animals which I consider to be the males and complemental males of the two species of Ibla, if classed by their own characters, would, from the reasons formerly assigned, form a new genus, nearer to Ibla than to the parasites of Scalpellum: so, again, the assumed males of the three latter species of Scalpellum would form two new genera, both of which would be more closely allied to Scalpellum, than to the parasites of Ibla. With respect to the parasites of the first three species of Scalpellum, they are in such an extraordinarily modified and embryonic condition, that they can hardly be compared with other cirripedes; but certainly they do not approach the parasites of Ibla, more closely than the parasites of Scalpellum; and in the one important character of the antennae, they are identical both with the parasitic and ordinary forms of Scalpellum. That two sets of parasites having closely similar habits, and belonging to the same subclass, should be more closely related in their whole organization to the animals to which they are respectively attached, than to each other, would, if the parasites were really distinct and independent creatures, be a most singular phenomenon; but on the view that they differ only sexually from the cirripedes on which they are parasitic, this relationship is obviously what might have been expected.

The two species of Ibla differ extremely little from each other, and so, as above remarked, do the two males. In Scalpellum the species differ more from each other, and so do the males. In this latter genus the species may be divided into two groups, the first containing *S. vulgare*, *S. ornatum* and *S. rutilum*, characterized by not having a subcarina, by the rostrum being small, by the constant presence of four pair of latera, and by the peculiar shape / of the carinal latera; the second group is characterized by having a sub-carina and a large rostrum, and may be subdivided into two little groups; viz., *S. rostratum* having four pairs of latera, and *S. Peronii* and *villosum* having only three pairs of latera: now the males, if classed by themselves, would inevitably be divided in exactly the same manner, namely, into two main groups – the one including the closely similar, sack-formed males of *S. vulgare*, *ornatum*, and *rutilum*, the other the pedunculated males of *S. rostratum*, *Peronii*, and *villosum*; but this latter group would have to be subdivided into two little sub-groups, the one containing the three-valved male of *S. rostratum*, and the other the six-valved males of *S. Peronii* and *S. villosum*. It should not, however, be overlooked, that the two main groups of parasites differ from each other, far more than do the two corresponding groups of species to which they are attached; and, on the other hand, that the parasitic males of *S. Peronii* and *S. villosum* resemble each other more closely, than do the two hermaphrodite forms; but it is very difficult to weigh the value of the differences in the different parts of species.

Besides these general, there are some closer relations between the parasites and the animals to which they are attached; thus the most conspicuous internal character by which *Ibla quadrivalvis* is distinguished from *I. Cumingii*, is the length of the caudal appendages and the greater size of the parts of the mouth; in the parasites, we have exactly corresponding differences. Out of the six species of Scalpellum in their ordinary state, *S. ornatum* is alone quite destitute of spines on the membrane connecting the valves; and had it not been for this circumstance, I should even have used the presence of spines as a generic character; on the other hand, *S. villosum*, in accordance with its specific name, has larger and more conspicuous spines than any other species. In the parasites we have an exactly parallel case; the parasite of *S. ornatum* being the only one without spines, and the spines on the parasite of *S. villosum* being much the largest! This latter species is highly singular in having no caudal appendages, and / the parasite is destitute of these same organs, though present in the parasites of *S. rostratum* and *S. Peronii*. Again, *S. villosum* approaches, in all its characters, very closely to the genus Pollicipes, and the parasite in having prehensile antennae, with the disc but little pointed, and with spines at the further end, departs from Scalpellum and approaches Pollicipes! Will any one believe that these several parallel differences, between the cirripedial parasites and the cirripedes to which they are attached, are accidental, and without signification? yet, this must be admitted, if my view of their male sex and nature be rejected.

One more, and the most important, special relation between the parasites and the cirripedes to which they are attached, remains to be noticed, namely that of their prehensile larval antennae. I observed the antennae more or less perfectly in the males of all, and except in *S. villosum*, in all the species, though so utterly different in general appearance and structure, I found the peculiar, pointed, hoof-like discs, which are confined, I believe, to the genera Ibla and Scalpellum. In the hermaphrodite forms of Scalpellum, I was enabled to examine the antennae only in two species, *S. vulgare* and *S. Peronii* (belonging, fortunately, to the two most distinct sections of the genus), and after the most careful measurements of every part, I can affirm that, in *S. vulgare*, the antennae of the male and of the hermaphrodite are identical; but that they differ slightly in the proportional lengths of their segments, and in no other respect, from these same organs in *S. Peronii* – in which again the antennae of the male and of the hermaphrodite are identical. The importance of this agreement will be more fully appreciated, if the reader will consider the following table, in which the generic and specific differences of the antennae in the Lepadidae, as far as known to me, are given. These organs are of high functional importance; they serve the larva for crawling, and being furnished with long, sometimes plumose spines, they serve apparently as organs of touch; and lastly, they are indispensable as a means of permanent attachment, being / adapted to the different objects, to which the larva adheres. Hence the antennae might, *à priori*, have been deemed of high importance for classification. They are, moreover, embryonic in their nature; and embryonic parts, as is well known, possess the highest classificatory value. From these considerations, and looking to the actual facts as exhibited in the following table, the improbability that the parasites of *S. vulgare* and *S. Peronii*, so utterly different in external structure and habits one from the other, and from the cirripedes to which they are attached, should yet have absolutely similar prehensile antennae with these cirripedes, appears to me, on the supposition of the parasites being really independent creatures, and not, as I fully believe, merely in a different state of sexual development, insurmountably great.

The parasites of *S. vulgare* take advantage of a pre-existing fold on the edge of the scutum, where the chitine border is thicker; and in this respect there is nothing different from what would naturally happen

Generic characters of the larval prehensile ANTENNAE in the Lepadidae, as far as known from their imperfect state of preservation, and the number of species examined	Name of species	Length of, from end of disc to the further margin of the oblique basal articulation: Scale, fractions of the 1/6000ths of an inch
LEPAS: disc *large, thin, almost circular*, slightly elongated, with several long spines on the hinder margin; end segment with three very long, plumose spines on the upper *exterior* angle.*	*L. anatifera* [?] *L. australis, L. pectinata, L. fascicularis,*	62 111 51 60
DICHELASPIS: disc *small*, thin, circular, with several spines on the hinder margin; end segment, with two long spines on the upper *exterior* angle.	*D. Warwickii,*	54
CONCHODERMA: disc large, thin, *transversely* elongated, with several long spines on the hinder margin; end segment, with two excessively long, plumose spines on the upper *exterior* corner	*C. Virgata, C. aurita,*	82 —
ALEPAS: disc small, slightly elongated, with two or more spines on the hinder margin; end segment, with two long spines on the upper *inner* corner, and four shorter ones on the *exterior* corner.	*A. cornuta,*	60
IBLA (parasitic males of): disc, *hoof-like, pointed,* elongated, with a single spine on the hinder margin; end segment, with four short spines on the upper *exterior* corner.	*I. Cumingii, I. quadrivalvis,*	22 32–33
SCALPELLUM: disc *hoof-like*, generally *pointed*, and elongated, with a single spine on the hinder margin; end segment, with a notch on the inner* side, bearing two spines, longer than those on the *exterior* corner.	*S. vulgare, S. ornatum, S. Peronii*	39 36 30
POLLICIPES: disc small, *hoof-like*, not pointed, with a single spine on the hinder margin; end segment, as in *Scalpellum.*	*P. cornucopia,*	20

* In the diameter of the disc, the thin membranous border, which is present in the first three genera, is included; but I have some doubts, whether this border be not the first rim of cementing tissue, as all the specimens, of which measurements are here given, had been removed after attachment. In using the terms inner and outer sides of the end segment, it is

Length of, from end of disc to the inner margin of the basal articulation	Width of basal segment, in widest part	Disc, length of	Disc, width of	Ultimate segment, length of	Ultimate segment, width of. Scale, fractions of the 1/20,000ths of an inch
Scale same	Scale same	Scale same	Scale same	Scale same	
—	20	23	22	—	—
—	40	42	39	18	30
—	23	16	14	9	16
40	22	16	15	—	—
—	11	7–8	7–8	6	13–14
40	28	25	35	12	26
—	—	28	40	11	26
—	24	14	12	8	20
—	7–8	7	—	3–4	7–8
—	10	8	5	4	8
19	10	10–11	5–6	6	7
21	10	12	—	—	—
19	—	9	6	5	10
—	6	6	6	6	8

supposed, that this segment is stretched straight, instead of being bent rectangularly outwards, as in its natural position; and then there can be no doubt which is the inner and outer sides.

with an independent parasite; but in *S. ornatum* the case is very different, for here the two scuta are specially modified, *before the attachment of the parasites*, in a manner which it is impossible to believe can be of any service to the species itself, irrespectively of the lodgment thus afforded for the males. So again in *S. rutilum*, the shape of the scutum seems adapted for the reception of the male, in a manner which must be attributed to its own growth, and not to the pressure or attachment of a foreign body. Now there is a strong and manifest improbability in an animal being specially modified to favour the parasitism of another, though there are innumerable instances in which parasites take advantage of pre-existing structures in the animals to which they are attached. On the other hand, there is no great improbability in the female being modified for the attachment of the male, in a class in which all the individuals are attached to some object, than in the mutual organs of copulation being adapted to each other throughout the animal kingdom. /

It should be observed that the evidence in this summary is of a cumulative nature. If we think it highly, or in some degree probable – from the ordinary form of *Ibla Cumingii* having been shown on good evidence to be exclusively female – from the absence of ova and ovaria in the assumed males of both species of Ibla, at the period when their vesiculae seminales were gorged with spermatozoa – from the close general resemblance between the parts of the mouth in the parasites and in the Iblas to which they are attached – from the differences between the two parasites being strictly analogous to the differences between the two species of Ibla – from the generic character of their prehensile antennae – and from other such points – if from these several considerations, we admit that these parasites really are the males of the two species to which they adhere, then in some degree the occurrence of parasitic males in the allied genus Scalpellum is rendered more probable. So the absolute similarity in the antennae of the males and hermaphrodites both in *S. vulgare* and *S. Peronii*; and such relations as that of the relative villosity of the several species in this same genus, all in return strengthen the case in Ibla. Again, the six-valved parasites of *S. Peronii* and *S. villosum* are so closely similar, that their nature, whatever it may be, must be the same; hence we may add up the evidence derived from the identity of the antennae in the parasite and hermaphrodite *S. Peronii*, with that from the antennae in the male *S. villosum*, approaching in character to Pollicipes, to which genus the hermaphrodite is so closely allied; and to this evidence, again, may be added the singular coincident absence of caudal appendages in the male and hermaphrodite *S. villosum*. If these two six-valved parasites be received as the complemental males of their respective species, no one, probably, will doubt regarding the nature of the parasite of *S. rostratum*, in which the direct evidence is the weakest; but even in this case, the particular point of attachment, and the state of development of the valves, form a link connecting in some / degree, the parasites of the first three species with the last two species of Scalpellum, in accordance with the affinities of the hermaphrodites.

When first examining the parasites of *S. rostratum*, *S. Peronii*, and *S. villosum*, before the weight of the cumulative evidence had struck me, and noting their apparent state of immaturity, it occurred to me that possibly they were the young of their respective species, in their normal state of development, attached to old individuals, as may often be seen in Lepas; this, however, would be a surprising fact, considering that *S. rostratum* and *S. Peronii* are ordinarily attached, in a certain definite position, to horny corallines, and considering that the exact points of attachment in these three parasites (of which I have seen no other instance among common cirripedes), namely, between the scuta, would inevitably cause their early destruction, either directly or indirectly, by their living supports being destroyed. Nevertheless, I carefully examined a young specimen of *S. rostratum* only thrice as large as the parasite; and not having very young specimens of *S. Peronii* and *S. villosum*, I procured the young of close allied forms, namely, of *S. vulgare* (with a capitulum only $4/100$th of an inch in length), and of *Pollicipes polymerus* (with a capitulum of less size than that of one of the parasites), and there was not the least sign of anything abnormal in the development of the valves. In *S. vulgare*, at a period when the calcified scuta could have been only $1/100$th of an inch in length (and therefore considerably less than the scuta in the parasites), the upper latera must have been as much as $4/1000$ths of an inch in length, and the valves of the lower whorl certainly distinguishable.

To sum up the evidence on the sex of the parasites, I was not able to discover a vestige of ova or ovaria in the two male Iblas; and I can venture to affirm positively, that the parasites of *S. Peronii* and *S. villosum* are not female. On the other hand, in the two male Iblas, I was enabled to demonstrate all the male organs, and I most / distinctly saw spermatozoa. In the parasitic complemental male of *S. vulgare*, I also most plainly saw spermatozoa. In the parasites of *S. rostratum*,

S. *Peronii*, and S. *villosum*, the external male organs were present. I may here just allude to the facts given in detail under Ibla, showing that it was hardly possible that I could be mistaken regarding the exclusively female sex of the ordinary form of *I. Cumingii*, seeing how immediately I perceived all the male organs in the hermaphrodite *I. quadrivalvis*; and as the parasite contained spermatozoa and no ova, the only possible way to escape from the conclusion that it was the male and *I. Cumingii* the female of the same species, was to invent two hypothetical creatures, of opposite sexes to the Ibla and its parasite, and which, though cirripedes, would have to be locomotive! I insisted upon this alternative, because if the parasite of *I. Cumingii* be the male of that species, then unquestionably we have in *I. quadrivalvis* a male, complemental to an hermaphrodite – a conclusion, as we have seen, hardly to be avoided in the genus Scalpellum, even if we trust exclusively to the facts therein exhibited.

With respect to the positions of the parasitic males, in relation to the impregnation of the ova in the females and hermaphrodites, it may be observed that in the two male Iblas, the elongated movable body seems perfectly adapted for this end; in the males of the first three species of Scalpellum, the spermatozoa, owing to the manner in which the thorax is bent when protruded, would be easily discharged into the sack of the female or hermaphrodite; this would likewise probably happen with the complemental male of *S. rostratum*, considering its position within the orifice of the capitulum, between the mouth and the adductor scutorum muscle. The males of *S. Peronii* and *S. villosum* being fixed a little way beneath the orifice of the sack, below the adductor muscle, are less favourably situated, but the spermatozoa would probably be drawn into the sack by the ordinary action of the cirri of the hermaphrodite, and therefore would at least have as good a chance of / fertilizing some of the ova, as the pollen of many dioecious plants, trusted to the wind, has of reaching the stigmas of the female plants. Regarding the final cause, both of the simpler case of the separation of the sexes, notwithstanding that the two individuals, after the metamorphosis of the male, become indissolubly united together, and of the much more singular fact of the existence of complemental males, I can throw no light; I will only repeat the observation made more than once, that in some of the hermaphrodites, the vesiculae seminales were small, and that in others the probosciformed penis was unusually short and thin.

Viewing the parasitic males, in relation to the structure and appearance of the species to which they belong, they present a singular series. In *S. Peronii* and *S. villosum*, the internal organs have the appearance of immaturity; the shape of the capitulum is specially modified for its reception between the scuta of the hermaphrodite, and several of the valves have not been developed. This atrophy of the valves, is carried much further in *S. rostratum*. In Ibla, many of the parts are embryonic in character, but others mature and perfect; some parts, as the capitulum, thorax, and cirri, are in a quite extraordinary state of atrophy; in fact, the parasitic males of Ibla consist almost exclusively of a mouth, mounted on the summit of the three anterior segments of the twenty-one normal segments of the archetype crustacean. In the males of the first three species of Scalpellum, some of the characters are embryonic – as the absence of a mouth, the presence of the abdominal lobe, and the position of the few existing internal organs; other characters, such as the general external form, the four bead-like valves, the narrow orifice, the peculiar thorax and limbs, are special developments. These three latter parasites, certainly, are wonderfully unlike the hermaphrodites or females to which they belong; if classed as independent animals, they would assuredly be placed not in another family, but in another order. When mature they may be said essentially to be mere bags of spermatozoa. /

In looking for analogies to the facts here described, I have already referred to the minute male Lerneidae which cling to their females – to the worm-like males of certain Cephalopoda, parasitic on the females – and to certain Entozoons, in which the sexes cohere, or even are organically blended by one extremity of their bodies. The females in certain insects depart in structure, nearly or quite as widely from the order to which they belong, as do these male parasitic cirripedes; some of these females, like the males of the first three species of Scalpellum, do not feed, and some, I believe, have their mouths in a rudimentary condition; but in this latter respect, we have, among the Rotifera, a closely analogous case in the male of the Asplanchna of Gosse, which was discovered by Mr Brightwell[59] to be entirely destitute of mouth and stomach, exactly as I find to be the case with the parasitic male of *S. vulgare*, and doubtless with its two close allies. For any analogy to the existence of males, complemental to hermaphrodites, we must look to the vegetable kingdom.

[59] *Annals of Natural History*, vol. ii (2nd series, 1848), p. 153, Pl. vi. Mr Dalrymple has published a very interesting paper on the same subject in the *Philosophical Transactions* (p. 342), 1849; and there is another Memoir by Mr Gosse in the *Annals of Natural History*, vol. vi (1850), p. 18.

Finally, the simple fact of the diversity in the sexual relations, displayed within the limits of the genera Ibla and Scalpellum, appears to me eminently curious; we have (1st) a female, with a male (or rarely two) permanently attached to her, protected by her, and nourished by any minute animals which may enter her sack; (2nd) a female, with successive pairs of short-lived males, destitute of mouth and stomach, inhabiting two pouches formed on the undersides of her valves; (3rd) an hermaphrodite, with from one or two, up to five or six similar short-lived males without mouth or stomach, attached to one particular spot on each side of the orifice of the capitulum; and (4th) hermaphrodites, with occasionally one, two, or three males, capable of seizing and devouring their prey / in the ordinary cirripedial method, attached to two different parts of the capitulum, in both cases being protected by the closing of the scuta. As I am summing up the singularity of the phenomena here presented, I will allude to the marvellous assemblage of beings seen by me within the sack of an *Ibla quadrivalvis* – namely, an old and young male, both minute, worm-like, destitute of a capitulum, with a great mouth, and rudimentary thorax and limbs, attached to each other and to the hermaphrodite, which latter is utterly different in appearance and structure; secondly, the four or five, free, boat-shaped larvae, with their curious prehensile antennae, two great compound eyes, no mouth, and six natatory legs; and lastly, several hundreds of the larvae in their first stage of development, globular, with horn-shaped projections on their carapaces, minute single eyes, filiformed antennae, probosciformed mouths, and only three pair of natatory legs; what diverse beings, with scarcely anything in common, and yet all belonging to the same species!

GENUS: POLLICIPES

PLATE VII

POLLICIPES. Leach, *Journal de Physique*, tom. lxxxv, Julius, 1817[60]
LEPAS. Linn, *Systema Naturae*, 1767
ANATIFA. Brugière, *Encyclop. Méthod. (des Vers)*, 1789
MITELLA. Oken, *Lehrbuch der Naturgeschichte*, 1815
RAMPHIDIONA. Schumacher, *Essai d'un Nouveau Syst. &c.*, 1817 (ante Julium)
POLYLEPAS. De Blainville, *Dict. des Sc. Nat.*, 1824
CAPITULUM (secundum Klein). J. E. Gray, *Annals of Philos.*, tom. x, new series, August, 1825 /

Valvae ab 18 usque ad 100 et amplius: lateribus verticilli inferioris multis; lineis incrementi deorsùm ordinatis: subrostrum semper adest: pedunculus squamiferus.

Valves from 18 to above 100 in number: latera of the lower whorl numerous, with their lines of growth directed downwards: subrostrum always present: peduncle squamiferous.

Hermaphrodite; filamentary appendages either none, or numerous and seated on the prosoma and at the bases of the first pair of cirri; labrum bullate; trophi various; olfactory orifices generally highly prominent; caudal appendages uniarticulate and spinose, or multiarticulate.

Attached to fixed, or less commonly to floating objects, in the warmer temperate, and tropical seas.

It has been remarked, under Scalpellum, how imperfectly that genus is separated from Pollicipes; and we have seen under *Scalpellum villosum* that the addition of a few small valves to the lower whorl, would convert it into a Pollicipes, most closely allied to *P. sertus* and

---

[60] This is one of the rare cases in which, after much deliberation, and with the advice of several distinguished naturalists, I have departed from the Rules of the British Association; for it will be seen that *Mitella* of Oken, and *Ramphidiona* of Schumacher, are both prior to *Pollicipes* of Leach; yet, as the latter name has been universally adopted throughout Europe and North America, and has been extensively used in geological works, it appears to me to be as useless as hopeless to attempt any change. It may be observed that the genus *Pollicipes* was originally proposed by Sir John Hill (*History of Animals*, vol. iii, p. 170), in 1752, but as this was before the discovery of the binomial system, by the Rules it is absolutely excluded as of any authority. In my opinion, under all these circumstances, it would be mere pedantry to go back to Oken's *Lehrbuch der Naturgeschichte* for the name *Mitella* – a work little known, and displaying entire ignorance regarding the Cirripedia.

spinosus. It has also been shown, that the six recent species of Pollicipes might be divided into three genera, of which *P. cornucopia*, *P. elegans*, and *P. polymerus*, would form one thoroughly natural genus, as natural as Lepas and the earlier genera; *P. mitella* would form a second; and *P. sertus* and *P. spinosus* a third; but I have acted to the best of my judgement in at present retaining the six species together. As far as the valves of the capitulum are concerned, it would be very difficult to separate *P. mitella* from *P. sertus* and *P. spinosus*. /

Description. The number of valves in the capitulum has in this genus acquired its maximum. The number varies considerably in the same species, and even on opposite sides of the same individual, and generally increases with age. It is more important, that the number of the whorls in *P. cornucopia*, and in the two following closely allied forms, also increases with age. In *P. sertus* and *P. spinosus*, even the number of the whorls varies in different individuals, independently of age. The valves are arranged alternately with those above and below; they are generally thick and strong, making the capitulum somewhat massive; in some species they are subject to much disintegration; but in others, the apices of the several valves, especially of the carina and rostrum, are well preserved, and project freely: they are covered with membrane, which, differently from in most species of Scalpellum, either does not bear any spines, or only exceedingly minute points. In all the species there is a sub-rostrum and sub-carina, and often beneath these a second sub-rostrum and sub-carina. In medium-sized specimens there are at least twenty valves in the lowermost whorl. The carina is either straight or curved, but never rectangularly bent, and is always of considerable breadth. None of the valves are added to at their upper ends. The scuta have a deep pit for the adductor muscle. The valves lie either some little way apart, or more commonly close together. In *P. mitella* the scuta and terga are locked together by a fold, and the valves of the lower whorl overlap each other in a peculiar manner, resembling that in which the compartments in the shells of sessile cirripedes fold over each other.

The *peduncle* is of considerable length in some of the species, and rather short in others; it is, in every case, clothed with calcified scales. The scales in the first four species are placed alternately and symmetrically; they are formed and added to in the same manner as in Scalpellum; they differ in size according to the size of the individual, and consequently the lower scales / on the peduncle, formed when the specimen was young, are smaller than the upper scales; the lower scales are separated from each other by wide interspaces of membrane, owing to the continued growth of the peduncle by the formation of new layers of membrane, and the disintegration of the old outer layers. Each scale is invested by tough membrane (or has been, for it is often abraded off), in the same manner as the valves; each is furnished with one or more tubuli, in connection with the underlying corium. In *P. sertus* and *P. spinosus*, the scales are small, spindle-shaped, and not of equal sizes, and the rows are distant from each other, so that their alternate arrangement is not distinguishable; in these two species, new scales are formed round the summit of the peduncle, and the growth of each is completed whilst remaining in the uppermost row; but, besides these normal scales, such as exist in the other species of Pollicipes and in Scalpellum, new scales are formed in the lower part of the peduncle, which are generally of very irregular shapes, are often larger than the upper ones, are crowded together, and sometimes do not reach the outer surface of the membrane. This formation of scales in the lower part of the peduncle, independently of the regular rows round the uppermost part, is perhaps a feeble representation of the calcareous cup at the bottom of the peduncle in the genus Lithotrya. The prehensile antennae will be described under *P. cornucopia*.

Size. Most of the species are large: and *P. mitella* is the most massive of the pedunculated cirripedes.

The *mouth* is not placed far from the adductor muscle. The labrum is highly bullate. The mandibles have either three or four main teeth (Pl. X, fig. 1), with often either one or two smaller teeth inserted between the first and second. The maxillae (Pl. X, figs 13,14), have their edges either straight and square, or notched, or more commonly with two or three prominences bearing tufts of finer spines. The outer maxillae (fig. 17) generally have a deep notch / on their inner edges, but this is not invariable. The olfactory orifices in most of the species are highly prominent.

Cirri. The first pair is never placed far distant from the second. The posterior cirri have strong, somewhat protuberant segments; and between each of the four or five pair of main spines (Pl. X, fig. 27), there is a rather large tuft of straight, fine, short bristles. The second and third pair have the basal segments, either of the anterior rami, or of both rami, so thickly clothed with spines (fig. 25), as to be brush-like: in *P. mitella*, however, the third pair is like the three posterior pair in the arrangement of its spines, in this respect resembling the sessile Chthamalinae. The caudal appendages are either uniarticulate and spinose, or multiarticulate: it is remarkable that there should be this difference in such closely allied species as *P. cornucopia* and *P. polymerus*: the short, obtuse, obscurely articulated caudal appendage of the former species (fig. 22) makes an excellent passage from the uni-articulate (fig. 19) to the multiarticulate form, as in *P. mitella*.

The stomach, in those species which I opened, is destitute of caeca; the hepatic glands are arranged in straight lines; the rectum is unusually short. The prosoma is well developed.

In *P. cornucopia*, *P. elegans*, and *P. polymerus*, there are numerous filamentary appendages both on the prosomoa, and at the bases of the first pair of cirri: these appendages are occupied by testes, and I suspect stand in relation to the length of the peduncle and consequent great development of the ovaria. In order to give space for the filamentary appendages, the sack (generally roughened by small inwardly pointing papillae) penetrates more deeply than usual into the upper part of the peduncle. There are small ovigerous fraena in *P. sertus*, *P. spinosus*, and *P. mitella*: in the three other species, the fraenum or fold occupies the usual position on each side, and is large; but in one specimen carefully examined by me, I was unable to see any glands; and in another specimen, the / ovigerous lamellae were not attached to the fraena; hence I conclude that the fraena are functionless in these three species.

Affinities. I have already remarked on the close relationship between this genus and Scalpellum; there is also some affinity with Lithotrya.

Distribution. All over the world. The *P. cornucopia* ranges from Scotland to Teneriffe: the *P. polymerus* is found in opposite hemispheres in the Pacific Ocean, extending from California to at least as far as 32° south of the Equator.

Geological history. Having so lately given, in the *Memoirs of the Palaeontographical Society*, a full account of all the fossil species known, I will not repeat here the conclusions there arrived at. I will only state, that species of Pollicipes are found in all the formations, extending from the Lower Oolite to the Upper Tertiary beds.

1. POLLICIPES CORNUCOPIA

PLATE VII. FIG. 1

POLLICIPES CORNUCOPIA. Leach, *Encyclop. Brit. Supp.*, vol. iii, 1824
POLLICIPES SMYTHII, var. Leach, ibid.
LEPAS POLLICIPES. Gmelin, *Systema Naturae*, 1789
LEPAS GALLORUM. Spengler, *Skrivter Naturhist. Selskabet*, Bd. i, Pl. vi, fig. 9, 1790

P. capitulo, valvarum duobos aut pluribus sub-rostro verticillis instructo: valvis albis, aut glaucis: pedunculo, squamarum densis verticillis symmetricè dispositis.
Capitulum with two or more whorls of valves under the rostrum; valves white or grey; scales on the peduncle symmetrically arranged in close whorls.

Maxillae with three tufts of fine bristles, separated by larger spines: segments in the first cirrus less than half the number of those in the sixth cirrus: caudal appendages multi-articulate: filamentary appendages attached to the prosoma. /

Coast of Portugal; mouth of the Tagus. England,[61] Ireland, and the Firth of Forth in Scotland. Mediterranean (according to Brugière): Teneriffe: Mogador, Africa.

Capitulum, obtusely triangular, massive: valves close together, rather thick, with their exterior surfaces convex, naked, except in the lower parts, where united together by tough, greenish-brown membrane, destitute of spines. The edges of the orifice are widely bordered by membrane, coloured fine crimson red. The valves, in a specimen with a capitulum above three-quarters of an inch long, were 52 in number; in a specimen one-fifth of an inch long, only between 20 and 30. Two whorls of valves are distinct beneath the carina and rostrum. In one specimen in Mr Cuming's collection, with a capitulum 1·4 of an inch long, there were three whorls beneath the rostrum, and four beneath the carina. The scuta, terga, and carina are much larger than the other valves.

Scuta, oval, the basal and tergo-lateral margins sweeping into each other, and the apex pointed; internally (Pl. VII, fig. 1*a*) the pit for the adductor muscle is deep.

---

[61] This species is said by Montagu (*Test. Brit. Supplement*) to have been found attached to drift timber in the Firth of Forth, and to the bottom of a wrecked vessel towed into Dartmouth. According to Mr W. Thompson (*Annals of Nat. Hist.* vol. xiii, p. 436), it has been found attached to woodwork near Dublin.

Terga, larger than the scuta, internally (fig. 1*a*) slightly concave; carinal margin much curved and protuberant; basal angle blunt; scutal margin either curved with the upper part straight, or formed of two almost distinct lines, corresponding with the tergal margin of the scutum, and with one of the sides of the upper latus.

Carina, much curved, extending far up between the terga, internally deeply concave, widening much from the top to the bottom; basal margin highly protuberant, with a central portion either truncated and very slightly hollowed out, or bluntly and rectangularly pointed, with the apex itself rounded.

Rostrum, not one-third of the length of the carina, concave, triangular, with the basal margin slightly protuberant. / Of the other valves, including the sub-carina and sub-rostrum, the shape of their inner surfaces is subtriangular, with the basal margin convex; externally the umbones are pointed, and slightly curled inwards, so as to overlap each other like tiles: the smaller valves, however, of the lower whorls (fig. 1*a*) are more or less transversely elongated, so as to become almost elliptic instead of triangular. Of the latera, the upper pair, which corresponds to the interspace between the scuta and terga, is the largest, but barely exceeds in size the pair answering to the carinal latera in Scalpellum, which lie between the terga and carina: the next largest pair is the rostral, or that between the scuta and rostrum. Some, however, of the lower latera are of nearly equal size.

Peduncle, narrower, but generally longer than the capitulum; upper part encased with small calcareous scales, with their apices curved inwards, and overlapping each other. The inner surface of each scale is triangular, with the basal margin protuberant. The scales continue to grow or be added to, only in about the ten upper whorls, which form but a small part of the whole peduncle; in the lower part, the scales become further and further separated from each other. The surface of attachment, in full-grown specimens, is broad; but in two very young specimens, which I removed with great care after the action of potash, I found the peduncle ending in a filiform prolongation, such as often occurs in *Scalpellum vulgare* and in *Lepas fascicularis*. At the extremity of the pointed peduncle, there were seated the larval prehensile antennae, of which the following measurements are given to show how minute they are.

THE LEPADIDAE

	Inch.
Length, from apex of disc, to the further edge of the basal articulation	20/6000
Breadth of basal segment, in broadest part	6/6000
Hoof-like disc, length of	6/6000
Ultimate segment, entire length of	6/6000
Ultimate segment, breadth, in broadest part	6/2000

The disc resembles a broad, rounded hoof, very little longer than broad, and narrowed in at the heel; the / apex is not at all pointed, and bears some minute and thin spines. There is one large spine on the underside of the disc; and another on the basal segment, on the outside, in the usual position. The ultimate segment is long and thin; it has a notch on the inner side (the segment supposed to be stretched forward), bearing two or three long flexuous spines; and there are three or four other spines on the summit: altogether there is a close resemblance with the antennae in Scalpellum, excepting that the hoof-like disc is not here pointed.

Colours. Valves internally tinted, in parts, grey; peduncle, brown; corium of sack, purplish-brown, of peduncle, rich coppery brown; cirri, banded dorsally, and with the front surfaces of the segments, purplish-brown. Edge of the orifice of sack, fine crimsom red. The specimen here described had been dried for a few weeks, and was then moistened.

Dimensions. The largest specimen which I have seen, in Mr Cuming's collection, had a capitulum 1 and 4/10ths of an inch long; a fine specimen, from Teneriffe, was 9/10ths in length. In a specimen with a capitulum 1/20th of an inch long, and about the same in breadth, there were eighteen valves; so that, besides the principal valves, five pair of latera, the sub-carina, and sub-rostrum, were already developed, and on the upper part of the peduncle, there were many calcareous scales.

Filamentary appendages. The prosoma is well developed, with thirteen or fourteen pair of short, blunt filaments, placed close together in two longitudinal rows; those nearest the thorax are the longest; outside this double row, on each side, there is a row of papillae, indicating a tendency to the formation of two other rows of filaments. There is a pair of longer filaments, one on each side of the mouth, pointing upwards, and thinly clothed with long spines; at the bases of the first pair of cirri there is a second pair of filaments, shorter and bearing a few minute spines. The bottom of the sack is studded with small rounded papillae, with roughened summits. /

Mouth, not placed very far from the adductor muscle.

Labrum, highly bullate, equalling, in its longitudinal diameter, the rest of the mouth; upper part square, not overhanging the lower part; there are some small teeth on the crest.

Palpi, oval, outer and inner margins nearly alike, thickly clothed with spines.

Mandibles, with three very strong, yellow teeth; inferior point broad, coarsely pectinated. In one specimen, on one side, the third tooth was represented by two smaller teeth.

The *maxillae* bear three conspicuous tufts of fine bristles, separated by larger spines; the first tuft is placed close to the two, upper, large, but unequally sized spines; the second tuft is placed in the middle, and the third at the inferior angle. The two latter tufts stand on prominences; between the two upper tufts there are three pair, and between the two lower tufts four or more pair of rather strong spines (see fig. 13, Pl. X, in the allied *P. polymerus*).

Outer maxillae, with the inner edge divided in the middle by a conspicuous notch, and with the bristles above and below short, making two *equal* combs. On the exterior surface, the bristles are longer and more spread out. Olfactory orifices prominent, protected by a punctured swelling between the bases of the first pair of cirri.

Cirri, short and rather thick; the first pair is not far removed from the second. The segments of the three posterior pair are somewhat protuberant, bearing six pair of short, strong spines, graduated in length, between which there is a very thick, longitudinal brush of short, fine, straight bristles, of which the lower ones are the longest; some thick, minute spines arise from the upper lateral edges of the segments. The spines in the dorsal tufts are short, much crowded, and of nearly equal length (see fig. 27, Pl. X, in the allied *P. polymerus*). In a specimen in which the sixth cirrus had seventeen / segments, the first cirrus had, in the shorter ramus, eight segments, of which the lower four were thick and protuberant, with the spines doubly serrated. In this same specimen, the anterior ramus of the second cirrus had twelve segments, of which the five basal ones were highly protuberant, and thickly clothed with non-serrated spines. In the third cirrus the basal segments of the anterior ramus are highly protuberant. The basal segments in the posterior rami of both these cirri, are slightly protuberant, but otherwise resemble the segments in the three posterior pair.

The *caudal appendages* (Pl. X, fig. 22), in full-grown specimens, just exceed in length the lower segments of the pedicels of the sixth cirrus; they are nearly cylindrical, bluntly pointed, with five oblique imperfect articulations; the lower or basal articulations cannot be traced all round, being distinct only on the ventral surface. There is a row of short spines round the upper edge of each segment, with a little, short tuft on the point of the terminal segment. In a rather young specimen, however, with its capitulum one-fifth of an inch long, each appendage certainly consisted of a single segment, with spines only on the summit.

Penis purple, with excessively short and fine spines in tufts, chiefly near the extremity. In a specimen with a capitulum only one-fifth of an inch long, the penis consisted of a mere pointed papilla, not so long as the caudal appendage, and therefore equalling in length only the lower segment of the pedicel of the sixth cirrus.

Ovigerous fraena. I could see none, though there were two large lamellae in the sack. The ova were flesh-coloured, but they had been dried and then placed in spirits. The ova were wonderfully numerous, oval, much elongated, and 1/100th of an inch in length. /

2. POLLICIPES ELEGANS

POLLICIPES ELEGANS. Lesson, *Voyage de la Coquille*, tom. ii, p. 441, 1830, et *Illust. Zool.*, Pl. xxxix, 1831
POLLICIPES RUBER. G. B. Sowerby, *Zoolog. Proc.*, p. 74, 1833

P. capitulo, valvarum duobus aut pluribus sub-rostro verticillis instructo: valvis et pedunculi squamis rufo-aurantiacis: squamarum verticillis densis symmetricè dispositis.
Capitulum with two or more whorls of valves under the rostrum: valves and scales of peduncle reddish-orange; the latter symmetrically arranged in close whorls.

Maxillae with three tufts of fine bristles, separated by larger spines; segments in the first cirrus more than half the number of those in the sixth cirrus; caudal appendages multiarticulate; filamentary appendages attached to the prosoma.

Coast of Peru, Payta, attached to wooden posts, according to Lesson: Lobos Island, Peru, Mus. Cuming: West Coast of Mexico, Tehuantepec, on an exposed rock, according to Hinds.

The resemblance of this species is so close to *P. cornucopia*, that it is quite useless to do more than point out the few points of difference. Valves of the capitulum and scales of the peduncle, coloured (after having been in spirits), reddish-orange. In a specimen in which the capitulum was 1·3 of an inch in length, there were three whorls of valves below the carina; in this large specimen altogether there were about eighty valves; in medium-sized specimens, the number is about the same as in *P. cornucopia*. The upper latus (viewed internally), has an area about twice as large as that latus, which corresponds to the interspace between the carina and terga; whereas in *P. cornucopia* the upper latus is only slightly larger than this same valve. The apex of the basal internal / margin of the carina is here rounded, instead of being square, as is generally the case with *P. cornucopia*. The strong membranous margin of the orifice of the sack, in its upper part, is almost one-third as wide as the widest part of the terga, whereas in *P. cornucopia* it is only one-fourth of the same width. The peduncle apparently is rather longer, compared with *P. cornucopia*, and the calcareous scales on it perhaps a little larger in proportion.

In a very young specimen, with the capitulum barely exceeding 1/20th of an inch in length, I could distinguish the sub-rostrum, sub-carina, the upper, and some of the lower latera.

Filamentary appendages. These, in a medium-sized specimen, are arranged on the prosoma in four longitudinal approximate rows, there being twelve in each row; those in the two outer rows are only half the length of those in the two inner rows; those nearest the thorax are the longest; there are some papillae outside the outer rows. In a very large specimen with its capitulum 1·3 in length, these filaments were very much more numerous, and some were placed on the first segment of the thorax, and at the bases of several of the posterior cirri. Some of the filaments are bifid, trifid, and even branched. In all the specimens, at the bases of the first pair of cirri, there are, on each side, a pair of filaments (one below the other), pointing upwards, less than half as long as those on the prosoma: also on each side of the mouth, there is a longer and thicker filament, pointing upwards, with a few very minute scattered spines on it; the apices of these three pair of filaments, as well as of some of the others, are roughened with very minute pectinated scales. All these filaments were gorged with the branching testes.

Mouth. The parts are closely similar to those in *P. cornucopia*; in the mandibles, the interspace between the third tooth and the inferior angle, is slightly pectinated: in the maxillae, there are six or eight pairs of spines between the two upper tufts of fine spines. /

Cirri. These are in most respects similar to those of *P. cornucopia*. In a specimen in which the sixth cirrus had eighteen segments, the shorter ramus of the first pair had ten segments, of which the five lower segments were thick and clothed with doubly serrated spines. In the second cirrus the anterior ramus had fifteen segments, of which the four basal ones were highly protuberant, and thickly clothed with spines. These spines, and some on the third cirrus, and a few on the first cirrus, have peculiar bent teeth, presently to be described under *P. polymerus*. These singularly toothed spines are absent in *P. cornucopia*. From the above numbers, we see that the first and second pairs of cirri have more segments in proportion to the sixth pair, than in *P. cornucopia*; and in the second pair, a fewer proportional number of the basal segments are protuberant and thickly clothed with spines.

Caudal appendages, shorter than the lower segments of the pedicels of the sixth cirrus, with only four articulations; rather constricted near the base.

The *ovigerous fraena* consist of very long and prominent folds, thinning out to nothing towards the bases of the scuta, but not furnished, as far as I could see, with glands, and therefore not normally functional.

Diagnosis with P. cornucopia. The reddish-orange colour of the valves alone suffices. There is a very slight difference, in the larger proportional size of the upper latera, and in the outline of the basal margin of the carina. In the maxillae there is, in *P. elegans*, a greater width between the two upper tufts of fine spines. In the cirri, the segments in the first pair, are more than half as many as those in the sixth pair; in the anterior ramus of the second pair, only $4/15$ths of the segments are protuberant and brush-like, whereas in *P. cornucopia* $5/12$ths are in this condition. /

3. POLLICIPES POLYMERUS

PLATE VII. FIG. 2

POLLICIPES POLYMERUS[!]. G. B. Sowerby, *Proc. Zool. Soc.*, p. 74, 1833
POLLICIPES MORTONI[!]. Conrad, *Journal Acad. Nat. Sci. Philadelphia*, vol. vii, p. 261, Pl. xx, fig. 12, 1837

P. capitulo, valvarum duobus, tribus, aut pluribus sub-rostro verticillis instructo: valvis sub-fuscis: lateribus à supremo ad infimum gradatim quoad magnitudinem positis: carinae margine basali (introrsùm spectanti) ad medium excavato: pedunculi squamarum verticillis densis, symmetricè dispositis.

Capitulum with two, three, or more whorls of valves under the rostrum: valves brownish: latera regularly graduated in size from the uppermost to the lowest: carina with the basal margin (viewed internally), hollowed out in the middle: scales of the peduncle symmetrically arranged in close whorls.

Maxillae with three tufts of fine bristles, separated by larger spines; caudal appendages uniarticulate; filamentary appendages attached to the prosoma.

Upper California, St Diego and Barbara, 32° to 35°N., according to Conrad; Mus. Cuming: Low Archipelago, Pacific Ocean; Mus. Coll. of Surgeons: Southern Pacific Ocean, collected during the Antarctic Expedition, Mus. Brit.

Capitulum, but little compressed, broad, with the scuta and terga placed in a more oblique direction, with respect to the peduncle, than is usual, so that the line of orifice forms an unusually small angle with the basal margin of the capitulum. The capitulum is composed of several whorls of valves, which gradually decrease in size from above downwards. In a medium-sized specimen there were four whorls under the rostrum; in the lowest of these whorls, there were between eighty and ninety valves, and in the whole capitulum from one hundred and seventy, to one hundred and eighty. The valves in the lower whorls / are not of equal sizes. Viewed externally, the valves seem to touch and overlap each other; viewed internally (Pl. VII, fig. 2a) they are found to be just separated from each other by transparent membrane; none of the valves are articulated together. The outer surfaces of nearly all the valves, except in the two last formed whorls, are much disintegrated, and seem to be composed of alternate white and brown layers of shell. The membrane connecting the valves, as well as that of the peduncle (in specimens long kept in spirits), is brown; but in some dried specimens, there are indications of its having been coloured crimson (as in *P. cornucopia*), round the orifice and between the valves.

Scuta, irregularly oval, convex, narrow at the upper end; basal margin may be almost said to be formed of three short, unequal margins, corresponding with the rostrum, the rostral and the adjoining latus. The edge corresponding with the latter, is the best marked, and is generally slightly hollowed out, as if a piece had been broken off. The tergo–lateral margin is curved and protuberant. The umbo projects a little over the scutal margin of the terga.

Terga, projecting beyond the other valves to an unusually small degree, broadly oval; basal angle bluntly pointed, apex rounded, blunt; scutal margin, hollowed out to receive the upper part of the tergal margin of the scuta; carinal margin curved and protuberant; occludent margin consists of two short sides at right angles to each other. The whole valve in length and area is about equal to the scuta; internally, somewhat concave.

Carina, triangular, rather narrow, internally deeply concave, very slightly curled inwards; basal margin protuberant, with a large central portion considerably hollowed out.

Rostrum, triangular, of nearly the same shape as the carina, but only one-third of its length, internally very slightly concave, and with the basal margin various, being either truncated or angularly prominent in the middle. /

Latera. The upper pair (corresponding to the interval between the scuta and terga) is only a trifle larger than the latera immediately beneath; and these only a little larger than those lower down. In the lowest whorl, the valves are very minute, though still about twice as large as the scales on the peduncle, and of a different shape from them. The upper latera (viewed internally) are almost diamond shaped, owing to the prominence of the basal margin, but this varies considerably in degree. The latera in the next whorl are triangular, with the basal margins protuberant and arched, in a less and less degree in the lower whorls, until in the lowest, the valves are elongated transversely.

Microscopical structure. A valve placed in acid leaves a thick opaque mass, formed of three different kinds of tissue, one having a finely shaded appearance; a second with a largely hexagonal reticulated structure, and the third thin, transparent, and marked with arborescent lines, which I imagine to be tubes, as will be hereafter seen in Lithotrya. Near the exterior surface, there are many tubuli. It appears to me probable that the strong tendency which the valves in this species have to disintegrate, is connected with the unusual quantity of animalized tissue contained by them. Externally the valves are covered by a strong membrane, either white or yellow, or white streaked with yellow, and marked by lines of growth, and by longitudinal, sinuous, little ridges.

Peduncle, in the upper part, of rather less diameter than the capitulum; twice or thrice as long as it; tapering a little downwards; surface of attachment wide and flat. Calcareous scales, minute, symmetrically and closely packed together: each scale is much flattened, and its shape, including the imbedded portion, is that of a spear with its point broken off. The basal end of each scale is conically hollow, and from the layers of growth conforming to this hollow, there is a false *appearance* of an open tube running through the scale.

Attachment. The surface of attachment is wide: the / two cement ducts, after running down the sides of the peduncle in a sinuous course, within the longitudinal muscles and close outside the ovarian tubes, pass through the corium, and then separately form the most abrupt loops or folds. These are represented in Pl. IX, fig. 2, in which a space about $1/10$th of an inch square is given, as seen from the outside. At each of the bends, an aperture has been formed through the membrane of the peduncle, and cement poured forth. The manner in which the discs of cement *b* come out of the two ducts *a a*, and reach the external surface, is shown in the section, figure 2*a*′. The two tubes are firmly attached to the older layers of membrane, and are covered by the last formed layers. In a young specimen, the cement ducts were a little above $2/2000$ths of an inch in diameter, which had increased, in a medium-sized specimen, to $5/2000$. The cement glands are retort shaped, seated near each other, high up in the peduncle.

Size. The largest specimen which I have seen, was three inches in length including the peduncle; the capitulum was $9/10$ths of an inch long, and one in width.

Young specimen. I examined one with a capitulum $18/1000$ths of an inch long, measured from the lowest whorl to the tips of the terga; the width was only $13/1000$ths of an inch; in old specimens the width of the capitulum is greater than the length. The length of one of the scuta was $^{14}/_{1000}$ths of an inch, therefore, greater than the width of the entire capitulum, which is not the case with mature specimens. Besides the scuta and terga, the carina and rostrum, and three pair of large latera, there was a lower whorl formed of ten or twelve valves, giving altogether to the capitulum of this very small specimen, either twenty-two or twenty-four valves.

Shape of body, sack, colours, etc. From the position of the orifice of the capitulum, the animal's body is suspended to the scuta in a more transverse direction than is usual. The prosoma is well developed, and is distinctly separated from the three posterior thoracic segments, / by a band of thin membrane. The tunic of the basal part of the sack, where it enters the peduncle in a blunt point, is thickened and covered with roughened rounded papillae. The corium of the sack under the valves, is coloured (after spirits) so dark a brown as to be nearly black; the cirri and trophi are similar, but with a tinge of greenish-purple.

Filamentary appendages. Of these there were, on the prosoma of one specimen, twelve pairs, and in another specimen fourteen pairs, seated in two approximate rows; the middle filaments are the longest, equalling about half the diameter of the thorax: each is flattened, and tapers but little towards its summit, which is roughened with microscopical crests serrated on both sides; on the summit, also, there are a few bristles and some very short, thick, minute spines. These appendages are directed rather towards each other, and towards the thorax. I do not doubt that their numbers vary according to the size of the specimen. I believe that they are occupied by testes. Outside these filaments, on each side of the prosoma, there are two very irregular rows of papillae, intermediate in length between the filaments and the rounded swellings at the bottom of the sack. Beneath the basal articulation of the first cirrus, there is on each side, a short appendage, with a few bristles on its summit. Lastly, on each side of the middle of the mouth, on the prosoma, there is a longer appendage, dark coloured, furnished with a few scattered bristles on its sides and apex, and directed upwards and a little towards the adductor scutorum muscle.

Mouth. Labrum highly bullate, but with the uppermost part not more bullate than the lower part, and therefore not overhanging it; basal margin much produced; crest with some small blunt teeth and some bristles. The inner fold of the labrum is much thickened, yellow, punctured, and with a tuft of fine bristles on each side.

Palpi, approaching each other but not touching, club shaped, or with broad and square extremities, thickly fringed with serrated bristles. /

Mandibles with three unusually strong teeth, slightly graduated in size, with the inferior angle very coarsely pectinated; the lower edges of the main teeth are roughened.

Maxillae (Pl. X, fig. 13). Spinose edge about half the length of the mandibles; the two upper spines are unusually strong; close under, and almost hidden by them, there is a tuft of fine spines; in the middle there is a second similar tuft mounted on a prominence; and at the inferior angle there is a third tuft, also mounted on a rather wider prominence, not quite accurately figured. In the interspaces between these tufts there are three or four pairs of spines of the usual appearance and projecting just beyond the fine tufts; the upper of the two interspaces is rather narrower, but rather deeper, than the lower interspace. Apodeme very long, irregularly shaped, like an S, with a remarkable elbow near its attachment; apex slightly enlarged, thin and rounded.

Outer maxillae. On the inner margin there is a deep and conspicuous notch, above and beneath which, there is a compact row of serrated bristles; exteriorly the bristles are rather longer.

Olfactory orifices very prominent, pointing obliquely towards each other.

Cirri. Posterior cirri moderately long, much curled, with the segments (Pl. X, fig. 27) flattened and wide; the anterior surface hemispherically protuberant, supporting six pairs of spines, of which the lower ones approach each other; between these spines there is a large tuft of very fine spines, of which the central ones are the longest; there is an upper lateral group of very short strong spines; dorsal tufts composed of short, fine numerous spines. *First pair* seated close to the second pair, short, having in both rami eight segments, whereas in the same individual the second pair, which is nearly twice as long, had thirteen, and the sixth pair eighteen segments. Rami of the first pair nearly equal in length, with their segments, excepting the two upper ones, thickly paved with bristles, in the / midst of which a tuft of fine spines, as in the posterior cirri, may be distinguished; the dorsal tufts encircle the whole of each segment; the spine-bearing anterior surfaces are protuberant chiefly in the upper part, so that they are oblique. The posterior [?] ramus has its segments much wider than those on the other ramus; and among the common spines, in the third and fourth segments (counting from the bottom), there are some very strong spines with their upper ends coarsely and doubly pectinated, each tooth being upwardly bent into a rectangular elbow. In the fifth segment, some of the spines are doubly pectinated with simple teeth; and most of the spines are doubly serrated. The *second* (Pl. X, fig. 25) and *third cirri* have the five basal segments ($5/13$ths of the whole number in the second cirrus, and $5/14$ths in the third cirrus) of their anterior rami, extremely broad, protuberant, and paved with serrated bristles, among which (except on the actual lowest segment), there are some simply pectinated spines, and others with their teeth elbowed, exactly as in the first cirrus. The basal segments of the posterior rami of the second and third cirri, differ from the three posterior cirri only in the spines being slightly more numerous; but none of them are pectinated.

Pedicels, rather short; the upper segment resembles, in the arrangement of its spines, the segments of the posterior cirri; the lower segment is longer than the upper, and has *two* tufts of fine spines, between the two rows of long spines. In the second and third cirri, these two intermediate tufts on the lower segment of the pedicel, are not so distinctly separated from each other.

Caudal appendages, very small, uniarticulate, blunt and rounded; tips bearing a few, very short, thick spines.

Alimentary canal. Oesophagus, somewhat curved at the lower end, where it enters the stomach, which has no caeca; rectum, unusually short, extending from the anus only to the base of the fifth pair of cirri. Within the stomach, from top to bottom, there were thousands of a bivalve entomostracous crustacean. /

Generative system. Both ovaria and testes are largely developed; the former fill the long peduncle; the testes enter both the pedicels of the cirri, and the filamentary appendages on the prosoma; vesiculae seminales very large, reflected at their ends, extending across each side of the stomach. Penis rather small, coloured purplish, with numerous little tufts of bristles.

Variation. In some specimens in the British Museum, collected by Sir J. Ross, in the Southern ocean, and in another older set from an unknown source, several parts of the outer tunic of the animal's body presented the remarkable fact of being calcified, but to a variable degree; whereas in several specimens from California, there was no vestige of this encasement. Considering it most improbable that the calcification of the integuments should be a variable character, I most carefully compared the above-mentioned sets of specimens, valve by valve, trophi by trophi, and cirri by cirri, and found no other difference of any kind; therefore I cannot hesitate to consider both to be the same species. The first Southern specimen which I examined presented the following characters: on the prosoma there was a central longitudinal band, formed of a thin, brittle, brown-coloured calcified layer, which became irregularly rather narrow towards the thorax; on each side it sent out six or seven irregular rectangular plates, which surrounded and supported the bases of the two rows of filamentary appendages; and outside these, some of the papilliform projections also had their bases surrounded by small, calcified, separate rings. The thoracic segments corresponding with the second, fourth, fifth, and sixth cirri had, on each side, an elongated calcified plate; on the ventral surface of the thorax, between the first and second cirri, there were two minute plates. In all the cirri, excepting the first pair, the segments of the rami, and in the three posterior pairs, the segments of the pedicels, had their dorsal surfaces strengthened by oblong, quadrilateral, calcified shields, the upper margins of which are notched for the dorsal tufts of spine, / and the two lateral margins are also slightly hollowed out (these are represented in fig. 27). The lower segments of the pedicels of some of the cirri, had an additional calcified plate on the antero-lateral face.

These plates are of a faint brown or yellowish colour, and are conspicuous: the degree of calcification differs considerably; some are quite brittle and very thin, others half horny, and effervesce only slightly in acids. After having been placed in acid, there is no apparent difference between the parts before occupied by the calcified plates and the surrounding membrane; these plates, however, are not superficial, but consist of several of the laminae, which together compose the ordinary integument, in a calcified condition. Like the integuments of the body, and unlike the valves of the capitulum, these calcified plates are thrown off at each exuviation. Neither the exact shape nor number of the plates corresponded in different individuals, nor even on opposite sides of the same individual. The margins of the plates often have a sinuous corroded appearance; they are, moreover, often penetrated by minute rounded holes, that is, by minute, rounded, non-calcified portions. In one specimen from the Antarctic expedition, there were only here and there a single shield on the segments of the posterior rami, and no plate on the prosoma. Of two specimens in another and older set in the British Museum, from an unknown locality, both had shields on the segments of the cirri, but only one had the large plate on the prosoma. I may here mention that in one specimen, in which the calcified plates were most developed, and which was nearly ready to moult, there were, within the filamentary appendages on the prosoma, small irregular balls of calcareous matter, appearing to me as if calcareous matter had been morbidly excreted, and not like a provision for the future.

Range. This species, in the present state of our knowledge, seems to range further than any other of the genus, extending from Upper California (lat. 32° to 35°N.), across the Pacific, to at least 32°S., perhaps / much farther south, for it was collected during the Antarctic expedition, and 32° was the highest latitude traversed by that expedition.

Affinities. This species is closely related to *P. cornucopia* and *P. elegans*, but differs rather more from them, than these two do from each other. In the capitulum the chief distinctive characters are – the more perfect graduation in size, and the greater number (taking equal-sized specimens), of the whorls of latera – the darker colours – the central part of the basal margin of the carina in this species, being considerably excised – the peculiar form of the basal margin of the scuta – and lastly, the scutal margin of the terga being more hollowed out. In the animal's body, the most obvious distinctive character is the uniarticulate caudal appendage. This species agrees with *P. elegans*, in the presence of the singular elbowed teeth, on some of the spines in the first three pairs of cirri.

4. POLLICIPES MITELLA

PLATE VII. FIG. 3

LEPAS MITELLA. Linn, *Systema Naturae*, 1767
POLLICIPES MITELLA. G. B. Sowerby, *Genera of Shells*, fig. 2
POLYLEPAS MITELLA. De Blainville, *Dict. Sc. Nat.*, Plate, fig. 5, 1824
CAPITULUM MITELLA. J. E. Gray, *Annals of Philosoph.*, new series, vol. x, 1825

P. capitulo valvarum unico sub-rostro verticillo instructo: laterum pari superiore (introrsum spectanti) inferiorum magnitudinem ter aut quater superante: lateribus inferioribus utrinque obtegentibus: pedunculi squamarum verticillis densis, symmetricè dispositis.

Capitulum with only one whorl of valves under the rostrum: the upper, pair of latera, viewed internally, are three or four times as large as the lower latera, which overlap each other laterally: scales of the peduncle symmetrically arranged in close whorls. /

Maxillae, deeply notched: caudal appendages, multiarticulated: filamentary appendages, none.

Philippine Archipelago, Mus. Cuming: China Sea, Mus. Brit.: Amboyna and East Indian Archipelago, according to Rumphius and other authors: Madagascar, according to J. E. Gray.

Capitulum, compressed, consisting of the scuta, terga, carina, rostrum, and a large pair of upper latera, with a single lower whorl of smaller valves; these latter vary from 22 in very small specimens, to 26 in large specimens. The capitulum, therefore, is formed of at most 34 valves; but in the largest specimen seen by me, the capitulum being 2·3 of an inch in width, there were only 32 valves. In the smallest, namely, with a capitulum 0·15 of an inch in width, there were 30 valves. The valves are remarkably strong, and formed of white shelly matter; they are closely approximate, and overlap each other: the scuta and terga are articulated together by a fold; the apices of the valves are either worn and disintegrated, or they project freely like horns beyond the sack, to a much greater extent than in any other recent species of the genus: even a considerable portion of the scuta projects obliquely upwards. The exterior surfaces of the valves (when not worn) are covered by a strong yellow membrane, and the upper free parts are generally attached together for some little length by this same membrane. The valves are plainly marked by the zones of successive growth; and most of them are ribbed and furrowed slightly, from their umbones to their basal margins. The yellow external membrane, examined microscopically, is marked by, or rather formed of, numerous growth lines, crossed by longitudinal beaded ridges. The tubuli are not numerous, and of small diameter.

Scuta (Pl. VII, fig. 3a', a) triangular, with the apex more or less produced, according to the state of its preservation, and a little curved towards the terga; basal margin, and in some degree the tergo-lateral margin, arched, and slightly protuberant; occludent margin thickened, / slightly prominent, with the inner edge covered by the yellow membrane, like the exterior surface of the valve. The upper part of the tergo-lateral margin overlaps a little the edge of the tergum, and receives it in a furrow – the two valves being thus locked together. This furrow lies in the freely projecting, membrane-covered portion, and extends up to the apex; it is of variable depth. Internally the scuta are concave, and in some old specimens to a high degree. In these latter, the basal margin, towards the tergo–lateral side, is strongly sinuous; the prominences are formed by the terminations of the external longitudinal ridges, and correspond to the interspaces between the valves of the lower whorl. These ridges, which are interesting, from throwing light on similar ridges in some fossil species, are present, both on old and young specimens, and run from the apex of the valve, in a slightly curved line, to the tergo–lateral half of the basal margin, where, as we have just seen, they sometimes form prominences. They consist of three of even four obscure, almost confluent, ridges, of which the middle one is generally (but not always) the smallest: together they cover the whole of that part of the scutum, which is not overlapped along the basal margin by the rostrum and large upper latus; and they seem evidently due to the growth of the shell in this interspace having been freer. So, again, the three or four small, confluent, component ridges have the same relation to the interspaces between the small latera of the lower whorl.

Terga large, four sided, with the internal growing surface (fig. 3a' b), almost diamond shaped; basal angle blunt, rounded; exteriorly, from the apex to the basal angle there is a rather broad, very slight prominence, which bears the same relation to the carina and upper latus, as do the compound ridges on the scuta to the rostrum and upper latus. The upper part of the scutal margin forms a slightly-projecting, rounded shoulder, though variable in its degree of prominence, in relation to the variable depth of the recipient furrow in the scuta. / Externally, parallel to the occludent margin, and close below the prominent shoulder, just mentioned, there is a slight and variable depression, extending up to the apex of the valve. This depression is due to the prominence, variable in degree, of the tergal edge of the recipient furrow in the scuta.

Carina, triangular, strong, inwardly bowed, generally with a large upper portion freely projecting; exteriorly with a narrow, sharp, central ridge or keel, which is solid, the interior concavity not reaching so deep; inner growing surface (fig. 3*b'*, *b*) deeply concave, triangular. Basal margin square – that is, transverse to the longer axis of the carina, or it even rises (as is best seen in the growth ridges) a little towards the exterior keel. On each side of the central exterior keel, there is a narrow longitudinal ridge, corresponding with the interspace between the sub-carina and the next but one latus of the lower whorl; the latus next to the sub-carina is very small, and overlies the ridge itself. In a very large specimen, these lateral longitudinal ridges formed (as they likewise did on the rostrum) slight prominences on the basal margin. In one specimen the carina was straight.

Rostrum closely similar, in almost every respect, to the carina, even to the exterior, lateral, longitudinal ridges, and in their relation to the interspaces in the lower whorl. The valve is generally not so long, but rather wider, more inwardly bowed, and with the exterior solid keel less prominent than in the carina. The inner growing surface (fig. 3*b' d*) is less acuminated at its upper end.

Upper pair of latera. These are much larger than the remaining valves of the lower whorl; they are straight, triangular, and much acuminated, with their apices, when well preserved, extending far up, for fully three-fourths of the height of the scuta. They nearly equal in length the carina. The growing surface (fig. 3*b'*, *a*) is flat, triangular, in well-preserved specimens forming only a third or a quarter of the entire length of the valve. In the middle of the basal margin there is a very slight prominence, / corresponding with a slight external central ridge, formed as heretofore by the overlapping of two of the valves of the lower whorl. Basal margin nearly on a level with that of the scuta and with the basal points of the terga. The foregoing eight larger valves form the main cavity, in which the body of the animal is lodged.

Valves of the lower whorl. These, seen externally, seem to belong to more than one whorl, but internally their basal margins stand on a level. They vary in number, as already stated, from 22 to 26. I have seen an individual with a valve more on one side than on the other. They are of unequal sizes, but they are rather variable in this respect: the largest are not above half the size of the upper latera: three or four pairs, together with the sub-rostrum (*e*) and sub-carina (*c*), are always larger than the others: these two latter valves differ from the others only in being more concave. Seen externally, all these valves project considerably, and curl a little inwards, with their apices generally worn and truncated. Viewed internally (fig. 3*b'*), whilst the valves are in their proper places, the inner and growing surfaces of the smallest are seen to be triangular – of the larger, some are rhomboidal, and others quadrilateral with the upper side much longer than the lower. These latter valves overlap the upper parts of the little valves on both sides of them; the rhomboidal valves overlap a valve on one side, and are overlapped on the other; the triangular valves are overlapped on both sides.

The corium lining the capitulum is produced into narrow purple crests, which enter the interstices between the valves, more especially along the line separating the upper and lower whorls. There is, also, a distinct flattened, tapering, free projection of corium, which enters between the carina and sub-carina; and another between the rostrum and sub-rostrum.

Peduncle, much compressed, short, rarely as long as the capitulum; in one very large specimen it was extremely short, barely one-fifth of the length of the capitulum. / The attached portion, which is moderately pointed in young specimens, becomes extremely broad in old specimens. The calcified scales sometimes differ a little in size, in specimens of the same age: they are always compactly and symmetrically arranged: in old specimens they are much larger than in young ones: each scale has, at first, a transversely elliptic growing base, which ultimately becomes nearly circular. Exteriorly the tips of the scales are always disintegrated; they are sometimes club-shaped, owing to the scales having been re-added to after a period of reduced growth. The scales are fringed with brown disintegrating membrane.

Attachment. At the base of the peduncle, the two cement ducts running together, twist about in a singular manner, and at their bends pour forth cement. According to the age of the specimen, the ducts vary in diameter from $\frac{1}{2000}$th to $\frac{5}{2000}$ths of an inch. The two cement glands are small and difficult to find; they are retort-shaped, with two ovarian tubes entering each. They lie close together, in nearly the centre of the peduncle, and less than halfway down it. This proximity of the two cement glands, and their position low down the peduncle, are of interest in relation to the position of these same glands in the sessile cirripedes.

Size and colours. This is the largest and most massive species in the family. I have seen one specimen in the British Museum, from the Coast of China, 2·3 inches across the capitulum, and 1·5 in length, with the valves surprisingly thick. The relative width and length of the capitulum varies. The sack (in specimens long kept in spirits) is dirty purple, and exteriorly between the scuta, dark purple. The cirri, trophi, penis, caudal appendages, three posterior segments of the thorax, and the abdominal surface are dark brownish-purple.

Body. Thorax remarkably compressed and carinated prosoma pretty well developed. Extending from the base of the second cirrus, to nearly a central line on the thorax, there is on each side a rounded ridge: there is a second / transverse ridge, running from the base of the first cirrus to near the adductor scutorum muscle: these ridges seem formed merely to allow of the larger development of the testes.

Mouth. Labrum highly bullate; crest without any teeth, but with a few minute hairs. The inner fold of the labrum forming the supra-oesophageal cavity, is thickened, and shows a trace of a central line of junction, as in sessile cirripedes.

Palpi (Pl. X, fig.7), small; of a singular club-like shape, owing to the convexity of the outer margin; exterior spines long, all doubly serrated.

Mandibles (Pl. X, fig. 1), with five teeth, of which the second is very small; inferior angle coarsely pectinated.

Maxillae (fig. 14), with a deep narrow notch (bearing some fine spines) beneath the two upper great spines, which stand on a prominence; edge straight, bearing fourteen or fifteen pairs of spines: on the inferior angle there is an obscure tuft of shorter and finer spines: apodeme long, sinuous, and slender.

Outer maxillae (fig. 17), with the inner margin divided by a deep notch into two lobes, of which the upper one is rather short; both are clothed with a compact row of short bristles; exterior margin with longer bristles.

Olfactory orifices, large and prominent to an unusual degree.

Cirri, moderately long and curled; the four posterior pair are alike; each segment has its anterior face somewhat protuberant, and bears six pairs of long spines, with a rather large, narrow tuft of intermediate spines, some of which are finely and doubly serrated. The dorsal tufts consist of short, thick spines, with some fine longer ones. The first cirrus is seated near the second; its rami are slightly unequal in length; lower segments paved with bristles; one ramus is thicker than the other, and some of its segments have coarsely pectinated spines. Second cirrus has the five basal segments of its anterior ramus highly protuberant, and paved with bristles, of which some are coarsely pectinated; the basal segments of the posterior ramus are rather more thickly clothed with bristles than are the posterior cirri, but otherwise resemble them. The third cirrus, as already stated, is exactly like the three posterior pairs; and this is a very unusual circumstance. On the dorsal surfaces and sides of the pedicels of the posterior cirri, there are some scattered, short, thick, minute spines.

Caudal appendages, multi-articulate: in a medium-sized specimen, each contained eight segments, which reached halfway up the upper segment of the pedicel of the sixth cirrus. Lower segments flattened; the upper, tapering, and cylindrical; all have their upper margins furnished with stiff, little spines. In a young specimen (only 0·3 of an inch in length, including the peduncle), the caudal appendage contained only four segments, and the tip did not reach to the upper edge of the lower segment of the pedicel of the sixth cirrus.

Stomach, without caeca.

Generative system. Vesiculae seminales not reflexed at their broad ends; white, spotted with black. Testes, pear-shaped, borne on long footstalks: penis covered with minute bristles, in little tufts arranged in straight lines. The ovarian tubes fill up the peduncle to its base, but do not surround the sack; they are of small diameter, and simply branched. There is a very narrow ovigerous fraenum, with a straight edge, lying on each side under the line of junction between the scutum and upper latus.

Affinities. This species differs from all the others of the genus, in the third cirrus resembling exactly the three posterior pairs. In most of its characters – namely, in the symmetrical arrangement of the scales on the peduncle, in the considerable size of the valves of the lower whorl, in the general approximation of the valves, in the multiarticulated caudal appendages, in the form of the outer maxillae, in the prominent olfactory orifices, in the basal segments of the anterior ramus alone of the second cirrus being paved with bristles, there is more affinity to *P. cornucopia*, / *P. elegans*, and *P. polymerus* than to *P. sertus* and *P. spinosus*.

In the scuta and terga being articulated together, in the union of all the valves by stiff membrane, in the peculiar manner in which the valves of the lower whorl overlap each other, in the corium entering between some of the valves in filiformed appendages, in the near equality of size of the rostrum and carina, in the shortness of the peduncle in old specimens, in the position of the cement glands, and lastly in the characters of the third pair of cirri, this species presents a closer affinity to the sessile cirripedes, more especially to the Chthamalinae, than does any other species of any other genus among the Lepadidae. The movements, however, of the four opercular valves are not at all more independent of the other valves, than in the other pedunculated cirripedes; and the peduncle is furnished with all its characteristic muscles.

5. POLLICIPES SPINOSUS

PLATE VII. *FIG. 4*

ANATIFA SPINOSA. Quoy et Gaimard, *Voyage de l'Astrolabe*, Pl. xciii, fig. 17

P. capitulo valvarum uno aut pluribus sub-rostro verticillis instructo: laterum pari superiore vix inferioribus longiore: membranâ valvas tegente (post desiccationem) subfuscaâ flavescente: pedunculi squamis inaequalibus, non symmetricis: verticillis longiusculè distanbibus.

Capitulum with one or more whorls of valves under the rostrum: upper pair of latera only slightly larger than the lower latera: membrane covering the valves (when dried) light yellowish-brown: scales of the peduncle of unequal sizes, unsymmetrical, arranged in rather distant whorls.

Maxillae, with the edge square and straight: caudal appendages uniarticulate: filamentary appendages, none.

New Zealand. Mus. Jardin des Plantes, Paris: Mus. Cuming. /

Capitulum, flattened, triangular, broad, with the valves varying in number, in full-grown specimens of the same size, from 30 to above 60; the scuta, terga, and carina are very much larger than the other valves; the rostrum, however, is nearly half the size of the carina; the remaining valves are exceedingly small. In some specimens there is only one whorl under the carina; in other specimens there are distinctly two whorls. The scuta, terga, and carina stand pretty close together; they are moderately thick, and are covered, in chief part, by yellowish-brown membrane, which is destitute of spines.

Scuta, triangular, broad, basal margin slightly protuberant.

Terga, as large as the scuta, flat, regularly oval, basal point blunt and rounded.

Carina very slightly curved, triangular, internally rather deeply concave, basal margin straight. The inner and growing surface is four-fifths of the entire length of the valve. In half-grown specimens the apex projects a little outwards.

Rostrum, small, much curled inwards; the basal margin is much hollowed out; the inner surface is broadly triangular, more than twice as wide as high, and about one-fourth of the entire length of the valve. The remaining valves, about twenty-six in number, do not correspond on the opposite sides of the same individual, they are exceedingly small, with the sub-carina, sub-rostrum, and three pairs of latera a trifle larger than the lower latera, which are generally arranged in two whorls. In shape all the latera are nearly alike; they consist of flattened styles, with their inner surfaces transversely oval, and more or less elongated, the larger ones being most elongated.

Peduncle, broad, barely as long as the capitulum. The calcareous scales are irregularly shaped, minute, elongated and pointed, placed in separate transverse rows, and crowded together in each row. Only the scales in the uppermost row grow regularly: but some of the lower scales continue to be added to irregularly, and hence are / the largest. On the other hand, the lower part of the peduncle, from the first formed scales having been worn away, is often quite naked. From this cause, and from the continued and irregular growth of some of the lower scales, the rows in this part of the peduncle, generally become irregular. The surface of attachment is broad.

In a half-grown specimen, with a capitulum only 3/10ths of an inch long, all the lower valves were considerably larger in proportion to the scuta, terga, and carina, than in full-grown individuals.

Size and colours. Length of capitulum in the largest specimen, 7/10ths of an inch; breadth, slightly exceeding the length. Colours after having been long in spirits – upper part of the sack, thorax, pedicels of cirri, and penis, clouded with fine purple; cirri banded with the same; exterior convex surface of the outer and inner maxillae and palpi dark purple; prosoma yellow. The membrane of the peduncle and of the capitulum is dirty yellow, with bands of purple between some of the valves.

Filamentary appendages, none. Ovigerous fraena placed near the middle of the basal margin of the scuta; small, semi-oval, with an elliptical ring of bead-like glands; glands seated on long footstalks.

Mouth. Labrum far produced towards the adductor muscle; upper part highly bullate, nearly equalling the longitudinal diameter of the rest of the mouth, and very slightly overhanging the lower part; crest with very minute bead-like teeth.

Palpi, with their inner margins considerably excised, most thickly clothed with spines.

Mandibles, with three strong teeth, two unequal-sized small teeth being placed between the first and second, thus making five altogether; inferior angle broad, pectinated.

Maxillae, with its edge broad, straight, bearing about twenty pairs of spines, shorter than the large upper spines.

Outer maxillae, with the bristles in front, continuous, and without any notch; exterior surface with a prominence / clothed with long spines. Olfactory orifices slightly prominent.

Cirri. First cirrus placed near to the second; posterior cirri not much elongated, with their segments slightly protuberant, bearing four pairs of spines, of which the lower pair is small; spines slightly serrated. In the lower segments, these spines are exceedingly unequal in length, the inner spines on both rami, not being above one-fourth of the length of the outer corresponding spine in each pair. The tufts intermediate between these pairs, are not very large: on the lateral upper rims there are some strong, short spines: dorsal tufts with short, thick spines. First cirrus about three-fourths as long as the second cirrus, with numerous tapering segments, three or four of the lower ones being thick and protuberant: in the first cirrus there are eleven segments, and in the sixth cirrus, seventeen. Second cirrus, with the the anterior ramus slightly thicker than the posterior ramus: a few of the basal segments of both rami are protuberant, and thickly clothed with spines. In the third cirrus, the two rami are nearly equally thick, with some of the basal segments in both clothed, like a brush, with spines. In these brushes on the first, second, and third cirri, most of the spines are doubly toothed, each tooth being simply conical.

Caudal appendages, small, much flattened, straight on the exterior side, and curved on the inner side, with a row of short, rather thick spines on the crest, and a few on the exterior margin.

The *affinities* of this species will be given under the head of the following, *P. sertus*.

6. POLLICIPES SERTUS
PLATE VII. FIG. 5

P. capitulo valvarum uno aut pluribus sub-rostro verticillis instructo: laterum pari superiore vix inferioribus longiore: membranâ valvas tegente (post desiccationem) fusco rufescente obscuro: rostro dimidiam carinae longitudinem / aequante, superficiei internae altitudine latitudinem plus duplo superante: pedunculi squamis inaequalibus, non symmetricis: verticillis longiusculè distantibus.

Capitulum with one or more whorls of valves under the rostrum: upper pair of latera only slightly larger than the lower latera: membrane covering the valves (when dried) dark reddish-brown: rostrum half as long as the carina, with its inner surface more than twice as high as broad: scales of peduncle of unequal sizes, unsymmetrically arranged in rather distant whorls.

Maxillae with two tufts of fine bristles, separated by larger spines: caudal appendages uniarticulate: filamentary appendages none.

New Zealand; Mus. Cuming.

Capitulum, much flattened, broad, subtriangular. Valves exceedingly various in number; in the largest specimen with a capitulum 8/10ths of an inch high, and 9/10ths of an inch wide, there were only thirty-one valves, and these formed only a single whorl under the carina and rostrum; whereas, in another specimen, which was barely 6/10ths of an inch in length, there were fifty-two valves, and these formed two or three distinct whorls under the carina. Scuta, terga, carina, and rostrum, much larger than the other valves. All are moderately thick, placed rather distant from each other, covered with thick membrane which abounds with tubuli, arranged in rows; surface apparently smooth, but with a very high power, extremely minute spines can be seen at the extremities of almost all the tubuli. Little bunches of reddish fibrous matter are imbedded in the membrane, like tufts of seaweed floating in water.

Scuta, triangular, basal margin curved, protuberant; the upper part of the tergo–lateral margin is, also, slightly protuberant.

Terga, large, oval, basal angle broad, square; lower part of carinal margin straight, upper part narrowed in; the apex is covered with membrane and projects freely. /

Carina, triangular, internally deeply concave, either straight, and with the apex free, or inwardly and considerably curved; basal margin nearly straight.

Rostrum, about half the length of the carina; either straight or inwardly curved; it projects freely for full half its length; inner growing surface triangular, more than twice as high as wide; basal margin very slightly hollowed out. The *sub-carina* and *sub-rostrum* are larger than the largest of the latera; their inner surfaces are transversely elongated, rounded at both ends, and slightly concave; externally they are pointed, and project outwards; sometimes the sub-carina, and sometimes the sub-rostrum is the largest.

Latera, small, with their inner surfaces transversely elongated, the larger being the most elongated. Externally they are acuminated, and directed upwards; they project but very little beyond the thick membrane in which they are imbedded. Neither the number, size, nor shape of the latera agree on opposite sides of the same individual; and it would appear that, occasionally, some of them cease to grow, and disappear. In the large specimen with only thirty-one valves, the three pairs of latera, corresponding to the upper, rostral, and carinal latera in Scalpellum, were larger in a marked manner than the others; but in the specimen with fifty-four valves, this could hardly be said to be the case. In this latter specimen, some of the valves in the lowermost whorl were exceedingly minute.

Peduncle, broad, about as long as the capitulum; surface of attachment wide; calcareous scales minute, placed in transverse rows, which become less and less regular in the lower part. The scales do not stand very close together; they are of unequal sizes and irregular outline; generally spindle shaped; calcareous matter is added regularly only to the scales in the uppermost row, and irregularly to some of the lower scales. The latter, consequently, are the largest, and often much elongated; they are sometimes of singular and irregular shapes.

Colour. The membrane covering the valves and / forming the peduncle (after having been long kept dry, and not having been in spirits), is dark reddish chocolate-brown; corium of sack dark purple; cirri banded with dark purplish-brown, with the lower parts of the trophi similarly coloured.

Filamentary appendages, none, but on the prosoma there are scattered some small papillae, which are roughened by finely spinose scales, like combs; these papillae certainly seem to represent the filaments in *Pollicipes cornucopia* and its two allies.

Ovigerous fraena, seated in the same position as in *P. spinosus*, but rather longer, with an elliptical *tuft* of glands on the crest.

Mouth, not placed far from the adductor muscle.

Labrum, moderately bullate, with the upper part not overhanging; no teeth on the crest. *Palpi*, short, broad, blunt.

Mandibles, with three main teeth, with either one or two smaller teeth inserted between the first and second, making four or five altogether; inferior angle rather narrow, pectinated with long and fine spines.

Maxillae, rather broad, with two long upper spines; beneath which there is a very small prominence bearing a minute tuft of fine bristles; beneath this, there are eleven pairs of rather long and strong spines; and the inferior angle is formed by a rather broad, upraised, and obliquely rounded prominence, bearing a broad tuft of fine spines.

Outer maxillae, with the inner surface continuously clothed with short spines; exteriorly there is a slight prominence with long hirsute spines.

Olfactory orifices barely prominent.

Cirri. First pair placed near the second; the segments of the three posterior pairs are slightly protuberant, and bear three or four pairs of finely serrated spines; intermediate tufts long, the middle spines being the longest; spines on the upper lateral edges long and strong; dorsal tufts rather short. *First cirrus*, long, multiarticulate, having fourteen or fifteen segments, whilst the sixth cirrus had nineteen segments; rami unequal in length by / about two segments; basal segments protuberant, brush-like. *Second* and *third cirri* with five basal segments of both rami protuberant and brush-like; but the anterior rami in both cirri are broader than the posterior rami. Spines on the protuberant segments of both rami of both cirri, coarsely and doubly pectinated.

Caudal appendages (Pl. X, fig. 19), minute, uniarticulate, club-shaped, with the enlarged ends directed inwards, or towards each other; summits sparingly clothed with very short spines.

Penis, small.

Affinities. This species makes a very close approach in the general form and relative sizes of all the valves, and in the variability of the number of the whorls, to *P. spinosus*; there is a still closer and more important resemblance, in the inequality and manner of growth of the calcareous scales on the peduncle. These species differ, in the colour of the membrane covering the valves, and in the greater development of both rostrum and sub-rostrum in *P. sertus*. The rostrum of the latter is longer than half the length of the carina, and its inner surface is more than twice as high as wide; and the sub-rostrum is twice as large as any of the latera – all points of difference from *P. spinosus*.

In the characters of the mandibles, and more especially of the outer maxillae; in the length of the first pair of cirri; in both rami of the second and third cirri having their basal segments brush-like, with pectinated spines; and in the shape of the caudal appendages, there is a close relationship to *P. spinosus*, and through this species to *Scalpellum villosum*. In the little prominence of the olfactory orifices, *P. sertus* differs from most of the allied forms, excepting *P. spinosus*. In the maxillae having two prominences bearing fine tufts of bristles, in the roughened knobs on the prosoma, and in the presence, in some individuals, of two or three whorls of valves under the carina and rostrum, there is a marked tendency in *P. sertus* to approach *P. cornucopia*, *P. elegans*, and *P. polymerus*. /

GENUS: LITHOTRYA

PLATE VIII, IX

LITHOTRYA. G. B. Sowerby, *Genera of Shells*, April, 1822
LITHOLEPAS. De Blainville, *Dict. des Scienc. Nat.*, 1824
ABSIA.[6a] Leach, *Zoological Journal*, vol. ii, July, 1825
BRISNAEUS et CONCHOTRYA. J. E. Gray, *Annals of Philosophy*, vol. x (new series), August, 1825
LEPAS. Gmelin, *Systema Naturae*, 1789
ANATIFA. Quoy et Gaimard, *Voyage de l'Astrolabe*, 1832

Valvae 8, si inter eas parvum (saepe rudimentale) rostrum et duo parva latera numerentur; incrementi lineis concinnè crenatis: pedunculus squamis calcareis parvis vestitus, in verticillis superioribus crenatis; aut calyci basali calcareo aut discorum ordini affixus.

Valves 8, including a small, often rudimentary rostrum and a pair of small latera: lines of growth finely crenated. Peduncle covered with small calcareous scales, those of the upper whorls crenated; attached either to a basal calcareous cup, or to a row of discs.

Body lodged within the peduncle: mandibles with three teeth, the interspaces being pectinated; maxillae various: olfactory orifices slightly prominent: caudal appendages multiarticulate.

Lodged in cavities, bored in calcareous rocks, or shells, or corals; generally within the tropics.

Description. The capitulum is not much compressed, a horizontal section giving an oval figure; it is placed obliquely on the peduncle, the scuta descending lower than the terga and carina. There are eight valves, of which the scuta, terga, and carina are large; the rostrum and a pair of latera are very small and often rudimentary. These three latter valves are essentially distinguished from the scales of the peduncle, the upper ones of which they / sometimes hardly exceed in size, by not being moulted at each period of exuviation. The latera

---

[6a] The description of Absis is so inaccurate, that I should not have recognized it, had not the *Lithotrya Nicobarica*, in a bottle in the British Museum, borne this name.

overlie the carinal half of the terga; I presume that they are homologous with the carinal latera in Scalpellum. Each successive layer of shell forming the valves is thick, and extends over nearly the whole inner surface; hence the carina and terga, and to a certain extent the scuta, either actually do project freely much beyond the sack, or would have done so, had not their upper ends been removed; for the upper and old layers of shell, in most of the species, either scale off or disintegrate and wear away. A rectangularly projecting rim, serrated by small teeth, is formed at the bottom of each fresh layer of growth along the external surfaces of each valve (see upper part of fig. 1b' Pl. VIII). This structure, as well as that of the crenated scales on the peduncle, is important, for by this means the animal, as we shall presently see, forms and enlarges the cavity in the rock or shell in which it is imbedded.

The scutum overlaps either about one-third or even one-half of the entire width of the tergum, and abuts against a prominent longitudinal ridge on its exterior surface. In *L. truncata* and *L. Valentiana*, this ridge on the tergum being folded over towards the scutum, forms a conspicuous furrow, receiving the tergal margin of the latter. In *L. Valentiana*, there is a second furrow on the carinal side of the tergum, receiving the upper end of the corium-covered or growing surface of the carina. Besides these provisions for holding together the valves, there are, apparently, others for a similar purpose; thus in each scutum, under the rostral angle, there is a roughened knob-like tooth, which touches the underside of the little rostrum, and no doubt serves to give attachment to the membrane uniting the three valves together. In some species, the adjoining basal margins of the scuta and terga, where touching each other, are inflected and roughened; again in *L. Rhodiopus*, the carinal angles of the terga are produced into points, and in *L. truncata* and *L. Valentiana* into prominent roughened knobs, which touch two corresponding / small knobs, on the upper part of the growing surface of the carina. Moreover, considerable portions of the inner surfaces of the scuta and terga, are roughened with minute sharp, imbricated points, apparently for the firmer attachment of the corium. The roughened knobs at the rostral angles of the scuta, no doubt are homologous with the teeth in a similar position on one or both scuta in Lepas, and in some fossil species of Pollicipes, as in *P. validus*. The other projections and roughened surfaces are peculiar to Lithotrya. The growth of all the valves is, as in Pollicipes, simply downwards.

The *scuta* are triangular, with their umbones or centres of growth at the apex; the tergal margin, as seen from within, is either nearly straight or much hollowed out, accordingly as the scuta simply overlap the terga, or are received in a furrow. In some of the species there is a distinct pit for the adductor muscle, and in others this cannot be distinguished.

Terga. These present great differences in shape; but all appear to be modifications (as seen internally) of a rhomboidal figure, which seems to be the normal form of the terga in the Lepadidae. Of the lower part of the valve, the whole exterior surface, with the exception of a narrow ridge running from the apex down to the basal angle, is hidden by the overlapping of the scuta, latera, and carina.

The *carin* in outline, is triangular, with the basal margin in some species extremely protuberant. In the first four species, the internal surface is concave, in *L. truncata* and *L. Valentiana* it is convex, with a central raised ridge, and consequently the upper freely projecting portion of the valve, has a prominent central crest or ridge; in *L. Nicobarica* and *L. Rhodiopus* there is only a trace of this ridge. The *rostrum*, as before stated, is always very small; it, as well as the latera, are most developed in *L. Nicobarica*, and least in *L. truncata* and *L. Valentiana*; generally only a few zones of growth are preserved, and from their being enlarged at their basal serrated rims, / the rostrum sometimes appears like a few beads of a necklace strung together.

The *latera* are remarkable from being placed over the carinal half of the terga, in an oblique position, parallel to the lower carinal margin of the terga. A section, parallel to the growth layers, varies in the different species from elliptic to broadly oval, and in *L. Nicobarica* it is triangular. Only a few layers of growth are ever preserved. In *L. truncata*, where the latera are represented by mere stiles (like strings of beads), and are even less in width than the rostrum, they are imperfectly calcified.

Microscopical structure of the valves. The shelly layers are white, and generally separate easily, so that in *L. dorsalis* it is rare to find a specimen with the upper part of the valves perfect. The valves are so translucent, that in the thin margins, even the tubuli could be sometimes distinguished. The valves are coated by strong yellow membrane, which, after the shelly matter in *L. dorsalis* had been dissolved in acid, separated into broad slips, answering to each zone of growth. On the lower margin of each slip, there is a row of closely approximate spines, generally slightly hooked, pointed, 1/650th of an inch in length, and 1/10000th of an inch in diameter; they arise out of a little fold; all are furnished with tubuli of the same diameter with themselves, running through the whole thickness of the shelly layers, and attached, apparently, by their apices, to the underlying corium. As the spines are very numerous, so are the parallel rows of tubuli. After the shelly layers had been dissolved, there was left in *L. dorsalis* (well seen in the latera), an extraordinary, conferva-like mass of branching, jointed, excessively thin tubes, sometimes slightly enlarged at the articulations, and appearing to contain brown granular matter: other portions of the valves, instead of this appearance, exhibited membranes or films with similar, branching, articulated tubes or vessels attached to them: I have not seen this appearance in any other cirripede. The yellow exterior enveloping / membrane, with its spines, is present in all the species of the genus; in *L. Rhodiopus* these spines are much larger than in *L. dorsalis*, and on the inner sides of the carina they are trifid and quadrifid and large enough to be conspicuous with a lens of weak power.

Peduncle. The most remarkable fact concerning this part, is that the outer tunic, together with the calcareous scales with which it is covered, is moulted at each successive period of exuviation and growth. I demonstrated this fact in *L. dorsalis* and *L. truncata*, by removing the old tunic and finding a new membrane with perfect calcified scales beneath; and as these two species (I obtained, also, pretty good evidence in *L. Nicobarica*), are at the opposite extremes of the genus, no doubt this fact is common to the whole genus. I know of no other instance, among Cirripedia, in which *calcified* valves or scales are moulted. I am not certain that the whole skin of the peduncle is thrown off in a single piece; though this almost certainly is the case with the uppermost and lowest portions. The animal's body is partly lodged within the peduncle, which is generally from one to three times as long as the capitulum, and in the upper part is fully as broad as it. The scales with which it is clothed, extend up in the triangular interspaces between the basal margins of the valves. The scales of the upper whorl, or of the two or three upper whorls (Pl. VIII, figs 1*b'* and 3*d*) are larger than those below; and these latter rapidly decrease in size, so as to become low down on the peduncle, almost or quite invisible to the naked eye. The scales in each whorl, are placed alternately with those in the whorls, above and below. All the upper scales are packed rather closely together; those in the uppermost row are generally nearly quadrilateral; those in the few next succeeding whorls, are triangular, with their basal margins protuberant and arched; the scales, low down on the peduncle, stand some way apart from each other, and generally consist of simple rounded calcareous beads, of which some of the smallest in *L. dorsalis* were only / $\frac{1}{400}$th of an inch in diameter. In the lowest part of the peduncle these scales, after each fresh exuviation, are apparently soon worn entirely away by the friction against the sides of the cavity; hence in most specimens this part of the peduncle is quite naked. This same part, however, is furnished with nail- or rather star-headed little projections of hard, yellow, horny chitine (fig. 3*e*). The star on the summit seems generally to have about five irregular points; one star which I measured was $\frac{7}{6000}$th an inch in total width, the footstalk being only $\frac{2}{6000}$th of an inch in diameter; the whole projected $\frac{10}{6000}$ths of an inch above the surface of the peduncle; from the footstalk a fine tubulus runs through the membrane to the underlying corium. These star-headed little points are often much worn down; in one specimen which was on the point of exuviation, there remained, in the lower part, close above the basal calcareous cup, only some hard, smooth, yellow, little discs, on a level with the general surface of the membrane – these being the intersected or worn down footstalks, with every trace of the calcareous beads gone. But in this same specimen, under the old peduncular membrane, there was a new one, studded with the usual circular calcareous beads, slightly unequal in size, generally about $\frac{1}{400}$th of an inch in diameter, and each furnished with a tubulus; but as yet none of the star-headed points of chitine had been formed. I believe that these latter are developed from the tubuli leading to the calcified beads, and, therefore, are formed directly under them. In *L. cauta* the lowest scales on the peduncle are a little larger than in *L. dorsalis*, giving a frosted appearance to it, and all of them are serrated (fig. 3*d*) round their entire margins. Generally only the scales in the uppermost, or in the three or four upper rows are serrated, and this only on their arched and protuberant lower margins. The state of the serrated edge varies extremely in the same species, from elongated conical teeth to mere notches, according to the amount of wear and tear the individual has suffered since the last period / of exuviation; so also do the teeth or serrated margins on the valves of the capitulum. Each scale has a fine tubulus passing from the corium through the membrane of the peduncle to its bluntly pointed imbedded fang or base. The membrane is transparent, thin, and tender, to a degree I have not seen equalled in the other Lepadidae, except, perhaps, in Ibla. It is much wrinkled transversely.

Muscles of the peduncle. These consist of the usual interior and longitudinal – exterior and transverse – and oblique fasciae; the former are unusually strong; downwards they are attached to the basal calcareous cup or disc, and upwards they extend all round to the lower curved margins of the valves. They are, as usual, without transverse striae. Besides these, there are (at least in *L. dorsalis* and *L. Nicobarica*), two little fans of striaeless muscles, which occur in no other pedunculated cirripede; they are attached on each side of the central line of the carina, near its base; they extend transversely and a little upwards, and each fan converges to a point where the lower margins of the carina and terga touch; of these muscles, the upper fasciae are the longest. Their action, I conceive, must be either to draw slightly together the basal points of the terga, and so serve to open their occludent margins, or to draw inwards the base of the carina: these muscles apparently first shadow forth the posterior or carinal, transversely striated, opercular muscles of sessile cirripedes.

Basal calcareous cup or discs. I have seen this part in all the species, except *L. Valentiana*, and in this it probably occurs, considering its very close alliance with *L. truncata*. The size, form, and conditions of the cup or disc varies infinitely according to the age, size, and position of the individual specimen. We will commence with a full-sized animal, which has ceased to burrow downwards into the rock, in which case the discs usually grow into a cup, and become largely developed. In *L. dorsalis* alone, I have seen many specimens, so that / the following description and remarks, though applicable I believe to all the species, are drawn up from that alone. The cup (Pl. VIII, fig. 1a', 1c') is hardly ever regular in outline, and is either slightly or very deeply concave; I have seen one, half an inch in diameter; it is formed of several thick layers of dirty white, translucent, calcareous matter, with sinuous margins; externally the surface is very irregular, and is coated by yellow membrane presently to be described. The innermost and last formed layer sometimes covers the whole inside of the cup, and extends a little beyond its margin all round; but more generally it projects beyond only one side, leaving the other sides deserted. I have seen a *single* new layer extending beyond the underlying old layers, as much as one-sixth of an inch; and again I have seen a part of the cup, as much as a quarter of an inch in width, deserted and covered with serpulae. So irregular, however, is the growth, that after a period an old deserted portion will occasionally be again covered by a new layer, though of course without organic adhesion. Again it sometimes happens that the last formed layer, remaining central, is very much less than the older layers; in one such instance the innermost and last formed layer (fig. 1a') had a diameter of only a quarter of that of the whole cup, in the middle of which it was placed; the cup thus tends to become filled up in the middle. The cup, in its fully developed condition, is seated at the very bottom of the cavity in the rock. From the aggregate thickness of the several component layers forming the cup, the old and mature animal rises a little in its burrow; for instance, the bottom of the cup in one specimen which I measured, was 4/10ths of an inch in thickness.

In a younger condition, before the animal has bored down to the full depth, and whilst the cavity is only of moderate diameter, the lower part of the peduncle, instead of being attached to the inside of a cup, adheres to small, irregular, nearly flat, calcareous discs, overlapping each other like tiles (figs 1, 2a'). They are placed one / below the other, generally in a straight line, and are attached firmly to one side of the burrow. The discs are oval, or rounded, or irregular, and are commonly from 1/20th to 1/10th of an inch across: they usually form a quite straight ribbon, widening a little downwards: each little disc overlaps and extends beyond the one last formed, fully half its own diameter. I have seen one row of discs an inch in length, but the upper discs are always worn away by the friction of the calcified serrated scales on the peduncle. It is very important to observe that the lowest disc is not fixed (as was the case with the cup) at the very bottom of the burrow, but on one side, just above the bottom, which latter part is occupied by the blunt basal end of the peduncle.

In a valuable paper on *L. Nicobarica*, by Reinhardt, presently to be referred to, the disc is said to be attached on the carinal side (see fig. 2) of the peduncle; and this, I believe, is general. I have seen one instance in which, during the excavation of a new burrow, an old burrow was met with, and the row of discs turned down it, making, with their previous course, nearly a right-angle. In another similar instance, the discs, instead of turning down, became very large and broad, and so fairly formed a bridge across the old burrow (fig. 1) – becoming narrow again as soon as the animal recommenced burrowing into the solid rock. Sometimes, as it appears, the animal, whilst still small, from some unknown cause, stops burrowing downwards, and then a cup is formed at the bottom of the hole. As soon as the animal has got to its full depth, the burrow increases only in diameter, and during this process the linear row of discs is ground away and lost; a cup is then formed. the little discs can be deposited or formed only at each fresh exuviation; and as some of the burrows are above two inches in depth, and as on an average each disc does not extend beyond the underlying disc more than $\frac{1}{15}$th of an inch, an animal which has bored two inches in depth, must have moulted at least thirty times. I may here remark that I have / reason to believe, from some interesting observations made by Mr W. Thompson, of Belfast, that some sessile cirripedes moult about every fortnight.

Internal structure of the cup. When the cup is dissolved in acid, each shelly layer is represented by a rather tough, pale-brown membrane, itself composed of numerous fine laminae, which, under a one-eighth of an inch object glass, exhibit generally only the appearance of a mezzotinto drawing; but there often were layers of branching vessels (like moss-agate), less than the $\frac{1}{10000}$th of an inch in diameter, and of a darkish colour; these vessels are not articulated, but otherwise resemble the same peculiar structure in the valves of the capitulum. The exterior yellow membrane is marked, or rather composed of successive narrow rims, which, in fact, are the lines of termination of the laminae of membrane, which in a calcified state form the cup itself. In most parts, both on the borders and under the centre of the cup, but not everywhere, there are imbedded in the yellow membrane, elongated, irregular, top-shaped masses of bright yellow chitine, each furnished with a tubulus, which penetrating the calcareous laminae leads to the corium; the little apertures thus formed, are clearly visible in the layers of membrane, left after exposure to acid. In *L. Nicobarica*, the innermost shelly layer of the cup was punctured, like the surface of the shell in Chthamalus and many other sessile cirripedes, by the internal orifices of these tubuli. The top-shaped masses often have star-shaped summits; and they differ in no essential respects from those on the lower part of the peduncle, excepting that they are quite imbedded in the membrane covering the under surface of the cup, whereas those on the peduncle project freely. I found these top-shaped bodies in the outer membrane of the cups in *L. dorsalis, L. cauta,* and *L. Rhodiopus,* which alone I was enabled to dissolve in acid; and I mention this fact, as indicating the probable presence of the more important star-headed projections on the lower parts of the peduncle in these same species. The basal calcareous / cup resembles, in essential structure, the valves of the capitulum; the chief difference being that in the former there is a larger proportion of animal matter or membranous layers.

After the dissolution of the cups, in *L. dorsalis* and *L. Rhodiopus*, I most distinctly traced the two cement ducts; they included the usual darker chord of cellular matter; they were of rather small diameter, namely, 2/3000th of an inch. The two (in *L. dorsalis*) ran in a very irregular course, not parallel to each other, making the most abrupt bends. They passed through the membranous layers (as seen after dissolution), and running for short spaces parallel to the component laminae, were attached to them. In their irregular course, these cement ducts resemble those of *Pollicipes mitella*, but I could not perceive that any cement had been poured out at the abrupt bends. In one specimen of a basal cup, which I was enabled to examine whilst still attached to the rock, I found under the very centre (and of course outside the yellow membrane), a very small area of dark brown cement of the usual appearance. In several specimens of full-sized cups, I was not able to perceive any cement on the external surfaces of the upper and later-formed layers; hence I believe that the cup is cemented to the bottom of the hole only during the early stages of its formation; and this, considering its protected situation, would no doubt be sufficient to affix the animal. This probably accounts for the small size of the cement ducts, and for the facility with which, as it appears, the cups can be removed in an unbroken condition from the rock. In the case, however, of the small, flat, calcareous discs, which are formed whilst the animal is burrowing into the rock, these are attached firmly to the sides of the holes, in the usual manner, by cement. In this cirripede it would be useless to look for the prehensile antennae of the larva under the cup, for the animal, during the formation of the successive discs, must have travelled some distance from the spot on which the larva first attached itself. /

The membrane of the peduncle is continuous with the yellow membrane coating the external surface of the cup; and this latter membrane is continuous with those delicate laminae which, in a calcified condition, form the layers of the cup itself. In an exactly similar manner, in this and other cirripedes, the membrane of the peduncle, at the top, is continuous with that coating the valves, and is attached to the lower exterior edge of the last formed layer of shell.

When a new shelly layer is formed, both under the valves of the capitulum and inside the basal calcareous cup, it projects beyond the old layer, and is included within the old, as yet not moulted, membrane of the peduncle. Within the cup of *L. Nicobarica* I found a lately formed layer of shell, projecting ⅒th of an inch on one side of the cup, and by its protuberance distinguishable even through the old coat of the peduncle, which was nearly ready to be moulted. In an analogous manner, in the capitulum of *L. dorsalis* and *L. truncata*, I have found a new peduncular membrane bearing the usual, but then sharp, calcified scales, attached to the lower projecting edge of the last formed shelly layer, lying under the old peduncular membrane, which was attached to the penultimate layer of shell, and with its worn scales was just ready to be moulted.

The final cause of the moulting of the calcified scales, together with the membrane of the peduncle to which they are attached – a case confined to Lithotrya – I have scarcely any doubt is the reproduction of a succession of scales, sharply serrated for the purpose of enlarging the cavity in which the animal is lodged. The extreme thinness of the membrane of the peduncle has been noticed; this may be partly related to its protected condition, but partly, I think, to the necessity of its being formed in a very extensible condition; for the new coat, owing to the projection of the new shelly layers under the valves, and within the basal cup, is by so much shorter than the old peduncle, yet after exuviation it has to stretch to a greater length than the old membrane, to allow of the growth of / the cirripede. Owing to the thinness and fragility of this membrane, the basal attachment of the cirripede is, no doubt, chiefly effected by the unusually strong longitudinal muscles; and the necessity of a surface of attachment for these muscles, stronger than the external membrane of the peduncle, probably is one of the final causes of the basal calcareous disc and cup, and likewise for the unusual manner in which the valves of the capitulum are locked together by folds and small roughened projections. The basal discs and cup, however, apparently serve for several other purposes, namely, for raising the animal a little in its burrow (which is narrow and pointed at the bottom), at that period of growth when it has ceased to burrow downwards, but still increases in diameter; also for carrying the animal, as over a bridge, across any pre-existing cavity in the rock; and lastly, perhaps, for removing lower down, in the intervals of exuviation, the point of attachment for the longitudinal peduncular muscles.

Position of the animal in the rock, and its power of excavation. A specimen of rock, two or three inches square, in Mr Cuming's possession, is full of Lithotryas; the cavities extend in every possible direction, and several were parallel, but with the animals in reversed positions; the same thing is apparent in some specimens of Mr Stutchbury's, and it was evident that the positions occupied by the animals were entirely due to chance. In Mr Cuming's specimen of rock, a considerable portion of the external surface is preserved, and here it can be seen that many of the specimens have their capitulums directed from the external surface directly inwards. These individuals, which were of full size, must have preyed on infusoria inhabiting the cavities of the porous, calcareous rock. On the other hand, I have seen some young specimens of *L. dorsalis* with their valves not at all rubbed, and others of full size with uninjured Balani and corallines on the tips of the valves, and again a specimen of *L. truncata* with minute pale-green seaweed on the summit of the capitulum – all which appearances induce me to / believe that in these cases, the valves had projected freely beyond the cavity in which their peduncles were lodged. I may here also mention that in Mr Cuming's specimen, above alluded to, the basal cups of five specimens touched and adhered to each other; I was not able to make out whether there had originally existed separate burrows, as I think is most probable, and that the walls had been wholly worn away, or whether the five specimens had fixed themselves on one side of a large pre-existing, common cavity. Young specimens seem to burrow to the full depth, before nearly acquiring the diameter which they ultimately attain. I measured one burrow, 1·2 of an inch in depth, which, at its mouth or widest part, was only 0·17 in diameter.

The several species occur imbedded in soft calcareous rocks, in massive corals, and in the shells of mollusca and of cirripedes. It has been doubted by several naturalists, whether the basal calcareous cup at all belongs to the Lithotrya, but after the foregoing microscopical observations on its structure, it is useless to discuss this point. So again it has been doubted whether the cavity is formed by the cirripede itself; but there is so obvious a relation between the diameters of specimens of various sizes, and the holes occupied by them, that I can entertain no doubt on this head. The holes, moreover, are not quite cylindrical, but broadly oval, like the section of the animal. The simple fact, that in this genus alone each fresh shelly layer round the bases of the valves, and therefore at the widest part of the capitulum, are sharply toothed; and secondly, that in this genus alone a succession of sharply serrated scales, on the upper and widest part of the peduncle, are periodically formed at each exuviation; and that consequently the teeth on the valves and scales are sharp, and fit for wearing soft stone, at that very period when the animal has to increase in size, would alone render the view probable that the Lithotrya makes or at least enlarges the cavities in which it is imbedded.

Although it may be admitted that Lithotrya has the / power of enlarging its cavity, how does it first bore down into the rock? It is quite certain that the basal cup is absolutely fixed, and that neither in form nor state of surface it is at all fitted for boring.[63] I was quite unable to answer the foregoing question, until seeing the admirable figures by Reinhardt[64] (Pl. VIII, figs 2 and 2a') of *L. Nicobarica*, still attached in its cavity. Subsequently I obtained from Mr Stutchbury several pieces of rock completely drilled with holes, many of small diameter, by *L. dorsalis*, and in these I found numerous instances of the linear rows of little discs, like those of *L. Nicobarica*, showing in the plainest manner, that each time a new disc is formed, that is, at each exuviation, the animal moves a short step downwards; and as the lowest of these little

[63] Mr Hancock, in his admirable account of his burrowing cirripede, *Alcippe lampas* (*Annals of Nat. Hist.*, p. 313, Nov., 1849), came to this conclusion regarding the cup of Lithotrya, and hence was led to think that this genus did not form its own burrows, but inhabited pre-existing cavities. I am much indebted to this gentleman, who has been so eminently successful in his researches on the boring powers of marine animals, for giving me his opinion on several points connected with the present discussion.

[64] I owe to the great kindness of Prof. Steenstrup the sight of this plate, published in the *Scientific Communications from the Union of Natural History*, Copenhagen, 30 January, 1850, No. I. Since this sheet has been set up in type, I have received from Prof. Steenstrup the memoir, in Danish, belonging to the figures in question; and the greater part of this has been translated to me by the kindness of a friend. My account of the means of burrowing is essentially the same as that published by Reinhardt; but the moulting of the scales on the peduncle, the presence of scales and of points of a different nature, the method of attachment by cement, the conversion of the discs into a cup, etc., seem not to have been known to this naturalist. Reinhardt states that the points on the peduncle will scratch Iceland spar, and that, apparently, they are formed of phosphate of lime: in the case of the closely allied *L. dorsalis*, I must believe that the scales or beads on the peduncle are formed of carbonate of lime, for they were quickly dissolved with effervescence in acetic acid; and the star-headed points, which are subsequently developed under the calcareous scales, appeared to me, under the compound microscope, to be formed of a horny or chitine substance. Reinhardt states that the basal point of the peduncle is arched a little under the lowest disc, and there forms for itself a slight furrow (as represented in the lateral view, Pl. VIII, fig. 2); but in the burrows examined by me, this furrow or depression did not really exist, the appearance resulting from the basal margin of the lowest disc, projecting beyond the wall of the cavity by the amount of its own slight thickness.

discs *in none of the burrows* was placed at the very bottom, we see that the lowest point of the peduncle must be the / wearing agent. In the peduncle of an individual of *L. dorsalis*, nearly ready to moult, I found, it may be remembered, beneath and round the basal disc, under the old membrane of the peduncle, a new membrane studded with calcified beads, but with the horny star-headed spines not yet developed, whilst on the old outer coat these latter had been worn down quite smooth, and the calcified beads worn entirely away. Here, then, we have an excellent rasping surface. With respect to the power of movement necessary for the boring action, the peduncle is amply furnished with transverse, oblique, and longitudinal striaeless muscles – the latter attached to the basal disc. In all the pedunculata, I have reason to believe that these muscles are in constant slight involuntary action. This being the case, I conceive that the small, blunt, spur-like portion of the peduncle, descending beneath the basal rim of the lowest disc, would inevitably partake slightly of the movements of the whole distended animal. As soon as the Lithotrya has reached that depth, which its instinct points out as most suitable to its habits, the discs are converted into an irregularly growing cup, and the animal then only increases in diameter, enlarging its cavity by the action of the serrated scales on the peduncle, and of the serrated lower edges of the valves of the capitulum. With respect to those reversed individuals attached with their capitulums downwards, I suppose that the larvae had crept into some deep cavity, perhaps made originally by a Lithotrya, of which the rock in the specimen in question was quite full, and had there attached themselves. Finally, it appears that in Lithotrya the burrowing is simply a mechanical action; it is effected by each layer of shell in the basal attached discs overlapping, in a straight line, the last formed layer – by the membrane of the peduncle and the valves of the capitulum having excellent and often renewed rasping surfaces – and, lastly, by the end of the peduncle (that is homologically the front of the head) thus roughened, extending beyond / the surface of attachment, and possessing the power of slight movement.

We will now proceed with our generic description.

Animal's body. This, as already stated, is partially lodged within the peduncle. The prosoma is rather largely developed.

The *mouth* is placed at a moderate distance from the adductor muscle.

The *labrum* is moderately bullate, with a row of blunt bead-like teeth, mingled with fine bristles, on the crest, which in the middle part is generally somewhat flattened.

The *palpi* are blunt, and even squarely truncated at their ends; they are of large size, so that, if they had been half as large again, or even less, their tips would have met.

Mandibles (Pl. X, fig. 2), with three nearly equal large teeth, and the inferior angle produced, broad, and strongly pectinated: in the interspaces between these teeth there are, in all the species, some very fine teeth or pectinations, which are seated a little on one side of the medial line. The mandibles are somewhat singular from the size of the transparent flexible apodemes (*a a*) to which the muscles are attached; these are oval and constricted at their origins: in *L. dorsalis* they are roughened with little points; in *L. cauta* and *L. truncata* they are large, of the same shape, but smooth.

Maxillae. These are larger, compared to the mandibles, than is usual with pedunculated cirripedes; they differ in shape in the different species, being either nearly straight on their edge, and notched or not (fig. 10), or notched with the inferior part forming a double prominence (fig. 12); the spines on the inferior angle, which is sometimes slightly produced, are always crowded together into a brush, and are finer than those on the upper parts. The apodemes are less straight than is usual, and at their origin take, in all the species, a rather abrupt bend; their extremity is enlarged into a little disc, which in / *L. dorsalis* is covered with strong points, but in the other species is, as usual, smooth.

Outer maxillae. The inner margin is slightly concave, and in *L. truncata* alone, the bristles are hardly continuous, being interrupted in the middle part. The olfactory orifices are only very slightly prominent. The spines on all the trophi are more or less doubly serrated.

Cirri. The three posterior pair are elongated, with their anterior surfaces not at all protuberant. The segments bear from three to five pair of spines, with a row of three or four small intermediate spines; there are, as usual, some little lateral upper rim spines; the dorsal tufts contain some thick and thin spines mingled. *First* cirrus is short, and placed not quite close to the second pair; the basal segments are broad and thickly paved with bristles. The *second* pair is rather short compared with the *third* pair; a varying number of the basal segments in both rami of both these cirri are protuberant, and are thickly paved with bristles; such segments are more numerous and are broader on the anterior rami than on the posterior rami. In *L. cauta* alone, none of the basal segments in the posterior rami of the second and third cirri are thickly paved with bristles. The pedicels of the first three pair are irregularly covered with spines; those of the three posterior pair have the spines arranged in a regular double line. Most of the spines are doubly serrated.

Caudal appendages (Pl. X, fig. 23 and 24), multiarticulate, with thin elongated segments fringed with short spines; in length generally exceeding the pedicel of the sixth cirrus, and in *L. Nicobarica* equalling half the entire length of this cirrus.

Stomach, destitute of caeca; oesophagus somewhat curled.

Filamentary appendages, none.

Ovaria filling up the peduncle and surrounding the sack, but not extending up to the bases of the scuta and terga; I saw the ova only in *L. truncata*; they were here oval and large, being nearly 9/400ths of an inch in length. /

Penis, elongated; vesiculae seminales extending into the prosoma. I noticed the ovigerous fraena only in *L. truncata*; here they were large, with an almost bilobed outline; the margin and whole lateral surface being covered with elongated cylinders, finely pointed, but not enlarged at their extremities, as are the glands observed in most of the other genera.

Colours. The posterior thoracic segments, the pedicels, the anterior and dorsal surfaces of the segments of the cirri, the caudal appendages, and the outer sides of the trophi are, in most of the species, more or less mottled with dark purple; parts of the interior surfaces of the valves in some of the species are coloured fine purple.

Geographical distribution. The species are found all round the world in the tropical seas; this fact may have some connection with the presence of soft coral-reef limestone and of massive corals in these seas. The presence, however, of *L. cauta* on the shores of New South Wales, shows that the genus is not strictly tropical.

Affinities. Lithotrya is a well-pronounced distinct genus; although there is a considerable difference in the shape of the valves between *L. dorsalis* and *L. Valentiana*, at the opposite extremes of the genus, the strict uniformity of the internal characters shows that there are no grounds whatever for any generic separation; moreover, *L. Rhodiopus* neatly blends together these extreme forms. Indeed it is not easy to imagine a better marked series of transitional forms, than those presented by the terga, in passing from *L. dorsalis* through *L. Nicobarica*, *L. Rhodiopus*, and *L. truncata*, to *L. Valentiana*. Lithotrya has most affinity to *Scalpellum villosum* or to *Pollicipes spinosus* and *P. sertus*; though the affinity is far from close. In these two species of Pollicipes, we have seen that large irregular calcified spines are formed at the base of the peduncle, whereas in the other Pedunculata the scales or spines are formed exclusively round the upper margin of the peduncle. Lithotrya, as has been remarked by Sowerby and other authors, exhibits some affinity to the sessile / cirripedes, as shown by the calcareous basis – by the manner in which the scuta and terga are locked together – by the two little fans of muscle attached to near the basal points of the terga – and perhaps by some of the characters of the trophi; nevertheless, this affinity is far from being well marked, and I think is hardly so plain as in *Pollicipes mitella*.

1. LITHOTRYA DORSALIS

PLATE VIII. *FIG. IA'*

LITHOTRYA DORSALIS. G. B. Sowerby, *Genera of Shells*, April, 1822
LEPAS DORSALIS. Ellis, *Nat. Hist. Zoophytes*, Pl. xv, fig. 5, 1786
LITHOLEPAS DE MONT SERRAT. De Blainville, *Dict. des Sc. Nat.*, Plate, fig. 5, 1824

L. scutis terga angustè obtegentibus: carinâ intùs concavâ: rostro, duorum aut trium squamarum subjacentium latitudinem aequante: lateribus, squamarum quinque subjacentium longitudinem aequantibus, superficie internâ angustè ellipticâ: pedunculi squamis superioribus verticillum secondum minus duplo superantibus.

Scuta, narrowly overlapping the terga: carina internally concave: rostrum as wide as two or three of the subjacent scales: latera with their internal surfaces narrowly elliptical, as long as five of the subjacent scales: upper scales of the peduncle less than twice as large as those in the second whorl.

Mandibles, with twice as many pectinations between the first and second main teeth, as between the second and third teeth. Maxillae without a notch, edge nearly straight, and spines very numerous: caudal appendages exceeding, by half, the length of the pedicel of the sixth cirrus.

Barbadoes, West Indies; Venezuela; Honduras; imbedded in limestone; Mus. Brit. Cuming and Stutchbury.

The state of preservation of the valves in different specimens varies greatly; generally only two or three, or / even only the last formed shelly layer, is preserved, the upper ones having scaled off; in a few young specimens, however, all the layers were perfect. The carina is generally better preserved than the other valves, and hence the upper part usually projects freely; in one specimen no less than ten zones of growth were preserved in the carina, whilst the other valves consisted of only three: the terga generally project rather more than the scuta. As each growth-layer is thick, if the scaling process had not taken place, all the valves would have projected greatly. The little teeth lie close together on the prominent serrated rims, on each zone of growth. The internal surfaces of the valves are roughened with small imbricated points. Exteriorly the valves are covered with yellow membrane, with rows, corresponding with each zone of growth, of very minute, yellow, horny spines, generally having their tips bent over, and so made hook-shaped. These spines are less than 1/600th of an inch in length.

Scuta, triangular; internally concave, with a large depression for the adductor muscle; there is the usual small roughened internal knob, or tooth, at the rostral angle of both the right- and left-hand valves. Tergal margin straight, overlapping about one-third of the entire width of the terga.

Terga, irregularly oval, with the scutal margin straight; basal point blunt, with the two sides placed at about an angle of 45° to each other; the lower part of the carinal margin, immediately over the latera (as seen internally), is slightly hollowed out. Exteriorly, towards the bottom of the valve, from the overlapping of the scuta, of the latera, and of the carina, only a narrow rounded ridge is exposed, which runs down to the basal angle at about one-third of the entire width of the valve, from the scutal margin. Internally the valve is slightly concave.

The *carina* slightly overlaps the terga; internally concave; generally with a large upper portion freely projecting; inwardly curved, without any central crest or ridge; valve nearly as wide as the middle part of the / terga; inner growing or corium-covered surface, with its basal margin, protuberant and arched.

Rostrum (Pl. VIII, fig. 1a', a, and greatly magnified 1b') very narrow; rarely more than two or three layers of growth are preserved; the sides are deeply sinuous, owing to each zone widening downwards; basal margin rounded; in width equalling about two and a half of the uppermost scales of the peduncle, and about half as wide as the latera.

Latera, small, placed obliquely, and parallel to the lower carinal margin of the terga; longer axis equal to five of the uppermost scales of the peduncle, and to nearly half the width of the base of the carina; growing surface (or a section made parallel to the growth-layers), is narrow, elliptic, pointed at both ends, but the carinal half rather thicker than the scutal half.

The *peduncle* varies in length, generally about twice as long as the capitulum, in one specimen above thrice as long. The upper part as wide as the capitulum, the lower part sometimes much attenuated. The calcified scales in the uppermost whorl (Pl. VIII, fig. 1b') are only slightly larger than those in the second whorl; the scales in the succeeding three or four whorls, are considerably larger than those below, which latter very gradually decrease in size, till, low down on the peduncle, they are barely visible to the naked eye. In this lower part, they may be called calcareous beads; they stand some way apart from each other; they are nearly hemispherical, smooth, translucent, and furnished with a conical fang; some of the smallest were $\frac{1}{325}$th and $\frac{1}{400}$th of an inch in diameter. The upper scales vary somewhat in the outline, the most usual shape being subtriangular, with the lower margin arched and protuberant; and this margin, in the two or three upper whorls, is crenated with teeth, which are conical and sharp, after exuviation, but soon become reduced to mere notches. The scales in the uppermost whorl are usually nearly quadrilateral; the imbedded portion, or fang of each scale, is, in all, produced into a blunt rounded point. / The basal calcareous cup (fig. 1a' and 1c') is well developed, and is sometimes even half an inch in diameter. Before the cup is formed, there is a row of small, flat discs (fig. 1, and like those in fig. 2a') attached to the sides of the burrow: but a full account of these parts of the peduncle, and of the burrowing habits of this species, has been given under the generic description.

Size and colour. Full average-sized specimens have a capitulum half an inch in width and height; the entire length, with the contracted peduncle, being about an inch and a half. Valves coloured dirty white, with the enveloping membrane, when preserved, yellow. The outer maxillae, palpi, pedicels of the cirri, anterior faces of the segments, dorsal tufts, caudal appendages, and penis, dark purple. Thoracic segments brown. There is a purple spot between the bases of the first pair of cirri.

Mouth. Labrum considerably bullate, equalling about half the longitudinal diameter of the mouth; inferior part produced so as to separate the mouth some way from the adductor muscle; crest with a row of blunt teeth and hairs; central part depressed and flattened.

Palpi, rather large, separated from each other by only half their own length; bluntly pointed, thickly clothed with spines.

Mandibles (Pl. X, fig. 2), with twice as many pectinations, namely 15, between the first and second main teeth, as between the second and third teeth, namely about seven; inferior angle strongly and coarsely pectinated; distance between the tips of the first and second main teeth, considerably less than between the tips of the second tooth and of the inferior angle; sides hirsute.

Maxillae (fig. 10), with the edge not quite straight, with the whole inferior part slightly projecting; spines very numerous, thirty or forty pairs; those close beneath the two upper great unequal spines, form a tuft and are rather thinner than the others, as are also those near the inferior angle; sides hirsute. /

Outer maxillae, rather pointed, with the inner edge slightly concave, continuously and thickly clothed with short spines; spines on the outer edge long; there are also some minute, short, thinly scattered spines or points on the sides. Bristles on all the trophi doubly serrated.

Cirri. The first pair is placed at a small distance from the second. The segments in the three posterior pairs, support five pairs of very long spines, with a row of (I believe) four small intermediate spines; on the lateral upper edges, there are some short blunt spines; anterior faces of the segments not protuberant; the dorsal tufts consist of thick serrated, and of thin spines. The whole integument is hirsute with minute pectinated scales. Two or three of the basal segments in the sixth cirrus are confluent. *First cirrus*, anterior ramus rather shorter and thicker than the posterior ramus; basal segments thickly paved with serrated spines; in the posterior ramus, the six terminal segments are not paved with bristles. *Second cirrus* has the seven basal segments of the anterior ramus very broad, and paved with bristles; the eight terminal segments having the usual structure; in the posterior ramus the three or four basal segments are similarly paved, but to a very much less degree, and the remaining thirteen have the usual structure. *Third cirrus* has the six basal segments of the anterior ramus very broad and paved, and the fourteen terminal ones of the usual structure; in the posterior ramus, the three or four basal segments are similarly paved, but to a very much less degree, and the seventeen terminal ones have the usual structure. The pedicel of the first cirrus has very few spines; those of the second and third cirrus are thickly and irregularly clothed with spines; and those of the three posterior pair have a double row with intermediate small spines. On the antero-lateral faces of the pedicels of the second, third, and fourth pairs of cirri, there is an elongated white swelling or shield. Moreover, on the posterior thoracic segments, there are similar white-coloured swellings, with the membrane more plainly marked with scales than in / other parts. The spines on the first three pairs of cirri are coarsely serrated.

Caudal appendages (Pl. X, fig. 23), with numerous tapering segments, almost equalling one and a half times the length of the pedicel of the sixth cirrus. Each segment is elongated and somewhat constricted in the middle, with its upper edge (fig. 24) crowned with short spines; in a full-sized specimen there were seventeen segments.

2. LITHOTRYA CAUTA

PLATE VIII. FIG. 3

L. scutis terga amplè obtegentibus: carinâ intus concavâ: rostro squamarum subjacentium latitudinem vix aequante: lateribus, squamas subjacentes sesquitertio superantibus; superficie internâ latè ellipticâ: pedunculi squamis superioribus verticillum secundum paene quadruplo superantibus.

Scuta largely overlapping the terga: carina internally concave: rostrum hardly as wide as one of the subjacent scales: latera with their internal surfaces broadly elliptical, as long as two and a half of the subjacent scales: upper scales of the peduncle nearly four times as large as those in the second whorl.

Mandibles with an equal number of pectinations between the first, second, and third main teeth: maxillae notched, edge nearly straight: posterior rami of the second and third cirri, with their basal segments not paved with bristles: caudal appendages slightly exceeding in length the pedicels of the sixth cirrus.

New South Wales, Australia, imbedded in a Conia (unique specimen), Mus. Stutchbury.

Valves thin, white, translucent; upper layers of growth well preserved, excepting on the terga. A large portion of the carina projected freely. The teeth on the projecting margins of the growth-layers are broad, blunt, and often stand rather distant from each other.

Scuta (Pl. VIII, fig. 3*a*), triangular, internally concave, / with no distinct pit for the adductor muscle. The scuta largely overlaps the terga.

Terga (fig. 3*b*) approaching to rhomboidal; basal angle rectangular, almost central, and consequently the exterior longitudinal ridge, which is rounded, is likewise nearly central.

Carina, internally concave, with no trace of a central internal ridge in the upper free portion; the growing or corium-covered surface is transversely oval, and is as wide as the widest part of the terga.

Rostrum, exceedingly minute, enlarged at each zone of growth, not so wide as the immediately subjacent scale on the peduncle.

Latera (fig. 3*c*), in width equalling two and a half of the upper peduncular scales, or about one-fourth or one-fifth of the width of the carina; growing surface (or a section parallel to the layers of growth), broadly elliptic, pointed at both ends.

Peduncle, about twice as long as the capitulum; the scales of the uppermost whorl are quadrilateral (fig. 3*d*), and nearly four times as large as those in the second whorl; these latter are about twice as large as those in the third whorl, which are very little larger than the small, almost equal-sized, equally distant, round beads scattered over the rest of the peduncle, down to the basal cup. All these scales are dentated, the upper rows most plainly and only on their basal margins; the lower little beads are very slightly crenated round their entire margins; they are mingled with star-headed spines (fig. 3a) of yellow chitine. Basal calcareous discs thin, plainly marked exteriorly by concentric lines of growth, and covered by the usual yellow membrane, including the horny, spindle-shaped bodies.

Size and colours. The whole specimen, including the peduncle, was only one-fifth of an inch in length; the capitulum being 3/40ths of an inch in width. I do not know whether the specimen had attained its full size, but think this is probable, as a large-sized species would not have / made its habitation in one of the valves of so small a shell as a Conia. Shell white, exterior membrane, where preserved, yellow, and bearing small spines. Thoracic segments, the lower segments of the second, third, and fourth cirri, all the segments of the first cirrus and the trophi, slightly mottled with darkish purple.

Mouth. The teeth or beads on the crest of the labrum are blunt, few, not very small, and equidistant.

Palpi, bluntly pointed.

Mandibles, with the three main teeth nearly equal in size; the pectinations are equal in number, namely, only three between the first and second, and the second and third main teeth; the inferior angle is coarsely pectinated, with one central spine much longer than the others; the distance between the tips of the first and second main teeth, equals that between the second tooth and the inferior angle.

Maxillae, with the two upper spines very large; beneath them there are two small spines, and a considerable notch; the inferior part of the edge is nearly straight, bearing about thirteen pairs of spines, obscurely divided into two groups, the lower spines being smaller than the upper ones. The upper convex margin is hirsute with long hairs.

Outer maxillae, blunt, with the inner margin slightly concave; continuously, but thinly clothed with spines.

Cirri. The segments of the three posterior pairs bear four pairs of spines, with the usual intermediate fine spines; dorsal spines thin and thick mingled together. *First cirrus,* short, with the anterior ramus rather the thickest and shortest; all the segments thickly paved with bristles, except the two terminal segments, of which the ultimate one bears some serrated spines of most unusual length, namely, equalling within one segment the entire length of the ramus. I presume that these spines serve as feelers. *Second cirrus*; anterior ramus much thicker and considerably shorter than the posterior ramus; six basal segments paved with bristles, the two terminal segments having the usual structure; posterior ramus with all its nine / segments on the usual structure. *Third cirrus*, longer, to a remarkable degree, than the second cirrus, with its anterior ramus having the four basal segments paved, and the seven terminal ones on the usual structure; posterior ramus with twelve segments, of which none are paved. The pedicels of the second and third cirri thickly and irregularly clothed with spines. The upper segments of the pedicels of all the cirri are unusually long.

Caudal appendages, longer than the pedicels of the sixth cirrus, by barely one-third of their own length. Segments much elongated, seven in number; I may add for comparison that each ramus of the sixth cirrus contained, in this specimen, sixteen or seventeen segments.

General remarks. It is difficult to give obvious characters (excepting the smallness of the rostrum compared with the scales on the peduncle), by which this species can be externally discriminated from *L. dorsalis*, *L. Nicobarica*, and *L. Rhodiopus*; yet almost all the valves differ slightly in shape. In this species alone (the peduncle of *L. Rhodiopus* is not known), the lower, microscopically minute, bead-like scales of the peduncle are crenated, though obscurely, all round. In the animal's body, the diagnostic characters are strongly marked; the long spines on the terminal segment of the first cirrus – none of the segments in the posterior rami of the second and third cirri being thickened and paved with bristles – the pectinations being equal in number between the main teeth of the mandibles – are all characters exclusively confined to this species.

3. LITHOTRYA NICOBARICA

PLATE VIII. FIG. 2

L. NICOBARICA. Reinhardt, *Naturhist*; Selskabet, Copenhagen, No. I, Pl. I, fig. 1–3,[65] 1850 /

L. scutis terga angustè obtegentibus: carinae cristâ internâ tenui in parte superiore positâ: rostro conspicuo, squamarum sex subjacentium latitudinem aequante: lateribus, superficie internâ triangulâ, squamarum septem subjacentium latitudinem aequantibus.

Scuta narrowly overlapping the terga: carina with a slightly central internal ridge in the upper part: rostrum conspicuous, as wide as six of the subjacent scales: latera, with their internal surfaces triangular, as wide as seven of the subjacent scales.

Palpi square at their ends: mandibles with twice as many pectinations between the first and second main teeth, as between the second and third: maxillae slightly notched, with the inferior angle slightly prominent: caudal appendages more than twice as long as the pedicels of the sixth cirrus.

Timor; Brit. Mus. (given by Cuvier to Leach); Nicobar Islands, according to Reinhardt.

Capitulum as in *L. dorsalis*. The teeth on the prominent rims of the valves are small and approximate; but the specimen was much worn.

Scuta, triangular, slightly overlapping the terga; the line of junction between these valves slightly sinuous, the upper part of the tergal margin of the scuta being slightly hollowed out, and the corresponding upper portion of the margin of the terga being slightly protuberant. Internally, there is a considerable depression for the adductor muscle; and besides the usual knob at the rostral angle, there is a trace of a knob at the baso–tergal angle.

Terga, as seen internally, irregularly rhomboidal, ending downwards in a blunt point, of which the two sides (neither being sensibly hollowed out), stand at about an angle of 45° to each other. Scutal

---

[65] I am not at all sure that the proper title of the periodical in which this species has been described, is here given. I am greatly indebted to Prof. Steenstrup for sending me a separate copy of the paper in question, written in Danish. I believe I am right in identifying the specimen here described, from Timor, with the species from the Nicobar Islands, named by Reinhardt, *L. Nicobarica*.

margin, with the upper part (as above remarked), slightly protuberant: / near the bottom of this margin, there is a very slight projection, answering to the small knob at the baso–tergal angle of the scutum. Externally, towards the basal angle, the narrow strip not concealed by the overlapping of the latera and carina is square-edged, with the zones of growth on it straight.

Carina, internally concave in the upper free part, with a slight, central, internal crest, caused by the projection of each successive zone of growth. The inner growing surface is almost pentagonal in outline; with the basal margin square and truncated in the middle.

Rostrum (fig. 2a), rather conspicuous, many zones of growth being preserved. It equals in width six of the subjacent scales of the peduncle, but as these are rather smaller than elsewhere, the width equals about five of the ordinary uppermost scales; compared with the latera, it is nearly $5/7$ths of their width.

Latera, unusually large; as seen on their interior surfaces (or in a section parallel to the zones of growth), they are triangular, elongated transversely, with the carinal angle a rectangle. In width they equal the seven subjacent scales of the peduncle, and are more than half as long as the basal margin of the carina.

Peduncle, with the upper scales varying from circular to quadrilateral, thrice as large as those in the second whorl; beneath which, in the next three or four whorls, the scales rapidly decrease in size; and beneath these the whole peduncle is studded with equal-sized, rounded, calcareous beads, so minute as to be quite invisible to the naked eye. This specimen was nearly ready to moult, and perhaps in consequence of this, even the upper scales were most obscurely serrated on their lower margins, and all the others quite smooth: there were some much worn horny spines close to the bottom of the peduncle. Basal calcareous cup slightly concave, of moderate size; its diameter, in the one specimen examined, was $3/10$ths of an inch; it was composed of several layers. In the specimen figured (2a') by Reinhardt, instead of a cup, there is a / straight row of small discs, which are attached to the walls of the cavity, as explained in the generic description.

Mouth. Palpi with their ends square and truncated; thickly clothed with long spines.

Mandibles, with fully twice as many pectinations (viz. from 16 to 20), between the first and second main teeth, as between (viz. 8 to 10) the second and third main teeth. Inferior angle, coarsely pectinated. The distance between the tips of the first and second teeth, is considerably less than between the tip of the second tooth and the inferior angle.

Maxillae, with the edge very slightly irregular; beneath the two great upper spines there is a slight notch, with some small spines: inferior angle slightly prominent, with a brush of moderately fine spines; besides these, there are about seventeen pairs of large spines; sides very hairy.

Outer maxillae, with the inner margin slightly concave, and with the spines continuous.

Cirri. The segments in the three posterior pairs support three or four pairs of long spines, with a single row of moderately long intermediate spines; the dorsal tufts consist of a few rather thick, and some long and thin spines. The front of the segments is not protuberant; the whole surface is hirsute with minute comb-like scales. *Second cirrus*, with the anterior ramus having its eight basal segments highly protuberant and thickly clothed with spines, the upper nine having the usual structure; the posterior ramus has four or five basal segments thickly clothed with spines, and the twelve upper ones with the usual structure. *Third cirrus*, with the anterior ramus having six segments highly protuberant and thickly clothed with bristles, and the fifteen upper ones on the usual structure; in the posterior ramus, only three or four of the basal segments are paved with bristles. The spines on the first three pairs of cirri, are coarsely and doubly serrated.

The *caudal appendages* are more than twice as long as the pedicels of the sixth cirrus, and equal half the / length of the whole cirrus. In a specimen in which the sixth cirrus contained twenty-two segments, the caudal appendages actually contained twenty. The segments are thin, with their upper edges clothed with serrated spines. The slip of membrane on each side, whence this organ springs is united, for a little space, to the lower segment of the pedicel of the sixth cirrus.

Size and colour. Width of the capitulum rather above $4/10$ths of an inch; length, including the peduncle (contracted by spirits), nearly one inch. Valves, as usual, dirty white, partly invested by yellow membrane, furnished with a few minute yellow horny spines. Pedicels of the first four cirri, caudal appendages, penis, the two posterior thoracic segments, the segments of the cirri, and the trophi, clouded, banded, or spotted, with blackish purple.

Affinities. This species, in the characters derived from the valves, comes perhaps nearest to *L. Rhodiopus*; in the characters derived from the animal's body, it is nearest to *L. dorsalis.*

4. LITHOTRYA RHODIOPUS

PLATE VIII. *FIG. 4*

BRISNAEUS RHODIOPUS. J. E. Gray, *Annals of Philosoph.*, vol. x (new series), 1825
BRISNAEUS RHODIOPUS. J. E. Gray, *Spicilegia Zoolog.*, Pl. xvi, fig. 17, 1830

L. scutis terga ample obtegentibus: carinae cristâ internâ tenui, in parte superiore positâ: lateribus, superficie internâ symmetricè et latè ovatâ, carinae latitudinis plus quam tertiam partem aequantibus: tergorum basali apice tenui, et angulo carinali producto: rostro et pedunculo ignotis.
Scuta largely overlapping the terga. Carina with a slight central internal ridge in the upper part. Latera with their internal surfaces symmetrically and broadly oval, more than one-third of the width of the carina. / Terga with the basal points narrow, and the carinal angle produced. Rostrum and peduncle unknown.

Mandibles, with four times as many pectinations between the first and second main teeth, as between the second and third; distance greater between the tips of the first and second teeth, than between the tip of the second tooth and the inferior angle. Maxillae widely notched, with the inferior part forming two obscure prominences.

Habitat unknown. Imbedded in a massive coral. Brit. Mus.

The specimens are in a rather bad condition, and have been disarticulated. They are of rather small size; the rostrum and peduncle are lost, and animal's body much injured.

Valves white, thin, translucent; teeth on the projecting rims small, narrow, standing further apart than their own width. The upper layers have undergone but little disintegration or scaling off, and consequently the carina and terga project freely. The valves, where not rubbed, are covered by bright yellow membrane, which is thickly clothed with rows of spines; these are small on the exterior surfaces, but are very large and hooked in certain parts, as near the tergal margins of the scuta, and on the carinal margins of the terga, and especially on the inner face of the upper free part of the carina. Here the hooked spines (fig. 4*d*) are trifid or quadrifid, and are very conspicuous.

Scuta, as seen externally, triangular; they overlap half the width of the terga; on their internal faces (fig. 4*a*), in the upper projecting part, there is a strong ridge, against which the scutal margin of the terga abuts. There is a deep and conspicuous pit for the adductor muscle.

Terga, as seen externally, nearly triangular. The ridge which leads from the apex to the basal angle, is rounded, central, and extremely prominent; but does not form a furrow, or include the overlapping margin of the scuta. The basal angle is narrow, spur-like, and slightly hollowed / out on both margins. The growing corium-covered surface (fig. 4*b*) is transversely elongated, with the occludent margin rounded, and the carinal angle much produced, but not forming a roughened knob.

Carina (fig. 4*d*), concave within, with a slight central ridge in the upper free portion. The inner growing surface is concave, almost pentagonal, with a just perceptibly raised central rim in the upper part, and with two minute prominences on each side, against which the produced carinal angles of the terga abut.

Rostrum, lost.

Latera (fig. 4*c*), growing surface (or a section parallel to the growth-layers), symmetrically oval, more than one-third as wide as the basal margin of the carina. Several zones of growth preserved.

Peduncle, lost, but a few scales accidently adhering to one of the valves, show that they are crenated in the three or four upper whorls. No basal calcareous cup was preserved, but by clearing out the base of one of the holes in the coral, in which a specimen had been imbedded, I found a little flat disc about the size of a pin's head; it was composed of two or three layers, and was externally coated by yellow membrane, including the usual spindle-shaped bodies and tubuli. The cement ducts were also discovered after dissolution in acid. So that there could be no doubt regarding the nature of the little disc.

Mouth. Labrum with a row of little blunt teeth.

Palpi, blunt, rather expanded at their ends, with the extreme margin much arched and furnished with two rows of long spines; there is a fringe of short spines on the straight inner side.

Mandibles. There are nine pectinations between the first and second main teeth, and only two between the second and third teeth; the inferior angle is coarsely pectinated, with one central spine twice as long as the others. The distance between the tips of the first and second main teeth, is greater than between the tip of the second tooth and the inferior angle. /

Maxillae (Pl. X, fig. 12). These may be described as having their edge formed into three prominences; or, as having a very wide notch under the two upper great spines, and with the whole inferior part forming two prominences. There are, altogether, about twelve pairs of spines, of which two stand singly on the inferior side of the wide notch under the two upper great spines. The spines on the inferior angle are rather smaller than those above; sides hirsute.

Outer maxillae, with the inner margin slighly concave, and sparingly covered with bristles.

Cirri, imperfectly preserved; the three posterior pairs have segments of the usual character, bearing five pairs of very long spines, with the usual little intermediate, the minute lateral, and the dorsal spines. First cirrus lost; second and third with only their few basal segments preserved, sufficient, however, to show that at least two or three segments, in both the anterior and posterior rami of both cirri, were paved with bristles.

Pedicels, as in the other species.

Caudal appendages, lost.

This species comes very close, as far as the characters derived from the trophi serve, to the *L. truncata*, though readily distinguished from that species by the shape of the valves. On the other hand, the capitulum of this species is distinguished with difficulty from that of *L. Nicobarica* and *L. cauta*; no doubt this difficulty is much enhanced by the rostrum and peduncle having been lost.

5. LITHOTRYA TRUNCATA

PLATE IX. FIG. 1

ANATIFA TRUNCATA. Quoy et Gaimard, *Voyage de l'Astrolabe*, Pl. xciii, figs 12 to 15, 1834

L. scutis in profundam tergorum plicam insertis: carinae cristâ centrali prominente et rotundatâ in parte superiore: rostro et lateribus rudimentalibus, carinae latitudinis quindecimam fere partem aequantibus. /
Scuta locked into a deep fold in the terga: carina with a prominent central rounded ridge in the upper part: rostrum and latera rudimentary, about 1/15th of the width of the carina.

Mandibles, with nearly three times as many pectinations between the first and second teeth, as between the second and third teeth; distance between the tips of the first and second teeth equal to that between the tip of the second tooth and inferior angle. Maxillae widely notched, with the inferior part forming two prominences. Caudal appendages shorter than, or barely exceeding in length, the pedicels of the sixth cirrus.

Friendly Archipelago, Mus. Paris; Philippine Archipelago, Mus. Cuming; imbedded in coral rock.

Capitulum rather thick, with the five main valves having their free apices, diverging and truncated. The upper and old layers of shell do not here scale off so readily as in many of the foregoing species; and hence an unusually large proportional length of each valve projects freely above the sack; and the valves are of unusual thickness. The capitulum is very nearly as wide at its summit as at its base, owing to the divergence of the apices of the valves. The scuta and terga are articulated together by a conspicuous fold, which, when seen from vertically above (Pl. IX, fig. 1a'), appears like a deep wedge-formed notch in the terga. On the exterior surfaces of the valves, the teeth on the successive rims are approximate; on the inner surfaces, the rims are covered by strong yellow membrane, which is generally fringed with small horny spines.

Scuta, exterior surface convex, subtriangular, with the apex truncated: seen vertically from above, there is a small rectangular indentation or fold which receives the projecting scutal margin of the terga. The inner growing or corium-covered surface (fig. 1b, b') is triangular, with its tergal margin *largely* hollowed out. Along the occludent margin there is a slight ridge, which terminates at the / rostral angle, in both the right- and left-hand valves, in a rounded, knob-like, roughened tooth. The lower part of the tergal margin is slightly inflected and roughened, where it meets the corresponding lower part of the scutal margin of the terga. There is a deep pit for the adductor muscle. The interior surface of the valve above this pit is faintly coloured purple. The inner surfaces of both scuta and terga, are roughened with little points.

Terga, seen externally, are almost quadrilateral (owing to the apex being truncated), with the free margin facing the scutum, arched. Seen vertically from above, each shows a deep fold, which receives the lower part of the tergal margin of the scutum. In the foregoing species, a prominent ridge runs down the exterior surface of the terga from the apex to the basal angle, against which ridge, the margin of the overlapping scuta abuts: here this ridge, instead of projecting straight out, is oblique or folded over, and thus forms a furrow, receiving the margin of the scuta. The interior growing surface of the tergum (fig. 1b', c), presents so irregular a figure, that it can hardly be described; in area it quite equals the scuta; it is slightly concave; at the upper point of the carinal margin, there is a large, rounded, protuberant, roughened knob, which corresponds with a small knob on each side of the inner face of the carina; these knobs seem firmly united together by membrane. The scutal margin of the terga, in the upper part, forms a shoulder, largely projecting over the scuta; on its lower part, there is a small roughened projection. The occludent margin is arched and protuberant, with a slight fold above the knob on the carinal margin, just mentioned: this fold is caused by the protuberance of the central internal ridge of the carina, but is so small, that when the capitulum is seen from vertically above, it can hardly be distinguished. Finally, the basal half of the carinal margin, runs in the same line with the basal margin of the scuta.

Carina, moderately large; seen externally, the surface presents an elongated triangle, with the apex truncated; / on the internal face (fig. 1b', d) of the free part, there is (instead of being concave as is usual) a great central ridge, which projects between the diverging apices of the terga, as may be seen from vertically above; hence the thickness of the upper part of the carina, in a longitudinal plane, almost equals its breadth. The edge of this ridge is rounded. The inner or growing surface of the carina is tinted purple, and lies in a plane, oblique to the longer axis of the valve; it is triangular, with the apex cut off, and the basal margin rounded and protuberant; it is not concave. There is a central raised line or slight ridge on this inner surface, and on each side in the upper part there is a small, white, roughened knob, corresponding with the similar knobs on the carinal margins of the terga.

Rostrum (fig. 1*b'*, *a*), rudimentary; in one specimen it was about 1/30th of an inch in width; it is either as wide, or only half as wide, as the subjacent scale on the peduncle.

Latera, rudimentary, placed between the edges of the carina and the terga; rather smaller than the rostrum; almost cylindrical, slightly flattened, enlarged at each zone of growth, with one or two sharp teeth or spines on both faces; imperfectly calcified; in width barely 1/15th part of the carina.

Peduncle, short; the scales alone in the uppermost whorl are plainly toothed; they are transversely elongated, and almost quadrangular, and are nearly twice as large as those in the second whorl. Beneath this second whorl, there are two or three whorls, with scales, graduated in size; and the rest of the peduncle is covered by rather distantly scattered, minute, rounded or acutely pointed scales: the pointed scales are directed upwards, and are best developed under the carina. The basal calcareous cup, judging from two specimens, is thin, and not much developed.

Size and colour. The largest specimen was nearly 6/10ths of an inch across its capitulum. The calcareous / valves are dirty white. The sack is (after having been long kept in spirits) pale coloured, excepting a small purple space, between the scuta and another over the carina. The three posterior segments of the thorax and portions under the second and third cirri, the trophi, the pedicels and the anterior faces of the segments (especially of the basal segments in the second and third cirri), and a spot on their dorsal surfaces, and the penis are all coloured dark purplish-black. The prosoma is pale coloured.

Mouth. Crest of labrum with a row of bead-like teeth and hairs. *Palpi* bluntly pointed, with neither margin hollowed out.

Mandibles, with eight pectinations between the first and second main teeth, and three between the second and third teeth; inferior angle coarsely pectinated, with a central spine much longer than the others; the distance between the tips of the first and second main teeth, is about equal to that between the tip of the second tooth and of the inferior angle.

Maxillae. Under the two upper long spines (associated with some smaller ones), there is a slight and wide hollow; and the whole inferior edge obscurely forms two blunt points, with the spines on the lower projection smaller than the upper spines.

Outer maxillae, considerably concave in front, with the spines almost discontinuous in the middle part.

Cirri. First pair rather far separated from the second pair. The segments of the three posterior cirri bear three or four pairs of main spines, and are otherwise characterized like the foregoing species. *First cirrus*, with its anterior ramus much thicker than the posterior ramus, and of nearly equal length; all the segments, except the two terminal ones, thickly clothed with serrated spines. *Second cirrus* considerably shorter than the third cirrus: anterior ramus with the seven basal segments very protuberant, and paved with bristles, and the four terminal ones on the usual structure; posterior ramus, with the five basal segments paved (but much less thickly than in the / anterior ramus), and the nine terminal ones on the usual structure. *Third cirrus*, the anterior ramus, with the five basal segments, thick and paved, and eleven terminal segments on the usual structure: posterior ramus, with one basal segment paved, and sixteen other segments on the usual structure. In the posterior rami, however, of both the second and third cirri, it is difficult to draw any distinct line between the paved segments and the others.

Caudal appendages, short, either just exceeding in length the pedicels of the sixth cirrus, or equalling only the lower segment: segments flattened, cylindrical, six in number, there being, in the same individual, twenty-one segments in both rami of the sixth cirrus.

6. LITHOTRYA VALENTIANA

PLATE VIII. FIG. 5

CONCHOTRYA VALENTIANA. J. E. Gray, *Annals of Philosoph.*, vol. x (new series), 1825

L. scutis in profundam tergorum plicam insertis: tergorum opposito superiore margine, plicâ alterâ aequè profundâ instructo: carinae cristâ prominente centrali, marginibus quadratis, in parte superiore: rostro rudimentali: lateribus et pedunculo ignotis.

Scuta locked into a deep fold in the terga; the latter having a second equally deep fold on the opposite upper margin. Carina with a prominent, central, square-edged ridge in the upper part: rostrum rudimentary. Latera and peduncle unknown.

Animal unknown.

Red Sea, imbedded in an oyster-shell. British Museum.

General remarks. The two specimens in the British Museum are small, and in an imperfect condition, without the peduncle or the latera, and without the body of the animal. The capitulum so closely resembles that of *L. truncata*, that it is quite superfluous to do more than / point out the few differences. It is just possible, though not probable, that this form may prove to be merely a variety or younger state of *L. truncata*, in which case this latter name would have to be sunk. The difference, though one only of degree, in the form of the terga of the two species is conspicuous, and there is a slight difference in the carina, and again some dissimilarity in habits.

Description. The valves, as just stated, generally resemble those of *L. truncata*; scarcely any appreciable difference can be detected in the scuta; the apex, however, of the inner surface seems coloured a darker purple. The terga, as seen from vertically above (Pl. VIII, fig. 5*b*), have a fold or indentation on the upper or occludent margin, as large and as conspicuous as that receiving the margin of the scuta: this fold, as seen on the inner corium-covered surface (fig. 5*a*), descends below the roughened knob at the upper angle of the carinal margin, which is not the case with the slight fold in the same place in *L. truncata*; its presence seems caused by the edge of the central internal crest, in the upper part of the carina, being square (instead of round, as in *L. truncata*), and thus more deeply affecting the outline of the terga, between which it is inserted. The upper part of the scutal margin of the terga, as seen internally (fig. 5a), overlaps the scuta in a large *rectangular* projection. From the depth of the two opposite folds, namely, that caused by the tergal edge of the scuta and that by the crest of the carina, the inner face of the tergum is divided into two almost equal areas. The carina has its central crest square (fig. 5c, d) instead of being rounded as in *L. truncata*. The inner growing or corium-covered face is nearly at right angles to the longitudinal axis of the whole valve, instead of being oblique to it; it is convex or protuberant, with a central raised line, and two little knobs on each side of the upper part; the two lateral margins are slightly hollowed out, and the basal margin is not highly protuberant. The rostrum is excessively minute, barely above 1/20th of an inch in width; it is a little enlarged at each / zone of growth. Latera lost; no doubt they were rudimentary.

A fragment of a posterior cirrus, which adhered to one of the valves, shows that each segment supported four pairs of spines.

Width of the capitulum before disarticulation, probably was about 1/10th of an inch.

Species mihi no satis notae, aut dubiae

ANATIFA VILLOSA. Brugière, *Encyclop. Meth. Des. Vers.*, tom. i, p. 62, Pl. clxvi, 1789

On ships: Mediterranean.

ANATIFA HIRSUTA.[66] Conrad, *Journal of the Acad. of Nat. Sc.*, Philadelphia, vol. vii, p. 262, 1837

On fuci, Fayal, Azores.

The specimens, to which these names have been given by the above two authors, are described as small, and the *A. villosa* was suspected by Brugière to be young. The *A. hirsuta* is said by Conrad to have the valves minutely striated, granulated, and covered by a strong hirsute epidermis; the scuta, compared with the other valves, are very large; the entire length of this specimen was a quarter of an inch. The *A. villosa* is described as having smooth valves, and apparently the peduncle alone is hirsute. Now, in young individuals of *Lepas australis*,

[66] The *Anatifa hirsuta* of Quoy and Gaimard is the *Ibla quadrivalvis* of this work.

the peduncle is hairy, whilst in full-grown specimens it is quite smooth. Again, in some varieties of *L. fascicularis*, the thorax, prosoma, and cirri are hirsute, whereas they are generally quite smooth; hence I am inclined to suspect that *A. villosa* is the young, in a state of variation, / of *L. anatifera*; and that *A. hirsuta* bears a similar relation to *L. anserifera*. In Lamarck's *Animaux sans Vertèbres*, *Pollicipes villosus* of Sowerby is quite incorrectly given as a synonym to the above *A. villosa*.

ANATIFA ELONGATA. Quoy et Gaimard, *Voyage de l'Astrolabe*, Pl. xciii, fig. 6

This, I think, is certainly a distinct and new species, but I am unable to decide whether to place it in Lepas or Paecilasma. It is briefly described and pretty well figured in the above work. It was procured at New Zealand, but it is not stated to what object it was attached. The capitulum is much elongated, and one inch in length; the peduncle is from six to eight lines long. The carina is said to be very narrow; it is not stated whether it terminates downwards in a fork or disc; judging from the figure, it extends some way up between the terga, the basal ends of which are bluntly pointed. The scuta are almost quadrilateral. The peduncle is short, yellow, and tuberculated. The general appearance of the drawing makes me suspect that it is a Paecilasma.

CLYPTRA. Leach, *Zoological Journal*, vol. ii, p. 208, July, 1825

Leach has most briefly characterized a specimen in Savigny's Museum, from the Red Sea, under the above name of *Clyptra*. It has only four valves, and its peduncle is smooth; by the latter character it is distinguished from Ibla. Apparently this is a distinct and new genus.

Mr J. E. Gray, in *Proc. Zoolog. Soc.*, p. 44, 1848, quotes a description by Stroem (*Nym. Saml. Danske*, 295, n. iii, f. 20, 1788), namely, '*Lepas testâ compressâ 7-valvis, stipite lamellosâ*'. It is found attached to / *Gorgonia placomus*, in the North Sea. I suspect that this is the common *Scalpellum vulgare*, and that Stroem counted the valves only on one side, overlooking the rudimentary and concealed rostrum; and this would give seven for the number of valves. Had it not been for the expression 'stipite lamellosâ', I should have thought this might have been an unknown species of Dichelaspis.

SCALPELLUM LAEVIS. Risso, *Hist. Nat. des Product. de l'Europe Mérid.*, Tom. iv, p. 385, 1826

The chief characteristic of this species appears to be indicated by its specific name. It is found in the Mediterranean, attached to Cidarites. I am inclined to believe that it is distinct from *S. vulgare*.

SCALPELLUM PAPILLOSUM. King, *Zoolog. Journal*, vol. v, p. 334

Captain King has described this species, taken from the depth of 48 fathoms, on the coast of Patagonia, in Lat. 44° 30′S. It is probably distinct, but is so imperfectly described, that not even the number of the valves is given.

POLYLEPAS (POLLICIPES), Sinensis. Chenu, *Illust. Conchyliolog.*, Pl. II, fig. 7

This species is said to come from China; it is nearest to *P. spinosus*, but is, I think, distinct. /

[377/8]

EXPLANATION OF THE PLATES

PLATE I

Fig. 1. *Lepas anatifera*, nat. size. *Var.*, with a row of square, dark-coloured marks on the scuta and terga.

Fig. 1a. *Lepas anatifera*, external view of carina, magnified thrice.

Fig. 1b. *Lepas anatifera*, lateral view of carina, magnified thrice; var. *dentata*.

Fig. 1c. *Lepas anatifera*, internal view of right-hand scutum, to show the tooth at the umbo.

Fig. 2. *Lepas Hillii*, nat. size.

Fig. 3. *Lepas pectinata*, magnified thrice.

Fig. 3a. *Lepas pectinata*, var. (*spirulae*), tergum, magnified thrice.

Fig. 4. *Lepas anserifera*, nat. size.

Fig. 5. *Lepas australis*, nat. size.

Fig. 5a. *Lepas australis*, carina, external view of, magnified twice.

Fig. 6. *Lepas fascicularis*, nat. size, with its peduncle, together with those of three other specimens, imbedded in a vesicular ball of their own formation, of which a slice has been cut off to show the internal structure. The specimen is in the College of Surgeons. /

Fig. 6a. *Lepas fascicularis*, carina of, nat. size.

Fig. 6b. *Lepas fascicularis*, var. *villosa*.

Fig. 6c. *Lepas fascicularis*, var. *villosa*, carina of.

Fig. 6d. Part of the membrane from one side of the peduncle of *Lepas fascicularis*, with the ball removed, showing one of the cement ducts, and the orifices through which the vesicular membrane forming the ball has been secreted; greatly magnified; viewed from the outside.

PLATE II

Fig. 1. *Poecilasma Kaempferi*, magnified two and a half times.

THE LEPADIDAE

Fig. 1a. *Poecilasma Kaempferi*, carina of.

Fig. 2. *Poecilasma aurantia*, magnified two and a half times.

Fig. 3. *Poecilasma crassa*, magnified twice.

Fig. 3a. *Poecilasma crassa*, carina of.

Fig. 4. *Poecilasma fissa*, magnified five times.

Fig. 5. *Poecilasma eburnea*, magnified five times.

Fig. 5a. *Poecilasma eburnea*, carina of, external view of.

Fig. 5c. *Poecilasma eburnea*, carina of, lateral view of.

Fig. 5b. *Poecilasma eburnea*, scutum, internal view of.

Fig. 6. *Dichelaspis Warwickii*, magnified five times.

Fig. 6a. *Dichelaspis Warwickii*, transverse section of the top of the peduncle, showing the deeply notched end of the inwardly bent carina; magnified five times.

Fig. 6b. *Dichelaspis Warwickii*, var., scutum and tergum.

Fig. 7. *Dichelaspis pellucida*, magnified five times.

Fig. 7a. *Dichelaspis pellucida*, basal end of carina of, much magnified.

Fig. 8. *Dichelaspis Lowei*, magnified nearly ten times.

Fig. 8a. *Dichelaspis Lowei*, fork of carina of, viewed internally. /

Fig. 9. *Dichelaspis Grayii*, magnified eight or nine times.

Fig. 10. *Dichelaspis orthogonia*, magnified six times.

Fig. 10a. *Dichelaspis orthogonia*, carina, lateral view of.

Fig. 10b. *Dichelaspis orthogonia*, basal end of carina, viewed internally, much magnified.

PLATE III

Fig. 1. *Oxynaspis celata*, magnified three times.

Fig. 1a'. *Oxynaspis celata*, with the skin of the encrusting horny zoophyte removed. (a), scutum; (b), tergum; and (c), carina.

Fig. 2. *Conchoderma virgata*, magnified twice.

Fig. 2a. *Conchoderma virgata*, carina, viewed externally.

Fig. 2b. *Conchoderma virgata*, summit of capitulum, showing the terga from vertically above.

Fig. 2c. *Conchoderma virgata*, var. *chelonophila*, magnified four times.

Fig. 2d. *Conchoderma virgata*, var. *Olfersii* (scutum).

Fig. 3. *Conchoderma Hunteri*, magnified five times.

Fig. 4. *Conchoderma aurita*, nat. size, with the rudimentary carina exhibited on the right hand.

Fig. 4a. *Conchoderma aurita*, summit of capitulum, viewed from vertically above, showing the ear-like appendages and rudimentary terga.

Fig. 4b. *Conchoderma aurita*, section near the bases of the ear-like appendages, showing their folds.

Fig. 4c. *Conchoderma aurita*, var., scutum.

Fig. 5. *Alepas minuta*, magnified five times.

Fig. 6. *Alepas cornuta*, magnified five times. /

PLATE IV

Fig. 1. *Anelasma squalicola* (copied from Lovèn). The ovigerous lamellae are seen within the edges of the aperture of the capitulum. Enlarged about one and a half times.

Fig. 2. *Anelasma squalicola* (from Lovèn), with the membranes removed from one side of the capitulum and of the peduncle, exhibiting the body.

    a. External membrane of the capitulum.

    a, a. Inner membrane of ditto, lining the sack, and separated from the external membrane by a double fold of corium.

    b. The ovigerous lamellae, the edge projecting beyond the orifice of the capitulum.

    c. Penis, succeeed by six pairs of rudimentary cirri.

    d. Probosciformed mouth.

    e. Orifice of the acoustic [?] sack.

    f. Ovigerous fraenum.

    g. Ovarian branching tubes filling up the peduncle.

    h. Outer integument of peduncle, lined by corium and muscles, continuous with the outer membrane *a* of the capitulum.

Fig. 3. *Anelasma squalicola*. Small portion of the outer integument of the peduncle, greatly magnified, exhibiting the natural lines of splitting, and showing / that it is composed of several distinct portions or layers, which are displayed by the corners having been turned over. Three of the branching filaments, filled with pulpy corium, are given; the others have been cut off. The membrane *a* extends under *b*, but not under the circular patches of membrane, *c, c*.

Fig. 4. *Anelasma squalicola*. Mandibles, seen from the side towards the maxillae.

Fig. 5. *Anelasma squalicola*. Mandibles, seen from the side towards the labrum.

Fig. 6. *Anelasma squalicola*. The right-hand, rudimentary cirrus, the third from the mouth.

Fig. 7. *Anelasma squalicola*. Maxillae. The thin horny apodeme, *a*.

Fig. 8. *Ibla Cumingii*, female, magnified four times.

Fig. 8*a'*. *Ibla Cumingii*, female, magnified about five times, with the right-hand valves and right side of the peduncle removed. The male *h* is seen attached in the sack. The peculiar form of the body, caused by the small development of the prosoma, by the distance of the first and second pairs of cirri, and by the distance of the mouth from the adductor muscle (a dark dotted circle opposite *i*), and lastly, the remarkable course of the oesophagus over the adductor muscle, together with the outline of the stomach, are here all exhibited.

a. Scutum; the end of the large rounded adductor muscle, which / was attached to the valve now removed, near its apex, is plainly seen.

b. Tergum.

c. On a line with this letter, is seen the largely bullate labrum, forming a blunt overhanging projection.

d. Palpus, close to the upper segment of the pedicel of first cirrus.

e. Orifice of the acoustic [?] sack, between the bases of the first and second cirrus.

f. Caudal appendages.

g. Branching ovarian tubes within the peduncle.

h. Male, on the same scale, lying in its natural position within the sack, with the lower part of its peduncle bent upwards, and imbedded in the corium and muscles of the female.

i. Adductor scutorum muscle.

Fig. 8*b'*. *Ibla Cumingii*. Internal view of the scutum and tergum, and of the upper part of the outer integument of the peduncle, with its horny spines magnified about three times.

Fig. 8*c'*. *Ibla Cumingii*. A small portion of the outer integument of the peduncle, greatly magnified, showing the horny persistent spines; two of the spines have been torn out.

Fig. 9. *Ibla quadrivalvis*. Internal view of scutum and tergum, and of the upper part of the outer integument of the peduncle; magnified four times.

Fig. 9a'. *Ibla quadrivalvis*. Penis, supported on a long unarticulated projection; greatly magnified. /

PLATE V

Fig. 1. Male of *Ibla Cumingii*, magnified thirty-two times.

   a. Mouth.

   b. A slight double fold, formed by the basal edge of the labrum, and by a lower fold, which at *h* becomes well developed; the latter is a rudimentary representation of the double membrane and valves forming the capitulum.

   c. Eye.

   d, d. Torn membrane from the sack of the female, constricted round the body of the male.

   e. Terminal or basal point, with the prehensile larval antennae, represented on rather too large a scale.

   f. The imbedded portion of the male.

   g. Two pairs of cirri.

   h. The fold above alluded to, concealing a small portion of the slightly retracted thorax.

Fig. 2. The male of *Ibla Cumingii*, viewed from vertically above; magnified about sixty times. The dotted lower portion, represents the outline of the thorax and the positions of the cirri, which, from standing below the mouth, could not be well seen, when the summit of the mouth was in the proper focus.

   a. Labrum, largely bullate.

   b. Palpi.

   c. Mandibles.

   d. Maxillae.

   e. Outer maxillae; between which and the crest of the labrum, the orifice of the oesophagus can be obscurely seen.

   f. Anus.

   g. Rudimentary caudal appendages, under which is the pore leading from the vesiculae seminales.

   h. Posterior cirrus.

   i. Anterior cirrus. /

Fig. 3. Male of *Ibla Cumingii*. Labrum and palpi, as seen with the eye on a level with the summit of the mouth.

Fig. 4. Male of *Ibla Cumingii*. Posterior cirrus (*h* in fig. 2) much magnified.

Fig. 5. Male of *Ibla Cumingii*. Larval antennae; from the terminal point of the body (*e* in fig. 1), as seen with a ⅛th of an inch object glass.

Fig. 6. Male of *Ibla Cumingii*. Outer maxillae.

Fig. 7. Male of *Ibla Cumingii*. Mandibles, with the underlying articulated membrane, forming the side of the mouth.

Fig. 8. Male of *Ibla Cumingii*. Maxillae, with the apodeme.

Fig. 9. Complemental Male of *Scalpellum vulgare*, attached over the fold in the occludent margin of the scutum of the hermaphrodite.

    a. Orifice of the sack of the male.

    b. Spinose projections above the rudimental valves; at the bottom of the figure are represented, as seen through the whole thickness of the animal, the prehensile larval antennae.

    d. The depression for the attachment of the adductor scutorum muscle of the hermaphrodite; see fig. 15a'.

    e, e. A transparent layer of chitine, which forms a border to the occludent margin of the scutum of the hermaphrodite. This border supports long spines which are connected with the underlying corium by sinuous tubuli.

Fig. 10. The basal (normally anterior) portion of the above complemental male, greatly magnified, viewed dorsally from above, exhibiting the larval prehensile antennae, attached to the anterosternal surface of the animal. /

Fig. 11. One of the antennae of ditto, viewed laterally and on the outside.

Fig. 12. Ditto, ultimate segment of.

Fig. 13. Body of the above complemental male, consisting of the thorax supporting the four pairs of limbs, and of the terminal abdominal lobe.

Fig. 14. Small portion of the outer integument of the complemental male, as seen with a ⅛th of an inch object glass.

Fig. 15. *Scalpellum vulgare* (hermaphrodite), magnified three times.

    a, a. Complemental males.

    b. Rostrum, of which a separate enlarged figure b' is given.

Fig. 15a'. Scutum of the hermaphrodite *Scalpellum vulgare*, internal view of.

    a. Fold on the occludent margin.

    d. Pit for the adductor muscle.

PLATE VI

Fig. 1. *Scalpellum ornatum*. Female, magnified seven times.

Fig. 1a'. *Scalpellum ornatum*. Upper latus, viewed internally.

Fig. 1b'. *Scalpellum ornatum*. Scutum of full-grown specimen, viewed internally, much magnified.
    a. Depression for the adductor muscle.
    b. Depression for the reception of the male.

Fig. 1c'. *Scalpellum ornatum*. Scutum of half-grown specimen, viewed internally, much magnified, on same scale with fig. 1b'. The depression *b* for the reception of the male is here seen, in almost the first stage of formation. /

Fig. 1d'. *Scalpellum ornatum*. An imaginary section through the cavity *x* in which the male is lodged.
    a. Section of the shell of the scutum of the female.
    b. A layer of chitine homologous with the shell, and *partially* lining the scutum.
    c. The inner lining (of chitine) of the sack of the female.
    d. A double fold of corium.

Fig. 2. *Scalpellum rutilum*, magnified two and a half times.

Fig. 2a'. *Scalpellum rutilum*. Internal view of scutum, enlarged.
    a. Depression for the adductor muscle.
    b. Cavity for the reception of the male.

Fig. 2b'. *Scalpellum rutilum*. External view of carina.

Fig. 2c'. *Scalpellum rutilum*. Section across middle of carina.

Fig. 3. Complemental male of *Scalpellum Peronii*, greatly magnified.

Fig. 4. Complemental male of *Scalpellum villosum*, greatly magnified.
    a'. Natural size.

Fig. 4*a*, *b*, *c*. Ditto, valves separated.
    a. Scutum.
    b. Tergum.
    c. Carina.

Fig. 5. Complemental male of *Scalpellum rostratum*, a restored figure, greatly magnified. Scutum and rudimentary carina correct.

Fig. 6. *Scalpellum Peronii*, one and a half the natural size.
    a. Rostrum a little more enlarged, front view of.

Fig. 7. *Scalpellum rostratum*, magnified six times.
   a. Rostrum, front view of. /

Fig. 8. *Scalpellum villosum*, magnified one and a half the natural size.

Fig. 8a, b. *Scalpellum villosum*. a. Internal view of rostrum; b. Internal view of sub-rostrum.

PLATE VII

Fig. 1. *Pollicipes cornucopia*, one and a half nat. size.

Fig. 1a. *Pollicipes cornucopia*, internal view of valves.

Fig. 2. *Pollicipes polymerus*, one and a half nat. size.

Fig. 2a. *Pollicipes polymerus*, internal view of valves.

Fig. 3. *Pollicipes mitella*, nat. size.

Fig. 3a'. *Pollicipes mitella*, nat. size, internal views of
   a. Scutum, and of
   b. Tergum, showing articular fold.

Fig. 3b'. *Pollicipes mitella*. Internal view of other valves, in a small specimen, showing the manner in which the valves of the lower whorl overlap each other.
   a. Upper latera.
   b. Carina,
   c. Sub-carina, both viewed a little obliquely.
   d. Rostrum,
   e. Sub-rostrum, both viewed a little obliquely.

Fig. 4. *Pollicipes spinosus*, one and a half nat. size.

Fig. 5. *Pollicipes sertus*, one and a half nat. size.

PLATE VIII

Fig. 1. A piece of rock bored in two directions by *Lithotrya dorsalis*, with the calcareous basal discs in the upper cavity, serving as a bridge for crossing an old cavity. About twice natural size. /

Fig. 1a'. *Lithotrya dorsalis*, nearly twice nat. size, with the basal calcareous cup adherent; a, rostrum on same scale, seen externally.

Fig. 1b'. *Lithotrya dorsalis*, rostrum and the rostral corners of the two scuta, together with a small portion of the subjacent membrane of the peduncle, with its calcareous scales; viewed externally, greatly magnified, showing the inferior crenated edges of the scales.

Fig. 1c'. *Lithotrya dorsalis*, basal calcareous cup, one and a half the natural size; this is the largest specimen which I have seen.

Fig. 2. *Lithotrya Nicobarica*, magnified nearly twice; attached to the rock, copied from Reinhardt; a, rostrum on the same scale, with the other valves, seen externally; b, section of the row of discs; c, extreme point of the peduncle, extending beneath the row of discs.

Fig. 2a'. Rock bored by *Lithotrya Nicobarica*, showing the row of calcareous discs, copied from Reinhardt.

Fig. 3. *Lithotrya cauta*, magnified between seven and eight times; a, scutum; b, tergum.

Fig. 3c. *Lithotrya cauta*, latus, greatly magnified.

Fig. 3d. *Lithotrya cauta*, uppermost scales of the peduncle, greatly magnified.

Fig. 3e. *Lithotrya cauta*, star-shaped discs of hard chitine, supported on a peduncle of the same substance, taken from the lower exterior surface of the peduncle, very greatly magnified. /

Fig. 4. *Lithotrya Rhodiopus*, magnified five times, internal views of; a, scutum; b, tergum; c, latus; d, carina.

Fig. 5. *Lithotrya Valentiana*, magnified between three and four times; a, internal view of scutum and tergum, locked together; b, capitulum seen from vertically above; c, internal view of carina; d, section across the middle of the carina.

PLATE IX

Fig. 1. *Lithotrya truncata*, magnified four times.

Fig. 1a'. *Lithotrya truncata*, capitulum seen from vertically above, not so distinctly represented as in fig. 5b, Pl. VIII.

Fig. 1b'. *Lithotrya truncata*, internal views of valves; a, rostrum, with a few subjacent scales of the peduncle; b, scutum; c, tergum; d, carina.

Fig. 2. A portion (about 1/10th of an inch square) of the surface of attachment of the peduncle of *Pollicipes polymerus*, seen from the outside, greatly magnified, showing the small circular bb patches of cement, poured out from the cement ducts aa which lie within the peduncle.

Fig. 2a'. *Pollicipes polymerus*, a section, still more magnified, through the basal membrane of the peduncle, through one of the loops of the cement ducts *aa*, and through one of the circular patches *b* of cement. /

Fig. 3. Cement gland, duct, and ovarian tubes of *Conchoderma aurita*; *aa*, ovarian tubes, with ova in process of formation; *b*, cement gland; *c*, cement duct.

Fig. 4. *Conchoderma virgata*, enlarged, with one side of the capitulum and of the peduncle removed, to show the form and position of the body.

    a. tergum, edge of.
    b. mouth, with one of the palpi seen on the inner, upper corner.
    c. adductor scutorum muscle.
    d. orifice of acoustic [?] sack.
    e. scutum, occludent margin of.
    f. branching ovarian tubes within the peduncle.
    g. filamentary appendage on the prosoma.
    h. ditto, close to basal articulation of the first cirrus.
    i. ditto, on the pedicel of the first cirrus.
    j. ditto, on the pedicel of the third cirrus.
    k. ditto, on the pedicel of the fourth cirrus.
    l. ditto, on the pedicel of the fifth cirrus.
    m. edge of the carina.
    n. prosoma.

Fig. 5. Apex of one of the filamentary appendages of *Conchoderma aurita*, greatly magnified, exhibiting the included branching testes.

Fig. 6. Acoustic [?] sack of *Conchoderma virgata*, taken out of the acoustic meatus, with the diaphragm from the summit removed; greatly magnified. /

Fig. 7. Terminal part (magnified seven times), of the peduncle of an elongated specimen of *Scalpellum vulgare*, slit open, with the corium removed, showing the two cement ducts *aa*, and a row of circular patches *bb* of cement, by which the peduncle, along its rostral edge, is attached to the thin horny branches of the coralline. The larval antennae are seen at the terminal point, and the two cement ducts can be traced into them.

PLATE X

Figures all greatly magnified

Fig. 1. Mandibles of *Pollicipes mitella*: exhibiting the upper *a* and lower *b* articulations, and the three principal muscles; the short upper cut-off muscle runs to its attachment at the base of the palpus.

Fig. 2. Mandibles of *Lithotrya dorsalis*, exhibiting four *aa* roughened, thin, ligamentous apodemes for the attachment of the muscles.

Fig. 3. Mandibles of *Scalpellum Peronii*.

Fig. 4. Mandibles of *Ibla Cumingii*.

Fig. 5. Mandibles of *Lepas anatifera*.

Fig. 6. Palpus of *Lepas anatifera*.

Fig. 7. Palpus of *Pollicipes mitella*.

Fig. 8. Palpus of *Alepas cornuta*.

Fig. 9. Maxilla of *Lepas anatifera*.

Fig. 10. Maxilla of *Lithotrya dorsalis*, exhibiting the horny, rigid apodeme *a* buried in muscles, together with the two other principal bundles of muscles.

Fig. 11. Maxilla of *Ibla Cumingii*. /

Fig. 12. Maxilla of *Lithotrya Rhodiopus*.

Fig. 13. Maxilla of *Pollicipes polymerus*.

Fig. 14. Maxilla of *Pollicipes mitella*.

Fig. 15. Maxilla of *Poecilasma eburnea*.

Fig. 16. Outer maxilla of *Conchoderma virgata*; *a*, orifice of the olfactory cavity, the inner delicate chitine membrane of which is seen within, the specimen having been treated with caustic potash.

Fig. 17. Outer maxilla of *Pollicipes mitella*, showing the two principal muscles, and the prominent, tubular, *b* olfactory orifices.

Fig. 18. Caudal appendages, and basal segments of the sixth pair of cirri, of *Lepas anatifera*; *a*, anus; *b*, caudal appendages; *c*, lower segment of pedicel of sixth cirrus; *d*, upper segment of ditto; *e*, basal segments of the two rami.

Fig. 19. Caudal appendage (right-hand side) of *Pollicipes sertus*.

Fig. 20. Caudal appendage (right-hand side) of *Scalpellum Peronii*.

Fig. 21. Caudal appendage (right-hand side) of *Scalpellum vulgare*.

Fig. 22. Caudal appendage (right-hand side) of *Pollicipes cornucopia*.

Fig. 23. Caudal appendage (left hand) *Lithotrya dorsalis*; *a*, caudal appendage; *c*, lower segment of pedicel of sixth cirrus; *d*, upper segment of ditto; *e*, segments of one of the rami.

Fig. 24. Portion of caudal appendage of *Lithotrya dorsalis*, highly magnified.

Fig. 25. *Pollicipes polymerus*; anterior ramus of the second cirrus.

Fig. 26. *Lepas anatifera*; a segment of the sixth cirrus, showing the arrangement of the spines; *a*, main anterior / spines, of which there is a corresponding row on the opposite side; *c*, dorsal tuft.

Fig. 27. *Pollicipes polymerus*; a segment of the sixth cirrus, showing the arrangement of the spines; *a*, main anterior spines, of which there is a corresponding row on the opposite side; *bb*, calcareous shields on the dorsal surfaces, with tufts of fine spines near their upper edges.

Fig. 28. *Alepas cornuta*; sixth cirrus of; *a*, basal portion of one ramus, consisting of numerous segments; *k*, the other and almost rudimentary ramus.

Fig. 29. *Poecilasma fissa*; segments of the sixth cirrus, showing the arrangement of the spines; *a*, anterior spines; *c*, dorsal tufts. /

INDEX

Synonyms and doubtful species are printed in italics

Abortion, extreme, in the male of Ibla, 202
Absia, 332
Acari, development of, 18
Acoustic [?] organs, general description of, 53
Adductor scutorum muscle, 39
Affinities of the Lepadidae, 64
Alepas, genus, 156
  cornuta, 165
  minuta, 160
  parasita, 163
  squalicola, 170
  tubulosa, 169
Allman, Professor, on Cyclops, 38
Anatifa vel *Anatifera*, genus, 67, 99, 215
  crassa, 107
  dentata, 73
  elongata, 374
  engonata, 73
  hirsuta, 203
  laevis, 73, 77
  oceanica, 92
  obliqua, 264
  parasita, 163
  quadrivalvis, 203
  sessilis, 81
  spinosa, 324
  striata, 81, 86
  substriata, 77
  sulcata, 86
  tricolor, 77
  truncata, 361
  univalvis, 163
  villosa, 367
  vitrea, 92

Anelasma, genus, 169
Antennae, larval, 33
  in the Lepadidae, table of measurements, 286
  of Ibla Cumingii, 191
  of Lepas australis, 15
  of Scalpellum vulgare, 237
Appendages, caudal, 43
  in larva, 19
  filamentary, 38
Asplanchna, male of, 292
Attachment of cirripedes, 33
  of Scalpellum vulgare, 226
  of Pollicipes polymerus, 310

Balanidae, affinities of, 64
Bate, Mr C. S., on the metamorphoses of cirripedes, 9–16
Bopyrus, parasite allied to, 55
Branta, 137
  aurita, 141
  virgatum, 146
Brightwell, Mr, on the Asplanchna, 292
Brisnaeus, 332
  Rhodiopus, 363
Brugière, date of work of, 67
Buoyancy, means of, in Lepas fascicularis, 95
Burmeister, Professor, on the metamorphoses of cirripedes, 9, 13 /
Burrowing powers of, in Lithotrya, 337

Calentica, 215
  Homii, 274

Capitulum, general description of, 28
Capitulum, genus, 293
  mitella, 316
Carapace of the larva, 15
Caudal appendages, 43
  in larva, 19
Cement discs, in a straight row, in Scalpellum vulgare, 226
  in Pollicipes polymerus, 310
Cement ducts, 34
  in the larva, 20
Cement glands, incipient in larva, 24, 34
Cement, nature of, 36
Cement tissue, modified as a float in Lepas fascicularis, 95
Chitine, chemical nature of, 30
Chthamalinae, 2, 65
Cineras, genus, 137, 156
  bicolor, 146
  Cranchii, 146
  chelonophilus, 146 151
  megalepas, 146
  membranacea, 146
  Montagui, 146
  Olfersii, 146, 152
  Rissoanus, 146
  vittatus, 146
Circulation, 46
Cirri, general description of, 42
  of young cirripede, 22
Cirripede, immature whilst within the larva, 20
Cirripedes, sessile, affinities of, 64
  subfamilies of, 2
  useful as food, 66
Clyptra, 374
Coates, Dr, on Lepas fascicularis, 96
Conchoderma, genus, 136
  aurita, 141
  Hunteri, 153
  leporinum, 141
  virgata, 146
Conchotrya, 332
  Valentiana, 371
Cuming, Mr, obligations to, 181, 189
  on the cirripedes of the Philippine Archipelago, 65

on Balanus psittacus, 66
Cup, basal calcareous, in Lithotrya, 338

Dana, Mr J. D., on the ovaria in certain Crustacea, 26
  on the antennae of larval cirripedes, 15, 26
Dichelaspis, genus, 115
  Grayii, 123
  Lowei, 128
  orthogonia, 130
  pellucida, 125
  Warwickii, 120
Distribution, geographical, 65
Dosima, 67
  fascicularis, 92
Dujardin, on the larvae of Acari, 18

Encyclopédie Method., date of, 67
Entozoons, sexes of, 201
Epidermis of valves, 31
Exuviation, 61, 63
  of the larval eyes, 24
  of the larval integuments, 20
  of the membrane of peduncle in Lithotrya, 336
Eyes, in the Lepadidae, 49
  of the larva, first stage, 10
  last stage, 16, 24

Families of cirripedes, 2
Farre, Dr, on the acoustic organs in Crustacea, 54
Female organs of generation in the Lepadidae, 56
Filaments, 38
Forbes, Professor E., on the homology of the peduncle, 26
Fraena, ovigerous, 59

Ganglia, ophthalmic, 49 /
Generation, organs of, in the Lepadidae, 55
Glands, supposed salivary, 57
  on the ovigerous lamellae, 60
Goodsir, Mr, on the metamorphosis of cirripedes, 9, 16

Goodsir, Mr (*cont.*)
  on the supposed male of Balanus, 55
Gray, Mr J. E., on the genus Dosima, 99
  on the metamorphosis of cirripedes, 9
  on the inequality of the valves in Paecilasma, 101, 103
  on an unknown seven-valved Lepas, 374
  on the genus Scalpellum, 216
Growth, rate of, 63
Gymnolepas, 137
  Cranchii, 146
  Cuvierii, 141

Habitats, 65
Hancock, Mr, on the burrowing of cirripedes, 346
  on the larva of Lepas, 11
Hectocotyle, 200
Heptalasmis, 115
Hermaphroditism, peculiar kind of, 201
Heteroura androphora, 201
Homologies of the Cirripedia, 25–8
Ibla, genus, 180
  Cumingii (female), 183
    (male), 189
  Cuvieriana, 203
  quadrivalvis (hermaphrodite), 203
    (complemental male), 207
  general summary on its sexual relations, 281
Impregnation of the females and hermaphrodites in Ibla and Scalpellum, 290

King, Captain, on a new Scalpellum, 375
Kölliker, on the males of Cephalopoda, 200
Labrum, general description of, 40
Lamellae, ovigerous, 58
Larvae, general description of, 8
Larva of Ibla quadrivalvis, 210
Leidy, Professor, on the eyes of cirripedes, 2, 49
Lepas, genus, 67

anatifera, 73
anserifera, 81, 86
australis, 89
australis metamorphis of, 14
coriacea, 146
cornuta, 141
cygnea, 92
dilata, 92
dorsalis, 351
fascicularis, 92
fascicularis, peduncle, remarkable structure of, 95
Gallorum, 298
Hillii, 77
leporina, 141
membrancea, 146
mitella, 316
muricata, 85
nauta, 81
pectinata, 85
pollicipes, 298
scalpellum, 222
sulcata, 86
virgata, 146
Lerneidae, males of, 200
Leucifer, 28
Litholepas, 332
  de Mont Serrat, 351
Lithotrya, genus, 332
  cauta, 356
  dorsalis, 351
  Nicobarica, 354
  Rhodiopus, 363
  truncata, 366
  Valentiana, 371
  powers of burrowing, 337
Lovèn, Dr, on the habits of the *Alepas squalicola*, 178
  on the homologies of cirripedes, 26
Lowe, Rev. R. T., on the fishes of Madeira and Japan, 106
  on the cirripedes of Madeira, 65 /

Macgillivray, Professor, on Conchoderma, 140
  on Lepas anserifera, 81
Malacotta, 137
  bivalvis, 141

Male cirripedes, discussion on, 281
  of Ibla Cumingii, 189
  of Ibla quadrivalvis, 207
  of Scalpellum ornatum, 248
    Peronii, 270
    rostratum, 268
    villosum, 278
    vulgare, 231
  organs of generation in the Lepadidae, 55
Mandibles, general description of, 41
Martin St Ange, on the affinities of cirripedes, 1
  on a closed tube within the stomach, 45
  on the generative organs, 55
Maxillae, general description of, 41
Membrane, covering valves, 30
Metamorphoses, first stage, 9
  second stage, 13
  last stage, 14
Mitella, genus, 293
Mouth, general description of, 39
  of young cirripede, 22
  of the larva, first stage, 11
  last stage, 17
Muscles, 39
  without striae in Anelasma, and in embryonic cirripedes, 172

Nerves, general system of, 46
  of Ibla Cumingii, 188
Nomenclature of the parts of cirripedes, 3
  Rules of, 293

Octolasmis, 115
  Warwickii, 120
Oesophagus, general description of, 44
Orders of cirripedes, 2
Organs acoustic [?] general description of, 53
Organs acoustic, of the larva of Lepas, 15
  female, of generation, in the Lepadidae, 56

  male, of generation, in the Lepadidae, 55
  olfactory, general description of, 52
Otion, 137
  auritus, 141
  Bellianus, 141
  Blainvillianus, 141
  Cuvieranus, 141
  depressa, 141
  Dumerillianus, 141
  Rissoanus, 141
  succutifera, 141
Ova, 58
Ovaria, incipient in the larva, 20, 24
  in the Lepadidae, 57
Oviducts (supposed), 59
Owen, Professor, on certain Entozoic worms, 201
  on the Conchoderma Hunteri, 154
Oxynaspis, genus, 133
  celata, 134

Pamina, 137
  trilineata, 146
Peach, Mr, obligations to, 240
  on the movements of pedunculated cirripedes, 33
Peduncle, general description of, 31
  origin and homologies of, 21
Penis, general description of, 56
  of Ibla quadrivalvis, 206
Pentalasmis, vel *Pentalepas*, 67
  anseriferus, 81
  dentatus, 73
  dilatata, 81
  Donovani, 92
  fascicularis, 92
  Hillii, 77
  inversus, 86
  laevis, 73, 77
  radula, 86
  spirulae, 86
  spirulicola, 92
  sulcata, 86 /
Pentalepas vitrea, 92
Poecilasma, genus, 99
  aurantia, 105
  crassa, 107

Poecilasma, genus (*cont.*)
  eburnea, 112
  fissa, 109
  Kaempferi, 102
Pollicipes, 293
  cornucopia, 298
  elegans, 304
  mitella, 316
  Mortoni, 307
  obliqua, 264
  polymerus, 307
  ruber, 304
  scalpellum, 222
  sertus, 327
  sinensis, 375
  Smythii, 298
  spinosus, 324
  tomentosus, 274
  villosus, 274
Polylepas, 214, 293
  mitella, 316
  sinensis, 375
  vulgare, 222
Primordial valves, 22
Prosoma, shape of, 39
Proteolepas, 3, 26
Pupa, locomotive or last larval state, in cirripedes, 18

Ramphidiona, 293
Range, geographical, 65
Rate of growth, 63
Reinhardt on the burrowing of Lithotrya, 346
Reproduction, organs of, in the Lepadidae, 55
Rotifera, sexes of, 292
Rules of nomenclature, 293

Sack, description of, 31
  origin of, 15, 23
Scalpellum, genus, 215
  laevis, 375
  laeve, 222
  Sicilice, 222
  ornatum (female), 244
    (male), 248
  papillosum, 375

Scalpellum Peronii, 264
  (male), 270
  rostratum, 259
  (male), 262
  rutilum, 253
  (male), 258
  villosum, 274
  (male), 278
  vulgare, 222
  larva of, 9
  (complemental male), 231
  general summary on sexual relations, 281
Schmidt, Dr, on chitine, 30
  on the muscles in young crustacea, 172
Senoclita, 137
  fasciata, 146
Sexes, discussion on, in Ibla and Scalpellum, 281
Siebold, Dr C. Von, 201
Smilium, 215
  Peronii, 264
Spermatozoa in Scalpellum vulgare, 236
Sprengel, Ch. K., on compositous flowers, 203
Steenstrup, Professor, on the homology of the peduncle, 26
  on the non-hermaphroditism of cirripedes, 55
Stomach of larva, 19
  general description of, 44
Stroem on a seven-valved Lepas, 374
Syngamus trachealis, 20*

Testes in the Lepadidae, 55
Tetralasmis, 180
  hirsutus, 203
Thaliella, 215
  ornata, 244
Thompson, Mr W., on Lepas anatifera (var.), 74
  on the exuviations of sessile cirripedes, 63
  obligations to, 240
  Mr Vaughan, on the metamorphoses of cirripedes, 9, 10 /

Trilasmis, genus, 99
   eburnea, 112
Triton, genus, 156
   fasciculatus, 163

Upopi, or young acari, 18

Vesiculae seminales, 56
Valves, general description of, 28

Valves, chemical nature of, 30
   horny, colour changed by pressure, 184
   primordial, 22

Wagner, R., on the male organs of generation, 55

Xiphidium, 215 /

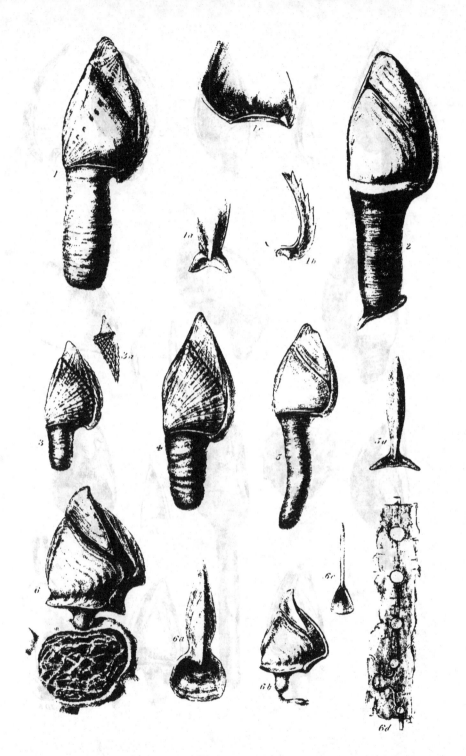

PLATE I
Lepas

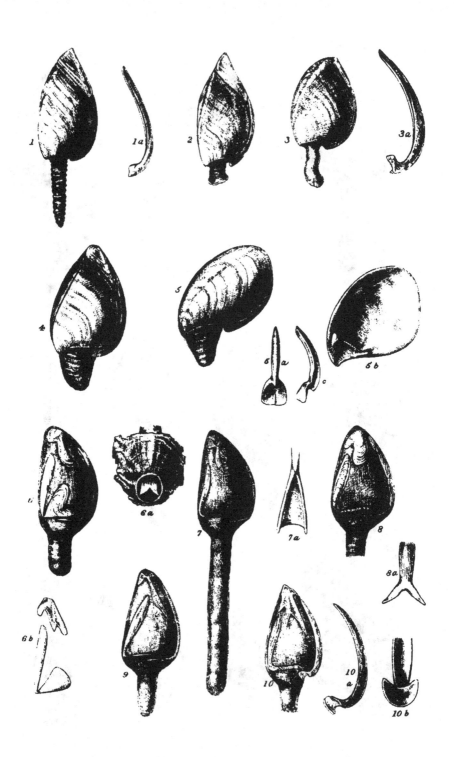

PLATE II

Poecilasma: Dichelapsis

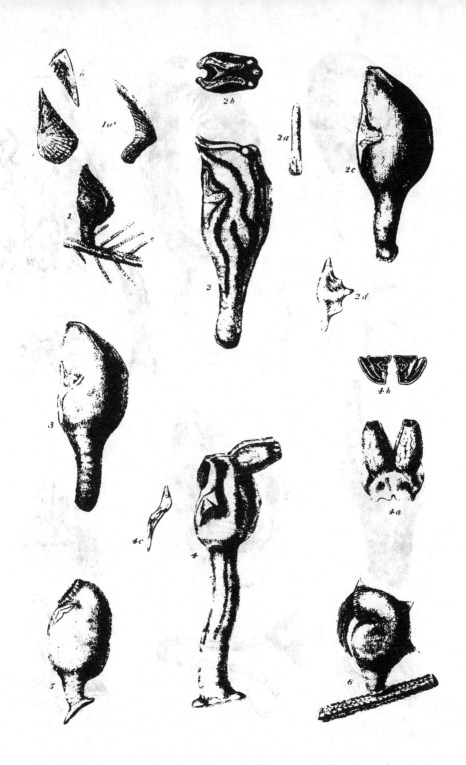

PLATE III

Oxynaspis: Conchoderma: Alepas

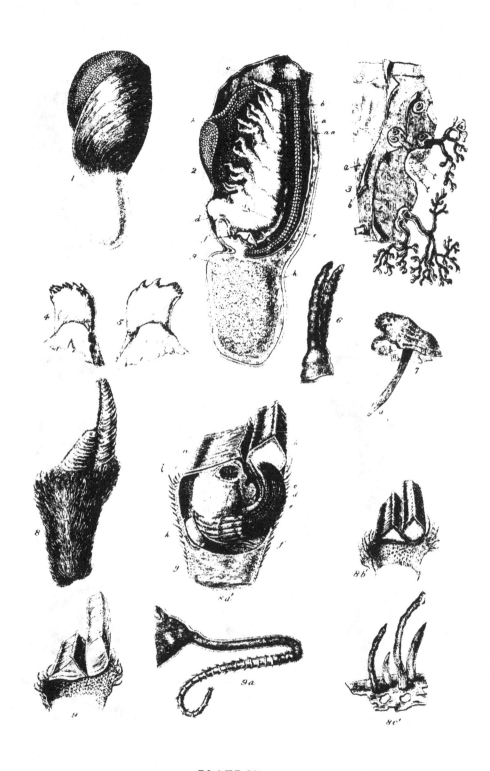

PLATE IV

Anelasma: Ibla

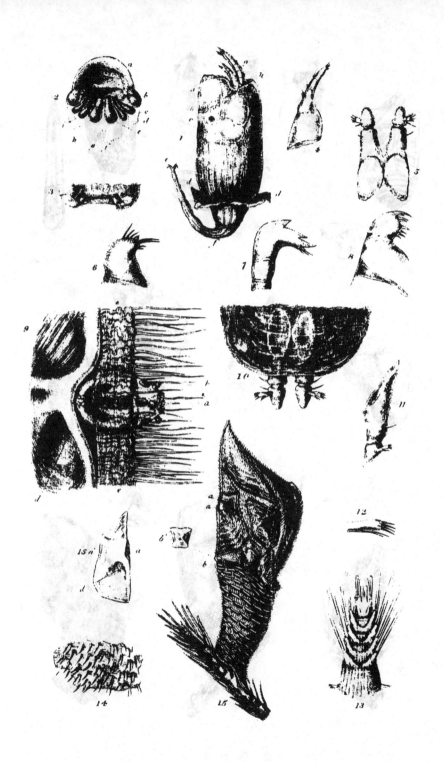

PLATE V

Ibla: Scalpellum

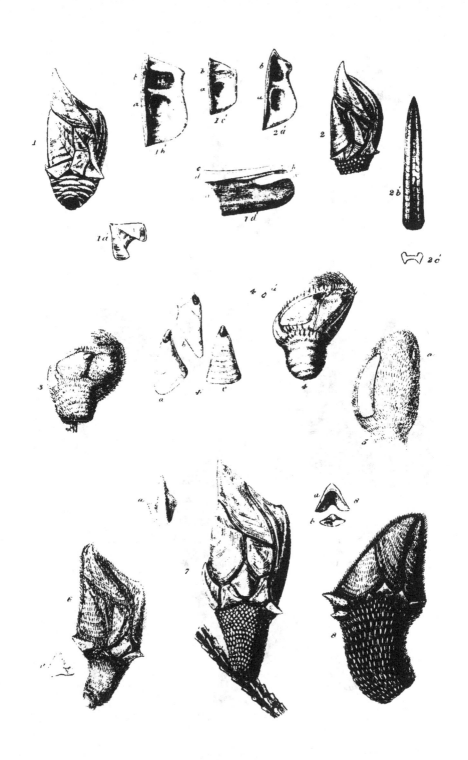

PLATE VI

Scalpellum

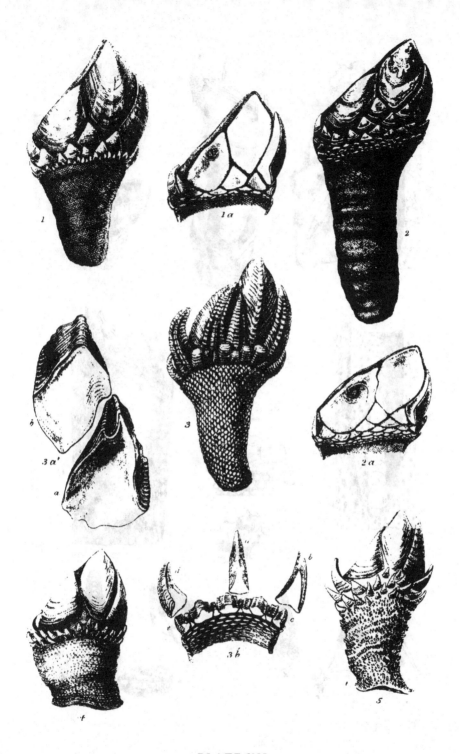

PLATE VII

Pollicipes

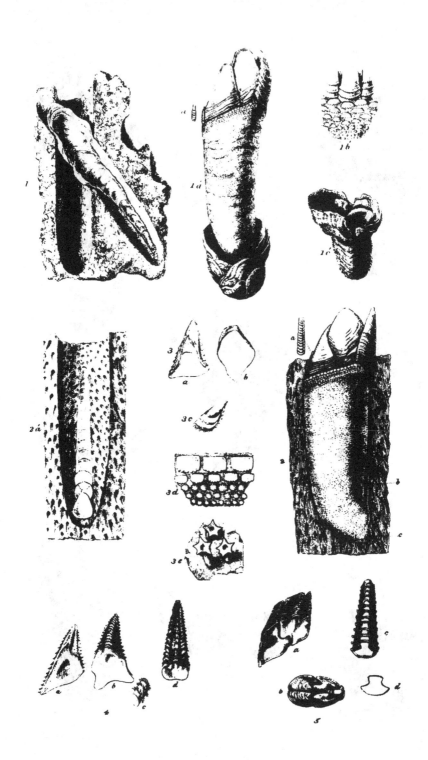

PLATE VIII

Lithotrya

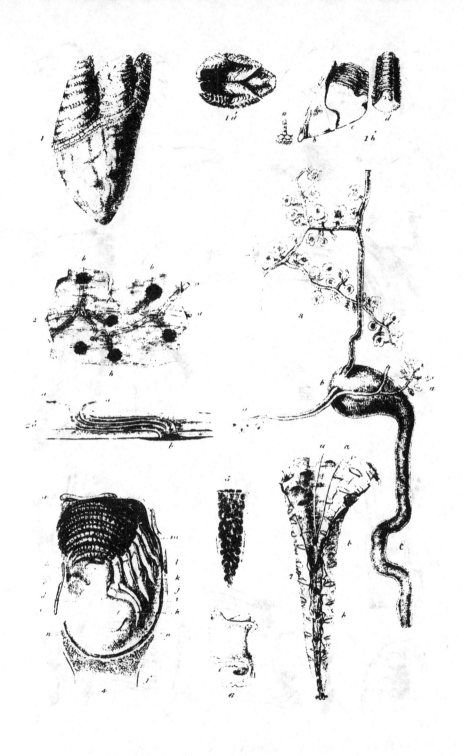

PLATE IX
Lithotrya, etc.

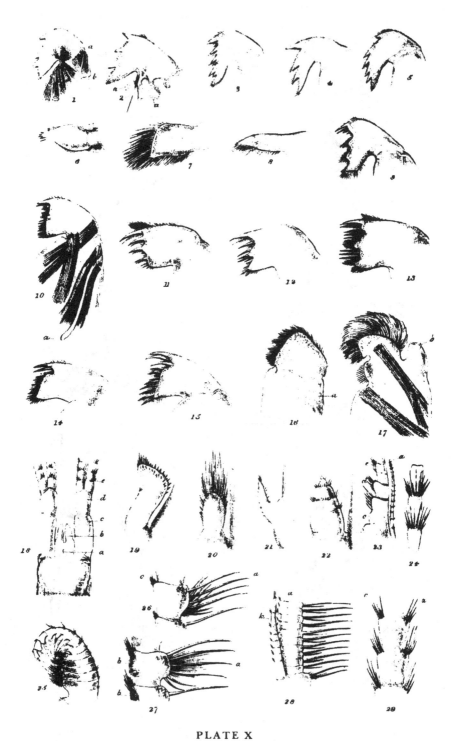

PLATE X

Mandibles, palpi, maxillae, outer maxillae, cirri and caudal appendages

Printed in the United States
by Baker & Taylor Publisher Services

Printed in the United States
by Baker & Taylor Publisher Services